MODERN ALGEBRA
AND TRIGONOMETRY

SECOND EDITION

MODERN ALGEBRA AND TRIGONOMETRY

J. VINCENT ROBISON

Associate Professor Emeritus
Department of Mathematics
Oklahoma State University

McGRAW-HILL BOOK COMPANY

New York   St. Louis   San Francisco   Düsseldorf   Johannesburg
Kuala Lumpur   London   Mexico   Montreal   New Delhi   Panama
Rio de Janeiro   Singapore   Sydney   Toronto

Library of Congress Cataloging in Publication Data
Robison, John Vincent.
  Modern algebra and trigonometry.
  1. Algebra. 2. Trigonometry, Plane. I. Title.
QA154.R655 1973   512'.13   72-6677
ISBN 0-07-053330-X

MODERN ALGEBRA
AND TRIGONOMETRY

1 2 3 4 5 6 7 8 9 0 KPKP 7 9 8 7 6 5 4 3 2

This book was set in Laurel by York Graphic Services, Inc. The
editors were Jack L. Farnsworth and Laura Warner; the designer
was Rafael Hernandez; and the production supervisor was
Joe Campanella.
The printer and binder was Kingsport Press, Inc.

CONTENTS

PREFACE

The primary aim of this edition of *Modern Algebra and Trigonometry* is the same as that of the first: to provide an adequate foundation for modern courses in analytic geometry and calculus. This is done (hopefully) by presenting, from a modern point of view, an orderly development of college algebra integrated with trigonometry.

Many students who take college algebra and trigonometry, however, are not mathematics majors. The needs of this rather large group have been kept in mind throughout the writing so that the book may be used for a terminal course in these subjects. While I have tried to be reasonably precise and rigorous in the development of the topics for the benefit of students who will take additional courses in mathematics, I have not emphasized rigor to the point of stifling interest for the student who does not plan to continue the study of mathematics.

There is enough material in *Modern Algebra and Trigonometry* for a five-unit semester course, or two three-unit quarter courses. By selection of the desired topics, the book can also be used for a three- or four-unit semester course or a five-unit quarter course. Most of the chapters are independent of each other, and some of them can be entirely omitted for short courses.

In selecting the content for this book, I have been guided to some extent by the recommendations of various curriculum study groups, such as the Committee on the Undergraduate Program in Mathematics (CUPM). The textbooks of the School Mathematics Study Group (SMSG) also influenced some of the writing.

It is impossible in a book of this length to include all the topics that belong to a study of algebra and trigonometry. Many that are important to the calculus, such as partial fractions and convergence of series, are not covered. I have included only those topics I consider to be most often taught in college algebra and trigonometry.

The arrangement of the chapters in this edition is a result of suggestions by many users of the first edition. Because most of the chapters are independent, the instructor will have much freedom in teaching them in a different order. For example, any or all of Chaps. 13 to 15 can follow Chap. 3.

Set operations and certain properties of sets are discussed in Chap. 1; and in Chap. 2 a brief axiomatic treatment of real numbers is presented. Simple inequalities and the fundamental operations also are included in Chap. 2.

Relations, functions, and graphs are introduced in Chap. 3. A relation is defined as a set of ordered pairs, and a function as a special kind of relation. The inverse of a function is then obtained by interchanging the coordinates of each of its ordered pairs. The frequently used distance formula is derived in this chapter, and a discussion of the linear function is included. Slopes and equations of lines, variation and proportion, and literal equations and formulas are also discussed in Chap. 3. The function concept, introduced in Chap. 3, provides the basis for developing most of the remaining topics.

Trigonometry as a study of periodic functions of real numbers is introduced in Chap. 5. The latter part of this chapter is devoted to the trigonometric functions of angles. The solution of triangles is deferred to a later chapter.

Matrices are considered in Chap. 13, where they are used in solving systems of linear equations. Determinants and Cramer's rule are topics included in this chapter.

In the study of permutations in Chap. 15, the cross product of sets is used to introduce the Fundamental Principle. Probability theory is considered after a short discussion of sample spaces and real set functions. Only finite sample spaces are considered.

In response to comments by users of the first edition, certain obviously meager discussions have been extended, and some sections of the book have been rewritten entirely for the sake of clarity.

Certain problems in the exercises call for proofs omitted in the text. Answers to odd-numbered problems are given in the answer section at the end of the book. If a problem (whether odd- or even-numbered) contains lettered parts, the answers to parts (*a*), (*c*), (*e*), etc., appear in the answer section.

Color has been used to highlight important points in the text and to clarify the illustrations. The marginal notes, placed to direct the student's attention to key definitions and discussions, have been arranged to aid him in reviewing the course material.

It is hoped that most of the errors occurring in the first printings of the first edition have been corrected and that the present book is relatively free of mistakes.

I would like to express my sincere thanks to Professor Robert W. Gibson, Department of Civil Engineering, Auburn University, for his many valuable criticisms and suggestions during the preparation of the manuscript.

J. VINCENT ROBISON

MODERN ALGEBRA
AND TRIGONOMETRY

1

SETS AND OPERATIONS

The theory of sets has probably had as great an impact on mathematics as any other idea of the past century. Set theory now permeates almost all of mathematics and provides the language and symbolism for constructing modern mathematics.

We shall begin with the fundamental idea of a set. In this and later chapters, we shall learn how to combine sets, how to describe relations between sets and between the elements in a set, how to construct new concepts which become tools for acquiring additional knowledge of mathematics. In particular, we hope to show how some of the traditional notions of algebra and trigonometry can be restated in this relatively new language of set theory.

1.1 CONCEPT OF SET

Sets

A basic notion of mathematics is the concept of *set*. We shall consider a set to be a collection of distinct things. The things may be physical objects or they may be mental concepts. Thus, we may speak of the set of chairs in the classroom; or the set consisting of the counting numbers 1, 2, 3, . . . , 20; or the set of all points on the line between the distinct points P and Q.

Integers

One of the most familiar sets is the set of *positive integers* 1, 2, 3, . . . used in counting. Another is the set of *integers* which consist of the positive integers and the negative integers . . . , -3, -2, -1, and *zero* (0). A third

Rational numbers

familiar set is the set of *rational numbers* which may be represented by fractions of the form p/q, where p and q are integers and q is not zero. A fourth

Real numbers

set is the set of *real numbers*. For the present, we shall consider real numbers as numbers that can represent distances measured on a line. The set of real numbers includes the set of rational numbers and the set of *irrational numbers*. Examples of irrational numbers are $\sqrt{2}$, $\sqrt{3}$, and π. We recall

1

from previous courses in mathematics that $\sqrt{2}$ represents the number of units in the length of the hypotenuse of a right triangle whose legs are each 1 unit in length. The number π is the ratio of the circumference of a circle to its diameter.

Elements of a set

Every object (or mental concept) that belongs to a set is called an *element* of the set. We may describe a set by stating the properties that every element of the set must possess and that no other object possesses. We use capital letters for the names of sets and lower-case letters for the names of elements of sets. We indicate that an element a belongs to the set A by writing

$$a \in A$$

and read it as "a is an element (or member) of the set A." If a does not belong to the set A, we write

$$a \notin A$$

and read "a is not an element of set A."

Describing sets

We describe sets in either of two ways. If the set has a limited number of elements, we may list the elements by name and enclose the list within braces { }. Thus, if the set B consists of all the positive integers less than 25 that are integer multiples of 4, we write

$$B = \{4,8,12,16,20,24\} \qquad \text{or} \qquad B = \{4,8,12, \ldots ,24\}$$

where the three dots indicate that some of the elements are not written but are understood to be in the set. We may also describe a set by enclosing within braces a description of the distinguishing properties of the elements of the set. For example, to describe the set A of all positive integers less than 6, we may write

$$A = \{\text{all positive integers less than 6}\}$$
or
$$A = \{x \mid x < 6, x \text{ a positive integer}\}$$

We read the first of these statements as "A is the set of all positive integers less than 6," and we read the second as "A is the set of all numbers x such that x is less than 6 and x is a positive integer." The bar ($|$) as used here is read "such that" and the symbol $<$ is read "is less than." We understand that the role of the letter x is that of a symbol which stands for an unspecified element of the set $\{1,2,3,4,5\}$. A colon (:) is sometimes used instead of the vertical bar for the words "such that."

EXAMPLE 1.   Use set notation and specify in two ways the set A of vowels in the English alphabet.

Solution. $A = \{a,e,i,o,u\}$
$A = \{x \mid x \text{ is a vowel in the English alphabet}\}$

1.2 ONE-TO-ONE CORRESPONDENCE

One-to-one correspondence

A second basic notion of mathematics is the concept of *one-to-one correspondence* between sets. Suppose $A = \{1,2,3\}$ and $B = \{a,b,c\}$. It is possible to pair each element of A with exactly one element of B in such fashion that each element of B is paired with exactly one element of A. Two such pairings can be displayed as follows:

$$
\begin{array}{ccc}
1 & \longleftrightarrow & a \\
2 & \longleftrightarrow & b \\
3 & \longleftrightarrow & c
\end{array}
\qquad
\begin{array}{ccc}
1 & \searrow\nwarrow & a \\
2 & \times & b \\
3 & \nearrow & c
\end{array}
$$

There are four other possible pairings. Can you find them? In each case, we say that sets A and B are in one-to-one correspondence.

> **DEFINITION.** A one-to-one correspondence between two sets A and B is a rule which pairs each element $a \in A$ with one (and only one) element $b \in B$ in such manner that each element of B is paired with exactly one element of A. **(1.1)**

To indicate that an element a_1 of set A is paired with the element b_1 of set B, and conversely, we use the symbol

$$a_1 \longleftrightarrow b_1$$

Equivalent sets

Two sets A and B are said to be *equivalent* if they can be put into one-to-one correspondence. For example,

$$A = \{1,2,7,9\} \qquad \text{and} \qquad B = \{a,p,b,q\}$$

are equivalent sets, and we write

$$A \longleftrightarrow B$$

We read this as "Set A is equivalent to set B" or as "Set A and set B can be put into one-to-one correspondence."

Identical sets

If every element of a set A is an element of the set B and every element of set B is an element of set A, then A and B are said to be *identical* sets. Thus the sets

$$A = \{5,7,14\} \qquad \text{and} \qquad B = \{7,14,5\}$$

are identical sets. We write

$$A = B$$

and read "Set A is identical to set B" or "Set A is the same as set B." An alternate definition of identical sets will be given in the next section.

1.3 SUBSETS

Subset

The set A is a *subset* of the set B if every element of A is also an element of B. To indicate that A is a subset of B we write

$$A \subseteq B$$

and read "A is contained in B" or "B contains A." Under this definition, every set is a subset of itself, that is, $A \subseteq A$. If

$$A = \{1,2,3\}$$
$$B = \{2,3,1\}$$
and
$$C = \{1,4,3,2,5\}$$

then A is a subset of B and B is a subset of A. Both A and B are subsets of C. Also, C is a subset of C.

A useful definition of equal sets can now be written concisely as follows:

DEFINITION. If A and B are sets, then $A = B$ if and only if $A \subseteq B$ and $B \subseteq A$. $\quad$ **(1.2)**

The expression "if and only if" occurs frequently in the statement of definitions and theorems. We use this expression to make two statements at the same time. For example, the preceding definition means two things:

If $A = B$, then $A \subseteq B$ and $B \subseteq A$.

If $A \subseteq B$ and $B \subseteq A$, then $A = B$.

Note that each of these statements is the converse of the other. Every "if and only if" statement can be written as two statements, one of which is the converse of the other.

Proper subset

The set A is a *proper subset* of the set B if every element of A is an element of B and B contains at least one element that is not an element of A. For example, if

$$A = \{1,2,3\} \qquad \text{and} \qquad B = \{1,2,3,4\}$$

then A is a proper subset of B. In this case, we write

$$A \subset B$$

DEFINITION. Set A is a proper subset of set B if $A \subseteq B$ and $A \neq B$. $\quad$ **(1.3)**

The symbol $\neq$ means "not identical," or "not equal."

1.4   EMPTY SET

When certain relations between sets are defined, conditions are frequently imposed upon the elements which we may later discover to be satisfied by *no* element. Hence, we would like to admit a very special set which we call the *empty* or *null* set. This set, denoted by $\varnothing$ or by $\{\ \}$, is the set which contains no elements. For example, the set of all positive integers between 8 and 9 is a set which contains no elements and is there-

$\varnothing = \{\ \}$

fore empty. Like most other sets, the empty set can be described in a variety of ways. Let $A = \{x \mid x + 3 = 2, x$ a positive integer$\}$. Since there is no positive integer x such that the sum of x and 3 equals 2, we conclude that A is the empty set, that is,

$$\{x \mid x + 3 = 2, x \text{ a positive integer}\} = \varnothing$$

Ø is a subset of every set

The empty set is a subset of every set. Since $\varnothing$ has no elements, it is certainly true that each element of $\varnothing$ is an element of any set A. Therefore,

$$\varnothing \subseteq A$$

It is important to note that if braces are used to denote the empty set, no symbol is enclosed within them. Thus, $\varnothing = \{\ \}$, but $\varnothing \neq \{\varnothing\}$.

EXERCISE SET 1.1

SPECIFY each of the following sets, using set notation, by listing the elements of the set enclosed within braces:

1. The positive integers (the counting numbers) less than 12.
2. The positive integers greater than 13 and less than 21.
3. The positive integers that are multiples of 4 and are less than 43.
4. The positive integers that are multiples of 4, are also multiples of 9, and are less than 100.
5. The rational fractions with numerator 3 and denominator a positive integer less than 9.

SPECIFY each of the following sets by enclosing within braces a description of the characteristic properties of the elements of the set:

6. $A = \{1,3,5,7,9\}$ 7. $B = \{3,6,9,12\}$ 8. $C = \{a,e,i,o,u\}$ 9. $D = \{a, an, the\}$

TELL which of the following are true, given the set $E = \{1,3,5,7,9\}$:

10. $7 \in E$ 11. $8 \in E$ 12. $9 \notin E$

TELL which of the following are true, given that $F = \{1,4,7,10\}$ and $G = \{1,4,7\}$:

13. $G \subset F$ 14. $F \subset G$
15. The set $\{1,7\}$ is a proper subset of F.   16. The set $\{1,4,7\}$ is a proper subset of G.
17. List three subsets of the set $\{a,b\}$. (Remember that the empty set $\varnothing$ is a subset of every set.)
18. List all subsets of the set $\{a,b,c\}$.   19. List all subsets of the set $\{1,2,3,4\}$.

ANSWER the following, given $A = \{a,b,c\}$, $B = \{1,2,3\}$, $C = \{b,c,a\}$, $D = \{3,2,1\}$:

20. Is $A = C$?   21. Is $A = B$?   22. Is $A \longleftrightarrow B$?   23. Is $B = D$?
24. Given that $A = \{1,4,9,15\}$ and $B = \{15,1,x,9\}$ and $A = B$, find x.

A *finite* set is defined to be the empty set or a set that can be put into one-to-one correspondence with the set of all positive integers less than some positive integer N. An *infinite* set is a set that is not finite. Thus, an infinite set has an "unlimited" number of elements.

25. Which of the following sets are finite? Infinite?

    (a) The set of all positive integers greater than 100.
    (b) The set of all positive integers less than 4,000,000.

(c) The set of all positive odd integers less than 348.
(d) The set of all former presidents of the United States.
(e) The set of all persons born after Dec. 31, 1960, and before Jan. 1, 1966.
(f) The set of all points on a line between the distinct points A and B.
(g) The set of all positive integers between 8 and 9.

1.5   UNIVERSAL SETS.   OPERATIONS

When each of the sets A, B, C, ... considered in a particular discussion is a subset of some particular set U, then U is called the *universal set*, or simply the *universe*.   The universe in any discussion is not necessarily unique.   For example, if the elements of the sets A, B, and C are positive integers less than 20, we can consider $U = \{1,2,3,\dots,19\}$ to be the universe.   But this would not rule out the set $\{1,2,3,\dots,35\}$ as a universal set, or any other set that contains A, B, and C as subsets.

Universe

Generally, the number of elements in the universe is not significant so long as the universal set is specified (or understood) and contains as a subset every set under consideration.   The concept of the universal set is necessary to the study of the algebra of sets.

The algebra of sets is similar in some respects to the algebra of numbers.   Recall that in the algebra of numbers we perform the operations of addition, subtraction, multiplication, and division on a pair of numbers to produce a single number.   We also perform the operations of raising to powers and extracting roots on a single number to produce a single number.

Similarly, in the algebra of sets we perform certain operations on a pair of sets to produce a single set.   We also perform an appropriate operation on a single set to produce a single set.

Operations on sets

A *binary operation* on the collection of all subsets of a universal set U is a rule that assigns a unique subset of U to any pair of subsets A and B of U, taken in the order A first and B second.   A *unary operation* on the collection of all subsets of U is a rule that assigns a unique subset of U to any subset A of U.   Thus, a binary set operation is a method of constructing a new set from two given sets, and a unary operation is a method of constructing a new set (or perhaps the same set) from one given set.

The next two sections are devoted to a discussion of these basic set operations.

1.6   INTERSECTION AND UNION OF SETS

New sets may be constructed from two or more given sets.   We now consider two basic sets that may be constructed from any two given sets A and B.

Intersection of sets

The *intersection* of the two sets A and B is the set of all elements that belong to both A and B. The intersection of A and B is denoted by $A \cap B$. Thus,

$$A \cap B = \{x \mid x \in A \text{ and } x \in B\} \qquad (1.4)$$

We read the symbol $A \cap B$ as "the intersection of A and B."

To help us visualize operations with sets, we often make use of what are called Venn diagrams. In a Venn diagram we let the elements of a set be represented by the points in a plane region. In Fig. 1.1 the shaded area represents the intersection of sets A and B.

EXAMPLE 1. If $A = \{9,10,11,12,13,14\}$ and $B = \{10,12,14,16,18\}$, find $A \cap B$.

Solution. $A \cap B = \{10,12,14\}$

The second basic set which may be constructed from any two sets is defined as follows:

Union of sets

The *union* of the sets A and B is the set of all elements which belong to A or to B (or to both). The union of A and B is denoted by $A \cup B$. Thus

$$A \cup B = \{x \mid x \in A, \text{ or } x \in B\} \qquad (1.5)$$

If $x \in A$, then $x \in A \cup B$, and if $x \in B$, then $x \in A \cup B$. That is, $x \in A \cup B$ if x is an element of A, or of B, or of both A and B.

The symbol $A \cup B$ is read "the union of A and B." In Fig. 1.2, the shaded area in each of the four Venn diagrams shown represents the union of the sets A and B in each case.

EXAMPLE 2. If $A = \{2,3,4,7\}$
and $\qquad\qquad B = \{4,7,8,11\}$
then $\qquad A \cup B = \{2,3,4,7,8,11\}$

Note that $A \cup B$ includes elements 4 and 7 only once. In general, the same symbol is not used twice or more in a given set notation. Thus "7" denotes just one number, and the set $\{7,7,7,8,8\}$ contains only two numbers.

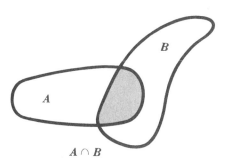

$A \cap B$

FIGURE 1.1
The shaded area represents $A \cap B$.

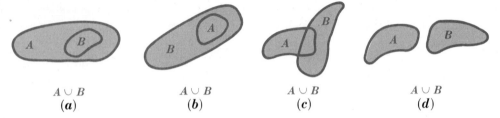

$A \cup B$
(a)
$A \cup B$
(b)
$A \cup B$
(c)
$A \cup B$
(d)

FIGURE 1.2   In each case the shaded area represents $A \cup B$.

1.7   COMPLEMENT OF A SET

The *complement* of a set A in a given universe U is the set of all elements that are not in A but are in U.   We denote the complement of the set A by A':

$$A' = \{x \mid x \notin A \text{ but } x \in U\} \tag{1.6}$$

For example, if $U = \{1,2,3,\ldots,9\}$ and if $A = \{1,2,3,\ldots,6\}$, then the complement of A is given by

$$A' = \{7,8,9\}$$

The Venn diagram, Fig. 1.3, will help in visualizing the complement of a set. The rectangle represents the universe U, and the shaded area represents the complement of set A.   The complement of a set A with respect to a given universe U is frequently denoted by $U - A$.   Thus,

$$A' = U - A$$

1.8   DISJOINT SETS

Two sets are *disjoint* (or mutually exclusive) if they have no elements in common.   We can state this definition precisely as follows:

DEFINITION.   Two sets A and B are disjoint if and only if $A \cap B = \varnothing$. $\tag{1.7}$

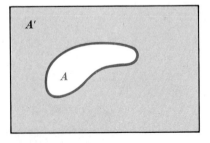

FIGURE 1.3
The shaded area represents the
complement of A.

For example, if $A = \{1,2,3\}$ and $B = \{4,5\}$, then A and B have no common element and therefore are disjoint; that is,

$$A \cap B = \emptyset$$

EXERCISE SET 1.2

1. Let $U = \{a,b,c,d,e,f,g,h,i,j\}$ be a universe and let $A = \{a,b,c,d\}$, $B = \{c,e,j\}$, and $C = \{a,d,f,g,h,i\}$. Enclose within braces the elements in each of the following sets:

   (a) $A \cap B$ — $\{c\}$ (b) $B \cap C$ (c) $C \cap A$ (d) $A \cup B$
   (e) $B \cup C$ $\{c,e,j,a,\}$ (f) $C \cup A$ (g) $B \cup \emptyset$ (h) $B \cap \emptyset$
   (i) $A \cap U$ $d,f,g,h,i\}$ (j) $A \cup U$ (k) $B \cap U$ (l) $C \cup (B \cap C)$
   (m) A' $\{e,f,g,h,i,j\}$ (n) C' (o) $(B \cup C)'$ (p) $(A \cap C)'$
   (q) $A \cap (B \cup C)'$ (r) $(U \cup \emptyset)'$

2. Let $U = \{0,1,2,3,4,5,6,7,8,9\}$, $A = \{0,1,2,3\}$, $B = \{2,4,6,8\}$, and $C = \{5,6,7,8\}$. Use set notation and write the elements in each of the following sets:

   (a) $B \cup C$ (b) $A \cap C$ (c) $A \cup \emptyset$ (d) $A \cap \emptyset$
   (e) $C \cup U$ (f) $C \cap U$ (g) A' (h) B'
   (i) C' (j) $(C \cup B)'$ (k) $(A \cap C)'$ (l) $(U \cap \emptyset)'$
   (m) $C \cap (A \cup B)$ (n) $(C \cap A) \cup B$ (o) $A \cup (B \cap C)$

3. Given $U = \{p,q,r,s,t,u,v,w\}$, $A = \{q,r,t,u\}$, $B = \{p,q,s,u\}$, and $C = \{t,u,v,w\}$:

   (a) Find the sets $A \cap (B \cup C)$ and $(A \cap B) \cup (A \cap C)$. Are they equal?
   (b) Find the sets $A \cup (B \cap C)$ and $(A \cup B) \cap (A \cup C)$. Are they equal?
   (c) Find $(A \cup B)'$ and $A' \cap B'$. Are these sets identical?
   (d) Find $(A \cap B)'$ and $A' \cup B'$. What can you say about the two sets?

4. Sketch Venn diagrams to illustrate each of the following statements:

   (a) $A \cap B = B \cap A$ (b) $A \cup (B \cap C) = (A \cup B) \cap (A \cup C)$
   (c) $A \cap (B \cap C) = (A \cap B) \cap C$ (d) $A \cap (B \cup C) = (A \cap B) \cup (A \cap C)$
   (e) $(A \cap B)' = A' \cup B'$ (f) $(A \cup B)' = A' \cap B'$

5. Sketch Venn diagrams to illustrate the following statements:

   (a) $(A')' = A$.   (b) $B \subseteq A$ if and only if $A' \subseteq B'$.
   (c) $A \cap B = B$ if and only if $B \subseteq A$.   (d) $A \cup B = A$ if and only if $B \subseteq A$.

1.9 SOLUTION SETS

Open sentence

An incomplete sentence such as "_____ is the capital of Oklahoma" is called an *open sentence*. It is neither true nor false as it stands. If we fill in the blank with the words "Oklahoma City," the statement is true. If we replace the blank with the word "Tulsa," the statement is false.

Open sentences in mathematics are generally in the form of equations or inequalities. For example, the statement

$$x + 3 = 8$$

Equation

is an open sentence called an *equation,* and the open sentence

$$x + 3 < 8$$

Inequality

is an *inequality.* In each statement, the letter x is a symbol that stands for an unspecified number. Replacing x by a numeral makes the statement

Variable

either true or false. Such a symbol is called a *variable.* The set of all eligible

Domain

replacements for a variable in an equation or inequality is called the *domain* of the variable.

Solution set

The *truth set,* or *solution set,* of an equation or inequality is the set of all replacements from the domain of the variable which make the open sentence a true statement. Therefore, before we can determine the solution set, we must know what the domain of the variable is; that is, we must know what replacements of the variable are eligible. For example, if the domain of x is the set of positive integers, the solution set of the equation

$$x + 5 = 8$$

is the set $\{3\}$, but the solution set of the equation

$$x + 5 = 3$$

is the empty set $\varnothing$. There is no positive integer x such that the sum of x and 5 equals 3.

Consider the equation

$$2x^2 - 5x + 3 = 0$$

If the domain of x is the set of positive integers, the solution set of the equation is the set $\{1\}$. On the other hand, if we take the domain of x to be the set of rational numbers, the solution set is the set $\{1, \frac{3}{2}\}$. Each element in this set makes the equation a true statement.

Open sentences may involve more than one variable. For example, the equation $y = 3x + 6$ and the inequality $z < 2x + y$ involve two and three variables respectively. The domains of the variables in such open sentences may be different. However, we will always consider them the same unless otherwise instructed.

1.10   SOME ALGEBRA OF SETS

We now consider certain properties of the set operations of union, inter-section, and complementation. In this discussion, let the universal set be denoted by U, and the empty set by $\varnothing$. Let A, B, C, ... be subsets of U. Then each of the following theorems can be proved by applying the basic definitions. Each theorem can be illustrated by a Venn diagram, but a Venn diagram is not a proof.

For compactness and because of the obvious relationships between the two theorems in each row, we state these theorems in parallel columns.

1. $U' = \emptyset$ $\emptyset' = U$

2. $(A')' = A$

3. $A \cup A = A$ $A \cap A = A$

4. $A \cup A' = U$ $A \cap A' = \emptyset$

5. $A \cup \emptyset = A$ $A \cap \emptyset = \emptyset$

6. $A \cup U = U$ $A \cap U = A$

Commutativity   7. $A \cup B = B \cup A$ $A \cap B = B \cap A$

Associativity   8. $A \cup (B \cup C)$ $A \cap (B \cap C)$

                     $= (A \cup B) \cup C$ $= (A \cap B) \cap C$

Distributivity   9. $A \cup (B \cap C)$ $A \cap (B \cup C)$

                     $= (A \cup B) \cap (A \cup C)$ $= (A \cap B) \cup (A \cap C)$

10. $(A \cup B)' = A' \cap B'$ $(A \cap B)' = A' \cup B'$

Theorems 7, 8, and 9 are called, respectively, the *commutative, associa-tive*, and *distributive* laws for the set operations $\cup$ and $\cap$.   A proof of the commutative property of the union of sets (the first part of Theorem 7) might appear as follows:

If $x \in A \cup B$, then $x \in A$ or $x \in B$, by the definition of union of sets. This means, of course, that $x \in B$ or $x \in A$.   Hence,

$$x \in B \cup A \qquad \text{by definition of } \cup$$

Therefore $A \cup B \subseteq B \cup A$ by definition of subset

We next show that $B \cup A$ is a subset of $A \cup B$.

If $y \in B \cup A$, then

	$y \in B$ or $y \in A$		why?
This means that	$y \in A$ or $y \in B$		why?
Hence	$y \in A \cup B$		why?
Therefore	$B \cup A \subseteq A \cup B$		why?
Since	$A \cup B \subseteq B \cup A$		
and	$B \cup A \subseteq A \cup B$		
we know that	$A \cup B = B \cup A$		why?

The associative and distributive laws can be proved in a similar manner. The student is urged to write proofs for all of these theorems.

1.11   SETS OF ORDERED PAIRS.   PRODUCT SETS

Let us denote by (x,y) a pair of numbers x and y, considered in the order

Coordinates of ordered pairs x first and y second.   We call x and y the *coordinates* of the ordered pair (x,y).

Any nonempty set of numbers can be made to yield a set of ordered pairs.   For example, let us form the set of all ordered pairs whose coordinates belong to the set

$$U = \{0,1,2,3\}$$

We proceed as follows:

1. Form all possible ordered pairs with the first coordinate 0:
   (0,0), (0,1), (0,2), (0,3)
2. Form all possible ordered pairs with first coordinate 1:
   (1,0), (1,1), (1,2), (1,3)
3. Form all possible ordered pairs with first coordinate 2:
   (2,0), (2,1), (2,2), (2,3)
4. Form all possible ordered pairs with first coordinate 3:
   (3,0), (3,1), (3,2), (3,3)

The set of all ordered pairs found in steps 1 to 4 above will be denoted by the symbol $U \times U$, which we read as "U cross U." Thus

$$U \times U = \{(0,0),(0,1), \ldots ,(3,2),(3,3)\}$$

Cartesian set

We call $U \times U$ the *cartesian set* of U, or the *product set* of U. If U is any given universe, then

$$U \times U = \{(x,y) \mid x \in U \text{ and } y \in U \} \qquad (1.8)$$

If X and Y are sets, the product set of X and Y is the set of all ordered pairs (x,y) such that $x \in X$ and $y \in Y$. We denote this product set by $X \times Y$ and write

$$X \times Y = \{(x,y) \mid x \in X \text{ and } y \in Y \} \qquad (1.9)$$

EXAMPLE 1. Given that $X = \{p,q,r\}$, $Y = \{1,2\}$, find $X \times Y$.

Solution. $X \times Y = \{(p,1),(p,2),(q,1),(q,2),(r,1),(r,2)\}$

EXAMPLE 2. If $A = \{1,2,3,4\}$ and $B = \{1,2\}$, find $A \times B$.

Solution. $A \times B = \{(1,1),(1,2),(2,1),(2,2),(3,1),(3,2),(4,1),(4,2)\}$

For a nonmathematical example of the cartesian product of two sets, let $B = \{\text{boys at a party}\}$ and $G = \{\text{girls at the same party}\}$. Then $C = B \times G = \{\text{possible couples at the party}\}$.

We shall make use of cartesian products of sets in Chap. 3 when we begin a study of one of the most important concepts in mathematics, the concept of a function. Cartesian sets form the basis for all plane graphing. Our study of permutations and combinations will also be based on this notion.

1.12 CARDINAL NUMBERS OF FINITE SETS

Let the sets A and B be equivalent finite sets. Then A can be put into one-to-one correspondence with B. We describe this property of A and B by saying that the sets A and B have the same *cardinal number*.

Cardinal number

All sets equivalent to the set $\{a\}$ have the same cardinal number, which we call *one*. All sets that can be put into one-to-one correspondence with the set $\{a,b\}$ have the same cardinal number, which we call *two*. All sets equivalent to the set $\{a,b,c\}$ have the cardinal number *three*, etc. We denote the cardinal numbers *one, two, three, four*, . . . by the symbols 1, 2, 3, 4, ...

DEFINITION. The cardinal number of a finite set is n if the set is equivalent to $\{1,2,3, \ldots ,n\}$. The empty set is said to have cardinality 0. **(1.10)**

Thus, the set $\{\triangle,\square,\bigcirc\} \longleftrightarrow \{1,2,3\}$ and has cardinal number 3. The set $\{8,13,17,21,32\} \longleftrightarrow \{1,2,3,4,5\}$ and has cardinal number 5.

Let A be a set whose cardinal number is x, and let B be a set of cardinal number y. If A and B are disjoint, we define the *sum* of the cardinal numbers x and y, denoted by $x + y$, to be the cardinal number of the set $A \cup B$.

Sum of cardinal numbers

EXAMPLE 1. The sets $\{a,b,c\}$ and $\{k,m,n,p\}$ have cardinal numbers 3 and 4 respectively. Find the sum of 3 and 4.

Solution. $3 + 4 =$ the cardinal number of the set $\{a,b,c\} \cup \{k,m,n,p\}$
$=$ the cardinal number of $\{a,b,c,k,m,n,p\}$
$= 7$

Product of cardinal numbers

We define the *product* of x and y, denoted by xy, to be the cardinal number of the set $A \times B$.

EXAMPLE 2. The set $\{a,b\}$ has cardinal number 2, and the set $\{r,s,t\}$ has cardinal number 3. Find the product 2(3).

Solution. $2(3) =$ cardinal number of $\{a,b\} \times \{r,s,t\}$
$=$ cardinal number of $\{(a,r),(a,s),(a,t),(b,r),(b,s),(b,t)\}$
$= 6$

We represent the cardinal number of the finite set A by the symbol $n(A)$. For example, since the cardinal number of the set $A = \{a,b,c,d\}$ is 4, we write in this case $n(A) = 4$. If $S = \{h,i,j,k,m,n\}$, then $n(S) = 6$, and so on.

The meaning of the important relation "is less than" applied to cardinal numbers of finite sets is explained as follows.

Suppose that x is the cardinal number of a finite set A and y is the cardinal number of a finite set B. We say that x is less than y, written

$$x < y$$

if and only if set A is equivalent to some proper subset of set B. Note that this definition applies to finite sets only.

We often use the statement "y is greater than x," written $y > x$, to mean the same thing as "x is less than y." Therefore,

$$x < y$$
and
$$y > x$$

mean the same thing and will be used interchangeably.

EXAMPLE 3. Given that $S = \{3,6,9,12\}$ and $T = \{a,b,c\}$ so that $n(S) = 4$ and $n(T) = 3$, show that $3 < 4$.

Solution. Since T is equivalent to a proper subset of S, say $\{3,6,9\}$, it follows from the definition that $3 < 4$.

To show that our definition of "is less than" does not hold for infinite sets, consider the infinite set $P = \{1,2,3, \ldots ,n, \ldots\}$ of positive integers and the infinite set $E = \{2,4,6, \ldots ,2n, \ldots\}$ of even positive integers. It is clear that E is a proper subset of P. The cardinal number of E, however, is not less than the cardinal number of P.

EXERCISE SET 1.3

1. Let $A = \{w,x\}$, $B = \{y,z\}$. Find (a) the set of all ordered pairs of $A \times B$, (b) the set of all ordered pairs of $B \times A$. Are these sets identical or equal?
2. Let $A = \{1,2,3\}$, $B = \{a,b\}$. Find (a) the set of all ordered pairs of $A \times B$, then (b) the set of all ordered pairs of $B \times A$. Is $A \times B = B \times A$ in this case?
3. Given $P = \{1,2,3\}$, $Q = \{1,2,3\}$. Find the set of all ordered pairs of $P \times Q$.
4. Given $X = \{x_1,x_2\}$, $Y = \{y_1,y_2,y_3\}$. Find the set of all ordered pairs of $X \times Y$.
5. If $R = \{3,5,7\}$, find $R \times R$.   6.   If $S = \{4,6,8\}$, find $S \times S$.
7. Find the sum of the cardinal numbers of the sets $\{1,2\}$ and $\{7,8,9,13\}$ by finding the cardinal number of their union.
8. Find the product of the cardinal numbers of the sets $\{3,7\}$ and $\{4,9\}$ by finding the cardinal number of their cartesian set.

We represent the cardinal number of a finite set A by the symbol $n(A)$. For example, if $A = \{a,b,c,d\}$, then $n(A) = 4$.

9. If $A = \{e,f,g,h\}$ and $B = \{m,n,o\}$, find $n(A \cup B)$.
10. If $R = \{1,2,3,4,5\}$ and $S = \{3,4,5,6,7\}$, find

(a) $n(R)$ (b) $n(S)$ (c) $n(R \cup S)$
(d) $n(R \cap S)$ (e) $n(R) + n(S) - n(R \cap S)$

11. If A and B are finite sets, show by using a Venn diagram that

$$n(A \cup B) = n(A) + n(B) - n(A \cap B)$$

12. If $P = \{0,1,2\}$, $Q = \{1,3\}$, find $P \times Q$ and $n(P \times Q)$.
13. If $T = \{4,8,12\}$, find $n(T \times T)$.

14. A certain company produces both tennis shoes and leather sandals. In a sample group of 588 persons the company found that each person used one or both products. If 384 used tennis shoes and 456 used leather sandals, how many used both? HINT: Let $A = \{$persons who used tennis shoes$\}$ and let $B = \{$persons who used sandals$\}$. Use the formula of Prob. 11 above and find $n(A \cap B)$.

15. At a dairy bar where many flavors of ice cream are sold it was found that of a group of 100 customers, 87 like vanilla or chocolate, 62 like vanilla, and 68 like chocolate. How many like both flavors?

16. Use a Venn diagram to show that if A, B, C are sets, then it is reasonable to conclude that

$$n(A \cup B \cup C) = n(A) + n(B) + n(C) - n(A \cap B)$$
$$- n(A \cap C) - n(B \cap C) + n(A \cap B \cap C)$$

17. Each of 400 families owns at least one of three makes of television sets and no family owns more than one of the same make. If 230 families own make A, 256 own make B, 240 own make C, and 360 families own at least two of the makes, how many families own all three makes of TV sets? HINT: Use the formula of Prob. 16 above and find $n(A \cap B \cap C)$.

2

THE REAL NUMBERS

The most important set considered in this book is the set of real numbers. We use these real numbers every day and are familiar with most of their properties, yet we might find it difficult to give a reasonably precise description of the system of real numbers. Our purpose now is to show how this system can be developed as a deductive system in which only a few of the properties of real numbers need to be taken as assumptions (or axioms). As we progress through this chapter, we will consider the meaning of the words *field*, *ordered*, and *complete*. When we have done that, we will describe the real numbers as a set of elements that form a complete, ordered field.

2.1 DEDUCTIVE SYSTEMS

In any deductive system, there are many technical terms, or symbols, used in stating the propositions of the system. Some of these terms are defined (described) by means of other terms whose meanings are assumed known. Those terms whose meanings are assumed are called *undefined elements* of the system. We shall take as undefined the term *real number*. We denote the set of all real numbers by $R = \{a,b,c, \ldots\}$. Certain relations among the undefined elements of the system and certain operations on them must be assumed at the beginning so that the propositions (theorems) of the system can be stated.

Also, in a deductive system, each proposition, after a beginning has been made, must be a logical consequence of the set of propositions and definitions which precedes it. Therefore, there must be one or more propositions at the beginning that cannot be deduced from preceding propositions. Those

Undefined elements

Axioms

Binary operation

Equality relation

propositions which are assumed to be true are called *axioms,* or *postulates.* In the next section we shall state the first six of the nine axioms for the set of real numbers R.

We define a *binary operation* on a set to be a rule which assigns to each pair of elements of the set (taken in a given order) a unique element of the set. The two basic binary operations on the set of real numbers are *addition,* denoted by the symbol $+$, and *multiplication,* denoted by the symbol $\times$. To each pair of real numbers a and b we assign unique real numbers x and y such that $a + b = x$ and $a \times b = y$. The number x is called the *sum* of a and b, and the number y is the *product* of a and b. The product of a and b is also denoted by $a \cdot b$, $a(b)$, $(a)b$, $(a)(b)$, and more frequently by ab.

The symbol $=$ (equals) is used to denote the relation of identity. Thus $a = b$ means that a and b are symbols standing for the same number. We assume that this familiar relation has the following properties for all elements a, b, c of the set R:

1. Equality is *reflexive:* $a = a$.
2. Equality is *symmetric:* If $a = b$, then $b = a$.
3. Equality is *transitive:* If $a = b$ and $b = c$, then $a = c$.

Finally, we shall assume that the equality relation satisfies the law of substitution:

4. Any quantity may be substituted for an equal quantity in any mathematical statement without changing the truth or falsity of the statement.

As a consequence of the substitution law we have that for all elements a, b, c, d of the set R:

5. Equality has the addition property: If $a = b$ and $c = d$, then $a + c = b + d$.
6. Equality has the multiplication property: If $a = b$ and $c = d$, then $ac = bd$.

2.2 FIELD AXIOMS FOR REAL NUMBERS

There are several methods of developing the real number system. In this book, we shall consider the real numbers first as a field, then as an ordered field, and finally as a complete ordered field. To begin this development, we assume the real numbers to be a set of elements with two binary operations on the set satisfying Axioms 1 to 6 (of this chapter).

Set of real numbers

Let $R = \{a,b,c, \ldots\}$ be a set of elements called real numbers on which the two binary operations, addition and multiplication, are defined. Assume for the present that the only known properties of the elements of R are given by Axioms 1 to 6, called the *axioms of a field.* Any set of elements that satisfy these six axioms is said to form a *field.*

AXIOM 1. THE CLOSURE PROPERTIES. For each $a, b \in R$,

$$a + b \in R \qquad ab \in R$$

This axiom is essentially the assertion that two binary operations exist on the set R. Thus, the sum or product of any two real numbers is a real number.

AXIOM 2. THE COMMUTATIVE LAWS. For each $a, b \in R$,

$$a + b = b + a \qquad ab = ba$$

Axiom 2 implies that the sum or product of any two real numbers is not affected by the order in which they are added or multiplied.

AXIOM 3. THE ASSOCIATIVE LAWS. For each $a, b, c \in R$,

$$(a + b) + c = a + (b + c) \qquad (ab)c = a(bc)$$

Axiom 3 states that the sum or the product of any three real numbers is not affected by the way in which they are grouped for addition or multiplication. A consequence of Axioms 1 to 3 is that the elements of any finite subset of the real numbers can be added or multiplied in any order. For this reason, parentheses are often omitted from sums or products of three or more numbers.

AXIOM 4. THE DISTRIBUTIVE LAWS. For each $a, b, c \in R$,

$$a(b + c) = ab + ac \qquad (b + c)a = ba + ca$$

A consequence of Axiom 4 is that the product of any given real number times the sum of two or more real numbers can be found by multiplying each number of the sum by the given number and then adding the products thus found. For example, $2(3 + 7 + 2) = 2(3) + 2(7) + 2(2)$.

We use parentheses ( ) or brackets [ ] to group two or more numbers as a single quantity. We also use braces { } for this purpose when it is clear that we are not defining sets.

AXIOM 5. THE IDENTITY ELEMENTS. There exists a real number called zero, denoted by 0, such that for each $a \in R$

$$a + 0 = a$$

There exists a real number different from zero, called one, denoted by 1, such that for each $a \in R$

$$a(1) = a$$

The real number 0 is called the *identity element for addition* and the real number 1 is the *identity element for multiplication*. Since 0 and 1 are in R, we have, from Axiom 2, $a + 0 = 0 + a = a$ and $a(1) = (1)a = a$.

AXIOM 6. THE INVERSE ELEMENTS. For each $a \in R$, there exists a real number, called the additive inverse (or the negative) of a, and denoted by $-a$, such that

$$a + (-a) = 0$$

For each nonzero $a \in R$, there exists a real number called the multiplicative inverse (or reciprocal) of a, and denoted by $1/a$, such that

$$a\left(\frac{1}{a}\right) = 1$$

By the commutative axiom, $a + (-a) = (-a) + a = 0$. Hence, a is the additive inverse of $-a$. In a similar way, we may show that a is the multiplicative inverse of $1/a$. Note that the sum of any real number and its additive inverse is the identity for addition. The product of any nonzero real number and its multiplicative inverse is the identity for multiplication.

Uniqueness of identity elements

The identity elements 0 and 1 of Axiom 5 are unique. Proof of the statement is not difficult. For example, to show that 1 is unique, suppose there exists another real number (say x) in R, such that for every $b \in R$

$$b(x) = b$$

Then $1(x) = 1$ since $1 \in R$

Therefore, $(x)1 = 1$ by Axiom 2

But $(x)1 = x$ by Axiom 5

Hence, $x = 1$ why?

Proof that 0 is unique is left as an exercise.

Uniqueness of inverse elements

The inverse elements $-a$ and $1/a$ of Axiom 6 are also unique. To prove that any real number a has only one additive inverse, suppose there is another, say x. Then

$$a + x = 0$$

Now,
$$-a = (-a) + 0 \qquad \text{by the identity axiom}$$
$$= (-a) + (a + x) \qquad \text{since } a + x = 0$$
$$= [(-a) + a] + x \qquad \text{by the associative law}$$
$$= 0 + x \qquad \text{why?}$$
$$= x \qquad \text{why?}$$

The proof that each nonzero real number a has only one multiplicative inverse is left as an exercise.

2.3   SOME PROPERTIES OF REAL NUMBERS

We will now derive certain properties of the real numbers directly from the field axioms of the preceding section.   It is our hope that the proofs of this section (and following sections) will not only establish some of the important properties of the real numbers but will emphasize the nature of algebra as a deductive science.   The student should supply reasons wherever they are omitted.

THEOREM. $0 + 0 = 0$ (2.1)

Proof.   Since $a + 0 = a$ for each $a \in R$, take $a = 0$ and obtain $0 + 0 = 0$.

THEOREM.   For each $a \in R$,
$$a(0) = 0 \qquad \text{and} \qquad 0(a) = 0$$ (2.2)

Proof.   Let $a(0) = x$, then $x \in R$ and there exists $(-x) \in R$ such that $x + (-x) = 0$, by Axioms 1 and 6.   Then

$$a(0) = a(0) + 0 \qquad \text{Axiom 5}$$
$$= a(0) + [x + (-x)] \qquad \text{substitution}$$
$$= [a(0) + x] + (-x) \qquad \text{associative law}$$
$$= [a(0) + a(0)] + (-x) \qquad \text{substitution}$$
$$= a(0 + 0) + (-x) \qquad \text{distributive law}$$
$$= a(0) + (-x) \qquad \text{by Th. (2.1)}$$
$$= x + (-x) \qquad \text{substitution}$$
$$= 0 \qquad \text{Axiom 6}$$

By the commutative law and the preceding, it follows that

$$a(0) = 0(a) = 0$$

THEOREM. If $ab = 0$, then either $a = 0$ or $b = 0$.   (2.3)

Proof. If $a = 0$, the theorem is proved. If $a \neq 0$, then

$$\frac{1}{a} \in R \qquad\qquad \text{Axiom 6}$$

Since $ab = 0$, $\qquad \dfrac{1}{a} \cdot (ab) = \dfrac{1}{a} \cdot 0 = 0 \qquad$ substitution and Th. (2.2)

and $\qquad\qquad \left(\dfrac{1}{a} \cdot a\right) b = 0 \qquad\qquad\qquad$ Axiom 3

Therefore $\qquad\qquad\quad (1)b = 0 \qquad\qquad\qquad\qquad$ Axiom 6

and $\qquad\qquad\qquad\quad b = 0 \qquad\qquad\qquad\qquad$ why?

THEOREM. If $a, b, c \in R$ and if $a + c = b + c$, then $a = b$.   (2.4)

Proof. Since $c \in R$, $(-c)$ is in R. Then $\qquad\qquad\qquad$ Axiom 6

$$(a + c) + (-c) = (b + c) + (-c) \qquad\qquad \text{why?}$$

and $\quad a + [c + (-c)] = b + [c + (-c)] \qquad$ by the associative law

Hence $\qquad\qquad\quad a + 0 = b + 0 \qquad\qquad\qquad$ why?

and $\qquad\qquad\qquad\quad a = b \qquad\qquad\qquad\qquad$ Axiom 5

THEOREM. If $a, b, c \in R$, $c \neq 0$, and if $ac = bc$, then $a = b$.   (2.5)

The proof is left as an exercise.

So far we have used the minus sign $(-)$ preceding an element of R to denote the additive inverse of that element. For example, $(-b)$ denotes the inverse of b under the operation of addition. There is another use of the minus sign which comes as a result of the following definition:

DEFINITION. For each $a, b \in R$,

$$a - b = a + (-b)$$ (2.6)

We call $a - b$ "the difference of a and b" or "b subtracted from a." Thus, the minus sign as used here indicates the familiar operation of subtraction.

THEOREM. For each $a, b \in R$,

$$(-a)b = -(ab)$$ (2.7)

Proof. $(-a)b + ab = [(-a) + a]b$ by Axiom 4

$\qquad\qquad\quad = (0)b \qquad\qquad\qquad$ why?

$\qquad\qquad\quad = 0 \qquad\qquad\qquad$ by Th. (2.2)

Therefore, $(-a)b$ is the additive inverse of ab. Since $-(ab)$ is, by Axiom 6, the additive inverse of ab,

$$(-a)b = -(ab)$$

THEOREM. For each $a, b, c \in R$,

$$a(b - c) = ab - ac$$

(2.8)

If $b \neq 0$, the product $b(1/b) = 1$, by Axiom 6. We now define the product $a(1/b)$ as follows:

DEFINITION. For each $a, b \in R$, $b \neq 0$,

$$\frac{a}{b} = a\left(\frac{1}{b}\right)$$

(2.9)

We call a/b "the quotient of a divided by b" or "the fraction a divided by b." We call a the numerator and b the denominator of the fraction a/b.

If a and b are integers, the real number a/b is a *rational number*. The following properties of rational numbers can now be established:

THEOREM. $b\left(\dfrac{a}{b}\right) = a$ provided $b \neq 0$ (2.10)

THEOREM. $\dfrac{a}{a} = 1$ provided $a \neq 0$ (2.11)

THEOREM. $\dfrac{0}{a} = 0$ provided $a \neq 0$ (2.12)

THEOREM. $\dfrac{a}{1} = a$ (2.13)

Integers as rationals

Theorems (2.12) and (2.13) imply that if a is any integer, then a is the same as the rational number $a/1$. Thus, every integer is a rational number. If I is the set of all integers and Q is the set of all rational numbers, then I is a proper subset of Q, that is,

$$I \subset Q$$

2.4   SIGNS RESULTING FROM OPERATIONS

The following theorems lead directly to certain rules of signs which are necessary if we are to operate effectively with real numbers.

THEOREM.   For each $a \in R$,   **(2.14)**

$$-(-a) = a$$

Proof. $-(-a)$ is the additive inverse of $(-a)$ why?

$\qquad\qquad$ a is the additive inverse of $(-a)$ why?

Hence $-(-a) = a$, since the additive inverse of $(-a)$ is unique.

THEOREM.   For each $a, b \in R$,   **(2.15)**

$$-(a + b) = -a - b$$

THEOREM.   For each $a, b \in R$,   **(2.16)**

$$-(a - b) = -a + b$$

Proof. $-(a - b) = -[a + (-b)]$ $\qquad\qquad$ why?

$\qquad\qquad$ $= -a - [(-b)]$ $\qquad$ by Th. **(2.15)**

$\qquad\qquad$ $= -a + b$ $\qquad\quad$ by Th. **(2.14)**

THEOREM.   For each $b \in R$,   **(2.17)**

$$(-1)b = -b$$

THEOREM.   For each $a, b \in R$,   **(2.18)**

$$(-a)(-b) = ab$$

Proof. $(-a)(-b) = -[a(-b)]$ $\qquad\qquad$ by Th. **(2.7)**

$\qquad\qquad$ $= -[(-b)a]$ $\qquad\qquad$ why?

$\qquad\qquad$ $= -[-(ba)]$ $\qquad\qquad$ by Th. **(2.7)**

$\qquad\qquad$ $= -[-(ab)]$ $\qquad$ commutative law

$\qquad\qquad$ $= ab$ $\qquad\qquad\quad$ by Th. **(2.14)**

THEOREM.   For each $b \in R$, $b \neq 0$,   **(2.19)**

$$\frac{1}{-b} = -\frac{1}{b}$$

THEOREM.   For each $a, b \in R,\ b \neq 0$,

$$\frac{a}{-b} = -\frac{a}{b} \quad \text{and} \quad \frac{-a}{b} = -\frac{a}{b}$$ (2.20)

THEOREM.   For each $a, b \in R,\ b \neq 0$,

$$\frac{-a}{-b} = \frac{a}{b}$$ (2.21)

Proof. $\dfrac{-a}{-b} = -a\left(\dfrac{1}{-b}\right)$ why?

$\qquad\quad = -a\left(-\dfrac{1}{b}\right)$ by Th. (2.19)

$\qquad\quad = a\left(\dfrac{1}{b}\right)$ by Th. (2.18)

$\qquad\quad = \dfrac{a}{b}$ why?

2.5   FURTHER PROPERTIES OF REAL NUMBERS

The theorems of this section provide methods for operating with fractions.   We shall not prove all the theorems, but we encourage the student to satisfy himself that they are indeed valid.   In all the theorems we shall consider $a, b, c, \ldots$ to be elements of R.

THEOREM. $\dfrac{1}{a} \cdot \dfrac{1}{b} = \dfrac{1}{ab}$ for $a \neq 0,\ b \neq 0$ (2.22)

THEOREM. $\dfrac{a}{b} \cdot \dfrac{c}{d} = \dfrac{ac}{bd}$ $b \neq 0,\ d \neq 0$ (2.23)

Proof. $\dfrac{a}{b} \cdot \dfrac{c}{d} = \left(a \cdot \dfrac{1}{b}\right)\left(c \cdot \dfrac{1}{d}\right)$ why?

$\qquad\quad = (ac)\left(\dfrac{1}{b} \cdot \dfrac{1}{d}\right)$ why?

$\qquad\quad = (ac)\left(\dfrac{1}{bd}\right)$ by Th. (2.22)

$\qquad\quad = \dfrac{ac}{bd}$ by Def. (2.9)

Product of fractions   This theorem can be extended to show that the product of a finite number of fractions is a fraction whose numerator is the product of all the numerators and whose denominator is the product of all the denominators of the fractions.

THEOREM. $\dfrac{ac}{bc} = \dfrac{a}{b}$ provided $b \neq 0,\ c \neq 0$ (2.24)

Proof. $\dfrac{ac}{bc} = \dfrac{a}{b} \cdot \dfrac{c}{c} = \dfrac{a}{b} \left[c \left(\dfrac{1}{c} \right) \right] = \dfrac{a}{b} (1) = \dfrac{a}{b}$

The student should supply the reasons in the preceding proof. Theorem (2.24) is called the *fundamental principle of fractions*. It states that a real number which is a fraction is unchanged when numerator and denominator are divided by the same nonzero number.

Fundamental principle

THEOREM. $\dfrac{a}{p} + \dfrac{b}{p} + \dfrac{c}{p} + \cdots + \dfrac{m}{p} = \dfrac{a + b + c + \cdots + m}{p}$

provided $p \neq 0$ (2.25)

Proof.

$$\dfrac{a}{p} + \dfrac{b}{p} + \dfrac{c}{p} + \cdots + \dfrac{m}{p} = a \left(\dfrac{1}{p} \right) + b \left(\dfrac{1}{p} \right) + c \left(\dfrac{1}{p} \right) + \cdots + m \left(\dfrac{1}{p} \right)$$

Def. (2.9)

$$= (a + b + c + \cdots + m) \left(\dfrac{1}{p} \right)$$ Axiom 4

$$= \dfrac{a + b + c + \cdots + m}{p}$$ Def. (2.9)

Sum of fractions

Thus, the sum of any finite number of fractions having the same denominator is a fraction whose numerator is the sum of the numerators and whose denominator is the common denominator of all the fractions.

THEOREM. $\dfrac{a}{b} + \dfrac{c}{d} = \dfrac{ad + bc}{bd}$ $b,\ d \neq 0$ (2.26)

Proof. $\dfrac{a}{b} + \dfrac{c}{d} = \dfrac{ad}{bd} + \dfrac{cb}{db}$ Th. (2.24)

$$= \dfrac{ad}{bd} + \dfrac{bc}{bd}$$ why?

$$= \dfrac{ad + bc}{bd}$$ Th. (2.25)

This theorem is useful in finding the sum of two fractions having unlike denominators.

THEOREM. $\dfrac{a}{p} - \dfrac{b}{p} = \dfrac{a-b}{p}$, for $p \neq 0$ (2.27)

THEOREM. $\dfrac{a}{b} - \dfrac{c}{d} = \dfrac{ad-bc}{bd}$, for $b,\ d \neq 0$ (2.28)

THEOREM. $\dfrac{\frac{a}{b}}{\frac{c}{d}} = \dfrac{a}{b} \cdot \dfrac{d}{c}$, for $b,\ c,\ d \neq 0$ (2.29)

Proof. $\dfrac{\frac{a}{b}}{\frac{c}{d}} = \dfrac{a}{b} \cdot \dfrac{1}{\frac{c}{d}}$ why?

            $= \dfrac{a \cdot 1}{b\left(\frac{c}{d}\right)}$ why?

            $= \dfrac{a \cdot d}{b\left(\frac{c}{d}\right)d}$ Th. (2.24)

            $= \dfrac{ad}{bd\left(\frac{c}{d}\right)}$ Axiom 2

            $= \dfrac{ad}{bc}$ why?

            $= \dfrac{a}{b} \cdot \dfrac{d}{c}$ Th. (2.23)

    The following examples illustrate some of the applications of the theorems considered in this and preceding sections.

EXAMPLE 1. $(-12) + (-12) = -(12 + 12) = -24$

EXAMPLE 2. $(-12) + 15 = (-12) + (12 + 3)$
                              $= [(-12) + 12] + 3$
                              $= 0 + 3$
                              $= 3$

EXAMPLE 3. $12 - 15 = 12 + (-15) = 12 + [-(12 + 3)]$
                          $= 12 + [(-12) + (-3)]$
                          $= [12 + (-12)] + (-3)$
                          $= 0 + (-3)$
                          $= -3$

EXAMPLE 4. $(-5)(+3) = (+3)(-5) = -[(3)(5)] = -15$

EXAMPLE 5. $(-4)(-7) = (4)(7) = 28$

EXAMPLE 6. $\dfrac{3/4}{-3/2} = \dfrac{3}{4}\left(-\dfrac{2}{3}\right) = -\left[\dfrac{3}{4}\left(\dfrac{2}{3}\right)\right] = -\dfrac{1}{2}$

EXERCISE SET 2.1

STATE which of the six axioms for the set of real numbers are used in each of the following:

1. $5 + 3 = 3 + 5$ commutative
2. $4 + 7 = 7 + 4$
3. $9 + 0 = 9$ identity
4. $0 + 5 = 5$
5. $-3 + 0 = -3$ "
6. $4(1) = 4$
7. $[-3 + (4)] + 7 = -3 + (4 + 7)$ assoc.
8. $4 + [(-2) + 1] = (4 + 1) + (-2)$
9. $0 + 0 = 0$ identity
10. $1(1) = 1$
11. $(-1)(1) = -1$ identity
12. $4 + (-4) = 0$
13. $(-3) + [-(-3)] = 0$ inverse
14. $(1)(7) = 7$
15. $4(\tfrac{1}{4}) = 1$ inverse
16. $-3\left(\dfrac{1}{-3}\right) = 1$
17. $\dfrac{1}{-5}(-5) = 1$ inverse
18. $\dfrac{1}{7}\left(\dfrac{1}{\frac{1}{7}}\right) = 1$
19. $x + y = 1(x) + 1(y)$ assoc.
20. $xy = yx$
21. $2(3 + 4) = 2(3) + 2(4)$ distributive
22. $4(x + y) = 4 \cdot x + 4 \cdot y$

FIND the value of each of the following:

23. $(-2) + 5$
24. $(7) + (-4)$
25. $(-3) + (-8)$
26. $-(-9) + (9)$
27. $7 - 2$
28. $(8) - (4)$
29. $(-7) - (-4)$
30. $(-3) - (-5)$
31. $0 + (-3) + (-5)$
32. $18 - (-15) + (-1)$
33. $(2)(-4)$
34. $(-3)(5)$
35. $(-7)(-3)$
36. $(-2)(-15)$
37. $(-3)(-4) + [5(0)]$
38. $(-2)(-4) - [(-1)(-4)]$

FIND the additive inverse of each of the following:

39. 7
40. 0
41. -3
42. y
43. 8
44. $-x$
45. $\tfrac{3}{2}$
46. $\tfrac{3}{5}$
47. $3x + 2y$
48. $2x - 7y$
49. $11a - 9b$
50. $-(7a + b)$
51. $-(3x - 5y)$
52. $-(2a - 6b)$

FIND the multiplicative inverse of each of the following:

53. 3
54. $\tfrac{1}{3}$
55. $\tfrac{2}{3}$
56. $\tfrac{4}{9}$
57. $1 + \tfrac{1}{2}$
58. $3 + \tfrac{7}{8}$

59. $5 - \frac{1}{5}$ **60.** $x + y$ **61.** $(2a - 3b)$
62. $-(4x/3y)$ **63.** $3a/7b$ **64.** $-3/x$
65. $2/(x + y)$ **66.** 3.7 **67.** $1/1.7$
68. $2/(3 + 0.7a)$

FIND the value of each of the following:

69. $\frac{1}{2} + \frac{3}{2}$ **70.** $\frac{5}{4} + \frac{3}{4}$ **71.** $\frac{8}{7} + \frac{5}{3}$
72. $\frac{4}{11} - \frac{2}{11}$ **73.** $\frac{3}{7} \cdot \frac{2}{5}$ **74.** $\frac{5}{8} \cdot \frac{2}{9}$
75. $(-\frac{1}{3})(\frac{5}{6})$ **76.** $(\frac{2}{3})(-\frac{1}{5})$ **77.** $(-2)(-7)(-\frac{1}{3})$
78. $(-\frac{2}{5})(-\frac{1}{9})$

PROVE each of the following if $a, b, c, \ldots$ are real numbers:

79. $a(b - c) = ab - ac$ **80.** $-(a + b) = -a - b$
81. $-(ac - ad) = [d + (-c)]a$ **82.** $(a - b) + b = a$
83. $(a - b) - c = -[b - (a - c)]$ **84.** If $a + b = c$, then $a = c - b$.

85. If $\dfrac{a}{b} = \dfrac{c}{d}$, then $ad = bc$.   **86.** If $ad = bc$, then $\dfrac{a}{b} = \dfrac{c}{d}$ $b, d \neq 0$

87. $\dfrac{a}{b}\left(\dfrac{b}{a}\right) = 1$ **88.** $\dfrac{1}{\dfrac{a}{b}} = \dfrac{b}{a}$

2.6   POSITIVE-INTEGER EXPONENTS

We use a very compact notation for the product of n factors each of which is a.   This notation is based on the following definition:

DEFINITION.   If a is any number and n is a positive integer, then

$$a^1 = a \tag{2.30}$$
$$a^{n+1} = a^n a$$

For example, by Axiom 3 and the definition,

$$a^2 = (a^1)a = aa \quad \text{(two factors)}$$
$$a^3 = (a^2)a = (aa)a = aaa \quad \text{(three factors)}$$
$$a^4 = (a^3)a = (aaa)a = aaaa \quad \text{(four factors)}$$

Continued applications of the definition to a^5, a^6, a^7, and so on, lead intuitively to the following:

$$a^n = aaa \ldots a \quad \text{(n factors)} \tag{2.31}$$

Equation (2.31) is frequently taken as the definition of a^n.   We read the symbol a^n as "the nth power of a" or as "a raised to the nth power."   We call the number a the *base* and the positive integer n the *exponent* of the power of the base.   Thus, in the expression $(\frac{3}{2})^5$ the number $\frac{3}{2}$ is the base and 5 is the exponent of the fifth power of the base.

Exponent

Proofs of the following theorems are often based on Definition (2.30) and the principle of mathematical induction which we will not consider here. At this point, we base our proofs on Eq. (2.31) and the associative and distributive laws. In each theorem, a and b are any numbers, and m and n are positive integers.

THEOREM. $a^m a^n = a^{m+n}$ (2.32)

Proof. $a^m = aaa \ldots a$ (m factors)

$a^n = aaa \ldots a$ (n factors)

$$a^m a^n = \overbrace{(aaa \ldots a)}^{m \text{ factors}} \ \overbrace{(aaa \ldots a)}^{n \text{ factors}}$$

$$= \overbrace{aaa \ldots a}^{m + n \text{ factors}}$$

$$= a^{m+n}$$

EXAMPLES. $(3^2)(3^4) = 3^{2+4} = 3^6$

$(-2)^3(-2)^4 = (-2)^{3+4} = (-2)^7$

THEOREM. If $a \neq 0$, then

(1) $\dfrac{a^m}{a^n} = a^{m-n}$ if m is greater than n

(2) $= 1$ if $m = n$ (2.33)

(3) $= \dfrac{1}{a^{n-m}}$ if m is less than n

Proof of (1). $\dfrac{a^m}{a^n} = a^m \left(\dfrac{1}{a^n} \right)$

$$= a^{m-n} a^n \left(\dfrac{1}{a^n} \right) \qquad \text{by Th. (2.32)}$$

$$= a^{m-n} \left[a^n \left(\dfrac{1}{a^n} \right) \right] = a^{m-n} \qquad \text{by Axiom 3}$$

The proof of (2) is left as an exercise.

Proof of (3). $\dfrac{a^m}{a^n} = a^m \left(\dfrac{1}{a^n} \right) = a^m \left(\dfrac{1}{a^m a^{n-m}} \right)$

$$= a^m \left(\dfrac{1}{a^m} \dfrac{1}{a^{n-m}} \right) = \left[a^m \left(\dfrac{1}{a^m} \right) \right] \dfrac{1}{a^{n-m}}$$

$$= \dfrac{1}{a^{n-m}}$$

EXAMPLES. $\dfrac{3^5}{3^2} = 3^{5-2} = 3^3$

$\dfrac{5^4}{5^4} = 1$

$\dfrac{7^2}{7^5} = \dfrac{1}{7^{5-2}} = \dfrac{1}{7^3}$

Proofs of the following theorems are left as exercises:

THEOREM. $(ab)^m = a^m b^m$ (2.34)

THEOREM. $\left(\dfrac{a}{b}\right)^m = \dfrac{a^m}{b^m}$ (2.35)

THEOREM. $(a^m)^n = a^{mn}$ (2.36)

EXAMPLES. $(-4y)^2 = (-4)^2 y^2$

$\left(\dfrac{2}{3}\right)^5 = \dfrac{2^5}{3^5}$

$(7^2)^2 = 7^{2(2)} = 7^4$

$(2a^2)^3 = 2^3 a^6$

2.7 IRRATIONAL NUMBERS

We have defined a rational number to be a real number that can be expressed in the form a/b where a and b are integers and $b \neq 0$. Thus, $\frac{3}{5}$, $-\frac{2}{7}$, $\frac{8}{1} = 8$, and $7.32 = \frac{732}{100}$ are rational numbers. The set of all rational numbers is a subset of the set of all real numbers.

Not all real numbers are rational, however, because some of them are not equal to the quotient of two integers. We call this important subset of the real numbers the set of *irrational numbers*. An example of an irrational number is the number whose square is 2, denoted by $\sqrt{2}$. Later in this section we show that $\sqrt{2}$ cannot be expressed as the quotient of two integers and is therefore irrational.

We recall from arithmetic the meaning of square roots and cube roots of numbers. Thus, $\sqrt{3}$ is a number such that $(\sqrt{3})^2 = 3$, and $\sqrt[3]{4}$ is a number such that $(\sqrt[3]{4})^3 = 4$, etc. In some cases we found that such roots are rational numbers. For example, $\sqrt{\frac{1}{4}} = \frac{1}{2}$ and $\sqrt[3]{27} = 3$. In other cases we could give only an approximation in the form of a decimal or rational number. We found that the rational number $\frac{1414}{1000} = 1.414$ could be used as an approximation of $\sqrt{2}$ and $\frac{1732}{1000} = 1.732$ as an approximation of $\sqrt{3}$.

Irrational numbers

We now show that $\sqrt{2}$ is not a rational number. To establish the proof we need the following theorem:

THEOREM. For any integer p, if p^2 is divisible by 2, then p is divisible by 2.   (2.37)

Proof.   If the integer p is divided by 2, the remainder is either 0 or 1. Hence p can be expressed in one and only one of the following two forms:

$$p = 2k \qquad \text{or} \qquad p = 2k + 1 \qquad \text{where } k \text{ is an integer}$$

Accordingly, then,

$$p^2 = 4k^2 = 2(2k^2) \qquad \text{and } p^2 \text{ is divisible by 2}$$
$$\text{or } \; p^2 = 4k^2 + 4k + 1 = 2(2k^2 + 2k) + 1 \qquad \text{and } p^2 \text{ is not divisible by 2}$$

Since we are given that p^2 is divisible by 2, it follows that p^2 must equal $4k^2$.

Therefore $p = 2k$ and p is divisible by 2.

THEOREM. $\sqrt{2}$ is not a rational number.   (2.38)

Proof.   The proof consists in showing that there is no rational number whose square is 2.

Suppose p/q is a rational number such that $(p/q)^2 = 2$. Suppose also that p/q is in lowest terms; that is, p and q have no common divisors except 1 and -1. Then

$$\frac{p^2}{q^2} = 2 \qquad \text{and} \qquad p^2 = 2q^2$$

Hence, p^2 is divisible by 2 and thus p is divisible by 2 as a consequence of Theorem (2.37). Write $p = 2r$, $r \in I$, and then $p^2 = 4r^2$. Substituting $2q^2$ for p^2, we obtain

$$2q^2 = 4r^2 \qquad \text{and} \qquad q^2 = 2r^2$$

Hence, q^2 is divisible by 2, and thus q is divisible by 2 as a consequence of Theorem (2.37).

Therefore, p and q have a common divisor 2, contrary to the supposition that the only common divisors of p and q are 1 and -1. This proves the theorem.

It can be proved that if the positive integer n is not the square of some integer k, then $\sqrt{n}$ is not rational and is therefore irrational. For example, $\sqrt{7}$, $\sqrt{8}$, $\sqrt{10}$ are irrational numbers. We shall consider other irrational numbers in later chapters.

2.8   REAL NUMBER LINE

A one-to-one correspondence between the set of real numbers and the set of points on a straight line has to be assumed. We can make the assumption more plausible, however, by actually showing how such a correspondence can be set up between the set of rational numbers and certain points on the line.

Construction of real line

Take a horizontal straight line and choose a point on the line. Call this point the *origin*. Choose a convenient unit of length and lay off on the line equal segments in both directions from the origin as in Fig. 2.1.

Assign the number 0 (zero) to the origin, the number 1 to the first point marked off to the right of 0, the number 2 to the second point, and so on. Assign the number -1, which is the additive inverse of 1, to the first point marked off to the left of 0. Assign the number -2, the additive inverse of 2, to the second point to the left, and so on. We then have a one-to-one correspondence between the set of integers and the set of points thus far located on the line. The segment from 0 to 1 is called the *unit segment*. The segment from 0 to 2 has length 2 units. The segment from 0 to 3 has length 3 units, and so on.

Unit segment

The rational numbers other than the integers can be made to correspond to certain other points on the line by the geometric construction used to divide a segment into any number of equal parts. Thus, if the segment from 0 to 1 is divided into four equal parts, the rational number ¼ may be represented by the point which is one-fourth of the distance from 0 to 1. The points corresponding to the rational numbers ¾, ³⁄₂, $-$⅞, $-$⁵⁄₂ are located on the line in Fig. 2.1.

The points on the line that correspond to certain irrational numbers may also be found geometrically. For example, construct the square whose base is the segment from 0 to 1, and let OP be the diagonal (see Fig. 2.2).

By the theorem of Pythagoras, $OP = \sqrt{1^2 + 1^2} = \sqrt{2}$. With O as center and OP as radius, describe an arc cutting the line at A. The segment $OA = OP = \sqrt{2}$, and point A corresponds to the irrational number $\sqrt{2}$.

Real numbers and the number line

To ensure a one-to-one correspondence between the set of all real numbers and the set of points on the line, which we now call the *number line* (or *number scale*), we make the following assumption:

Every real number corresponds to a point on the number line, and every point on the number line corresponds to a real number.   **(2.39)**

FIGURE 2.1   The real number line

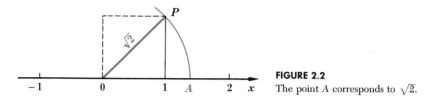

FIGURE 2.2
The point A corresponds to $\sqrt{2}$.

The real number x corresponding to a point on the number line is called

Graph of a number   the *coordinate* of the point, and the point is called the *graph* of the number. We refer to this point as the point (x) and adopt the convention that there is no verbal distinction between the points on the number line and the corresponding real numbers. Thus, we may speak of the *point* (3) as if it were the *number* 3, and we may speak of the number -5 as if it were the point (-5).

The graphic representation of the real numbers as points on a straight line may be used to separate the real numbers into three disjoint (mutually exclusive) subsets. The real numbers corresponding to points on the number

Positive numbers   scale to the right of 0 are called *positive numbers*. The additive inverses of the positive numbers correspond to points to the left of 0 and are called

Negative numbers   *negative numbers*. The real number 0 is neither positive nor negative.

2.9   THE ORDER AXIOMS

The six field axioms make no reference to the important order relation "less than." We have assumed, however, an intuitive understanding of this relation when we spoke of such things as "the set of all positive integers less than 6."

We use the following symbols in connection with the property of order for the real numbers:

$$< \qquad \text{"is less than"}$$
$$> \qquad \text{"is greater than"}$$

The precise meaning of these symbols is found in the following definition:

DEFINITION.   If a, $b \in R$, then $a < b$ if and only if $b - a$ is a positive number; $a > b$ if and only if $b < a$. **(2.40)**

Recall from Sec. 1.3 that here the expression "if and only if" means that if $b - a$ is positive, then $a < b$; conversely, if $a < b$, then $b - a$ is positive.

It can be proved that if $a < b$, then the point (a) lies to the left of point (b) on the number scale. Likewise, if $a > b$, which means $b < a$, then (b) lies to the left of (a). Thus, if a is positive, then $a > 0$, and if a is negative, then $a < 0$.

We now state the order axioms for the real numbers:

AXIOM 7. THE TRICHOTOMY AXIOM. If $a, b \in R$, then one and
only one of the following relations holds: **(2.41)**

$$a < b \qquad a = b \qquad a > b$$

AXIOM 8. CLOSURE FOR POSITIVE NUMBERS. If $a, b \in R$ and
$a > 0$ and $b > 0$, then   **(2.42)**

$$a + b > 0 \qquad ab > 0$$

Axiom 8 implies that the sum and the product of any two positive
numbers is positive.

THEOREM. If $a, b, c \in R$ and if $a < b$ and $b < c$, then $a < c$.   **(2.43)**

Proof. Since $a < b$ and $b < c$,

$$b - a > 0$$

and $\qquad\qquad\qquad\qquad\qquad\qquad c - b > 0 \qquad\qquad\qquad$ by definition

Hence, $\qquad\qquad\qquad (b - a) + (c - b) > 0 \qquad$ by the closure axiom

Then $\qquad\qquad\qquad\qquad\qquad c - a > 0 \qquad\qquad\qquad\qquad$ why?

Hence, $\qquad\qquad\qquad\qquad\qquad a < c \qquad\qquad\qquad\qquad$ by definition

THEOREM. If $a, b, c \in R$ and if $a < b$, then
$$a + c < b + c$$ **(2.44)**

The proof is left to the student.

THEOREM. If $a, b, c \in R$ and if $a > b$ and $c > 0$, then $ac > bc$.   **(2.45)**

The proof is left as an exercise.

THEOREM. If $a, b, c \in R$ and if $a > b$ and $c < 0$, then $ac < bc$.   **(2.46)**

The proof is left to the student.
The following examples illustrate uses of certain of these theorems:

EXAMPLE 1. Find the solution set of the inequality $6x - 3 > 21$, $x \in R$.

Solution. $(6x - 3) + 3 > 21 + 3$ by Th. (2.44)

and $6x > 24$ why?

therefore $x > 4$ by Th. (2.45)

Thus, the solution set is $\{x \mid x > 4\}$.

EXAMPLE 2.   Solve the inequality $4 + 6x > 17 - 4x$.

Solution. $(4 + 6x) + (4x - 4) > (17 - 4x) + (4x - 4)$ by Th. (2.44)

and $10x > 13$ why?

Hence, $x > \dfrac{13}{10}$ why?

The solution set is the set $\{x \mid x > {}^{13}\!/_{10}\}$.

EXERCISE SET 2.2

1. Use Definition (2.40) and Axioms 7 and 8 to prove the following. Assume a, b, c are real numbers.

   (a) If $a > b$ and $c > 0$, then $ac > bc$.
   (b) If $a > b$, then $a + c > b + c$.
   (c) If $a > 0$, then $-a < 0$.
   (d) If $a \neq 0$, then $a^2 > 0$.
   (e) If $a > b$, then $a^2 > b^2$, $a > 0$, $b > 0$.

2. Prove that $\frac{1}{4} < \frac{2}{5}$.
3. Prove that the product of two negative numbers is a positive number.
4. Prove that the product of a negative number and a positive number is a negative number.

FIND the solution set for each of the following inequalities:

5. $2x + 3 > 4$ 6. $5x - 7 < 1 - 3x$
7. $6 - 2x < x - 3$ 8. $3 + 7x > 2x - 15$

2.10   THE COMPLETENESS PROPERTY

The order properties discussed in the preceding section are possessed by both rational and real numbers. The real numbers, however, possess the important property of "completeness," a property not possessed by the rationals. Our discussion of this completeness property will be informal and intuitive.

Consider, for example, the set of rational approximations for $\sqrt{2}$ which we can obtain by the ordinary process of finding the square root of a positive number. Call this set S. Then

$$S = \{1, 1.4, 1.41, 1.414, 1.4142, \ldots\}$$

This set of rational numbers has two important properties which we describe as follows:

Upper bound   1.   The set S has an *upper bound.* In other words, there exists a real
number b such that if $x \in S$, then $x \leq b$. The symbol $\leq$ is read "is less than
or equal to," and $x \leq b$ means $x < b$ or $x = b$. Thus $\sqrt{2}$ is an upper bound
of the set S, but 1.5, 2, $5\!/\!2$, 7, and so on, are also upper bounds.

Least upper bound   2.   The set S has a *least upper bound.* A least upper bound is an upper
bound which is less than every other upper bound. Intuitively, we see that
$\sqrt{2}$ is the least upper bound of the set S; that is, $\sqrt{2}$ is an upper bound and
is less than or equal to every other upper bound.

These two properties of the set S are possessed by every subset of the
real numbers that has an upper bound. We state this in the form of an axiom:

AXIOM 9.   THE COMPLETENESS PROPERTY. Every nonempty
subset of the real numbers that has an upper bound has a least   **(2.47)**
upper bound.

The rational numbers do not satisfy the completeness axiom. For example, it can be shown that the least upper bound of the set of all rational
numbers x such that $x^2 < 2$ is not a rational number but is the real number $\sqrt{2}$.

The following examples illustrate the completeness property of certain
sets:

1.   The finite set $\{-3,2,5,7\}$ has the least upper bound 7.
2.   The infinite set $\{0.3,0.33,0.333, \ldots\}$ has the least upper bound $1\!/\!3$.
3.   The set $\{x \mid x^2 < 9\}$ has the least upper bound 3.
4.   The set $\{x \mid x^2 < 3\}$ has the least upper bound $\sqrt{3}$.
5.   The infinite set $\{1\!/\!2,2\!/\!3,3\!/\!4,4\!/\!5, \ldots\}$ has the least upper bound 1.
6.   The infinite set $\{1.1,1.01,1.001, \ldots\}$ has least upper bound 1.1.
7.   The set $\{x \mid x = 2 - 2/n, \ n = 1, 2, 3, \ldots\}$ has least upper
bound 2.
8.   The set $\{x \mid x = (1.001)^n, \ n = 1, 2, 3, \ldots\}$ has no least upper
bound.

The real numbers satisfy the completeness axiom as well as Axioms
1 to 8, and for this reason they are said to form a *complete ordered field.*
The rational numbers obey Axioms 1 to 8 but lack the completeness
property. For example, the set of all rational numbers x such that $x^2 < 3$
does not have a rational number for its least upper bound. On the other
hand, the set of all real numbers x such that $x^2 < 3$ does have a real number for its least upper bound, namely, the real number $\sqrt{3}$. Thus Axiom 9
is necessary to give the real numbers a property not possessed by the
rationals. The set of rationals and the set of irrationals are each a proper
subset of the reals. The set of reals is the union of the set of rationals and
the set of irrationals.

2.11 ABSOLUTE VALUE AND INEQUALITIES

A consequence of the order properties of the real numbers is that if $a \in R$, $a \neq 0$, then one and only one of the real numbers a, $-a$ is positive. We use the symbol $|a|$, read "the absolute value of a," to denote this positive value of a or $-a$.

Absolute value

> **DEFINITION.** For each $a \in R$, the absolute value of a satisfies the following equations:
>
> $$|a| = a \qquad \text{if } a \text{ is positive} \qquad\qquad (2.48)$$
> $$|a| = -a \qquad \text{if } a \text{ is negative}$$
> $$|0| = 0$$

For example, $|23| = 23$, $|-36| = -(-36) = 36$.

We often combine the two statements $a < b$ and $b < c$ by writing $a < b < c$. For example, $-2 < x < 3$ is equivalent to the two inequalities $-2 < x$ and $x < 3$. An inequality involving an absolute value is frequently equivalent to two inequalities. For example, $|x| < 3$ is equivalent to the two inequalities $-3 < x$ and $x < 3$, which we combine into $-3 < x < 3$. In general, if $|x| < a$, $a > 0$, then

$$-a < x < a$$

To prove this, we reason as follows:
If $x > 0$, then $x = |x| < a$. Since $-a < 0$ and $0 < x$, it follows that $-a < x$. From $-a < x$ and $x < a$, we have in this case

$$-a < x < a$$

If $x < 0$, then $-x = |x| < a$. Hence, $x > -a$. Since $x < 0$ and $0 < a$, it follows that $x < a$. From $x > -a$ and $x < a$, we have

$$-a < x < a$$

Note that $-a < x < a$ requires that x satisfy both inequalities $-a < x$ and $x < a$. Hence, the solution set of the inequality $|x| < a$ is

$$\{x \mid -a < x\} \cap \{x \mid x < a\}$$

which can be written in a slightly more compact form as $\{x \mid -a < x < a\}$.

EXAMPLE 1. Given $|2x - 4| < 8$, express the inequality without using the absolute value symbol, and find the solution set.

Solution. $$|2x - 4| < 8$$

means $$-8 < 2x - 4 < 8 \qquad \text{why?}$$

Then $-8 + 4 < (2x - 4) + 4 < 8 + 4$ why?

and $-4 < 2x < 12$ why?

Hence, $-2 < x < 6$ why?

The solution is $\{x \mid -2 < x\} \cap \{x \mid x < 6\}$, which we write as $\{x \mid -2 < x < 6\}$.

The symbol $\leq$ is read "is less than or equal to," and $a \leq b$ means $a < b$ or $a = b$. Thus, $6 \leq 7$, and $3 \leq 3$.

EXAMPLE 2. Solve the inequality $|2x - 9| \leq 5.$ $-5 \leq 2x - 9 \leq 5$

$4 \leq 2x \leq 14$

$2 \leq x \leq 7$

Solution. $|2x - 9| \leq 5$

means $-5 \leq 2x - 9 \leq 5$

Then $-5 + 9 \leq 2x - 9 + 9 \leq 5 + 9$ why?

and $4 \leq 2x \leq 14$ why?

Therefore $2 \leq x \leq 7$ why?

The solution set is $\{x \mid 2 \leq x \leq 7\}$.

Let us now consider the inequality $|x| > b,\ b > 0$. If $x > 0$, then

$$x = |x| > b$$

If $x < 0$, then $-x = |x| > b$

Hence $x < -b$

We see that x must satisfy $x < -b$ or $x > b$. There is no number that can satisfy both inequalities. The solution set is, therefore, the union of two sets:

$$\{x \mid x < -b\} \cup \{x \mid x > b\}$$

EXAMPLE 3. Solve the inequality $|3 - 2x| > 7$.

Solution. $3 - 2x > 7$ or $3 - 2x < -7$

Hence, $3 > 7 + 2x$ or $3 < -7 + 2x$ why?

and $-4 > 2x$ or $10 < 2x$ why?

Therefore $-2 > x$ or $5 < x$ why?

Consequently $x < -2$ or $x > 5$ why?

The solution is $\{x \mid x < -2\} \cup \{x \mid x > 5\}$.

EXERCISE SET 2.3

1. Find the absolute value of

> (a) -9 (b) -14 (c) $3-7$ (d) $5-9$
>
> (e) $-5-8$ (f) $-4-7$ (g) $(-3)(8)$ (h) $(-4)(-5)$
>
> (i) $8(-7)$ (j) $\left(\dfrac{-12}{5}\right)$ (k) $\left(\dfrac{3}{-4}\right)$ (l) $(-\frac{7}{8})(-1)$

2. Find $\{x \mid x \in R \text{ and } |4x| = 8\}$.   3. Find $\{x \mid x \in R \text{ and } |x-2| = 4\}$.

4. Find $\left\{x \mid x \in R \text{ and } \left|\dfrac{x}{3}\right| = 1\right\}$.

5. In each of the following find the set of values of x which satisfy the conditions:

> (a) $|x+3| = 7$ (b) $|2x-6| = 14$ (c) $|x-1| < 3$ (d) $|x-4| < 7$
>
> (e) $|2x-4| < 5$ (f) $|3x+2| < 10$ (g) $|2x-4| > 6$ (h) $|4x-2| > 4$
>
> (i) $|5x-1| \le 4$ (j) $|3x-7| \le 5$ (k) $|7x+2| \le 12$ (l) $|2x+3| \le 9$
>
> (m) $|4x-1| \le 15$ (n) $|7x+2| \ge 19$ (o) $|5+2x| \le 9$ (p) $|3-5x| \le 17$
>
> (q) $|2-3x| \ge 1$ (r) $|5+9x| \ge 4$

PROVE the following theorems:

6. $|ab| = |a| \cdot |b|$.   HINT: Case 1, $a > 0$, $b > 0$. Here $|a| = a$ and $|b| = b$. Hence, $|a| \cdot |b| = ab = |ab|$. There are three other cases to be considered.

7. $\left|\dfrac{a}{b}\right| = \dfrac{|a|}{|b|}$ $(b \ne 0)$ 8. $|a-b| \le |a| + |b|$ 9. $|a+b| \le |a| + |b|$

10. $|a-b| \ge |a| - |b|$ 11. $|a+b| \ge |a| - |b|$

2.12   ADDITION OF ALGEBRAIC EXPRESSIONS

Constant

A variable is a symbol that can be replaced by any element of the number system (see Sec. 1.9). A *constant* is an element of the number system. In any particular problem a variable may have many values, but a constant can have only one value in that problem. It is convenient to denote variables by the letters $x, y, z, w, \ldots$ and constants by the letters $a, b, c, \ldots$.

Expression

An *expression* is a constant, or a variable, or a finite number of indicated operations involving variables and constants when expressed by means of mathematical symbols. Thus, x, 3, $2x-5$, and $3x - \sqrt{7y} + 2/z$ are expressions. When an expression is written as the sum of other expressions, each of these is called a *term* of the given expression. If an expression has only one term, it is called a *monomial;* if it has more than one term, it is a *multinomial.* A multinomial having exactly two terms is a *binomial;* a multinomial of exactly three terms is a *trinomial.* For example, $3x^2$, $2x/z$, $(x + 2y)/3z$ are monomials. Note that the numerator of the third one is a binomial.

Monomial, multinomial

Binomial, trinomial

Factor

If an expression consists of the product of two or more expressions, each is a *factor* of the original expression. If the factors are letters, they are called

literal factors, and if they are numerals, they are called *numerical factors.*
If any term is equal to the product of two factors, then either factor is called
Coefficient   the *coefficient* of the other. Thus, in the term $\frac{3}{2} x^3yz^2$, the coefficient of z^2
is $\frac{3}{2} x^3y$. The coefficient of y is $\frac{3}{2} x^3z^2$ and the coefficient of x^3yz^2 is $\frac{3}{2}$, and
so on. If the coefficient is the numerical factor only, it is called the *numerical
coefficient.*

Similar terms   Two or more terms of an expression are *similar terms* (or *like terms*) if
(1) they have the same literal factors and (2) each literal factor of any term
has the same exponent as that factor in each of the other terms. Thus, $4x^2y^3$
and $-9x^2y^3$ are like terms, whereas $4x^2y^3$ and $-9x^2y^2$ are unlike terms.

The indicated sum of two expressions is sometimes called their *formal
sum.* For example, the formal sum of x and $3x + 2y$ is $x + (3x + 2y)$. The
commutative, associative, and distributive axioms often enable us to simplify
a formal sum into an equal expression which we call the sum. The following
examples illustrate the use of the axioms.

EXAMPLE 1. Find the sum of $3x^4y^2z$ and $8x^4y^2z$.

Solution. $3x^4y^2z + 8x^4y^2z$ given
$$= (3 + 8)(x^4y^2z) \qquad \text{distributive axiom}$$
$$= 11x^4y^2z$$

EXAMPLE 2. Simplify the following expression by combining like terms:
$3x^2y - 5z - 4x^2y + 7z + x^2y$.

Solution. $3x^2y - 5z - 4x^2y + 7z + x^2y$ given
$$= (3x^2y - 4x^2y + x^2y) + (-5z + 7z) \qquad \text{comm. and assoc. axioms}$$
$$= (3 - 4 + 1)(x^2y) + (-5 + 7)z \qquad \text{distr. axiom}$$
$$= 0(x^2y) + 2z$$
$$= 2z \qquad \text{identity axiom}$$

It is often convenient to arrange expressions to be added so that like
terms are in the same column, and then to find the sum of the terms in
each column.

EXAMPLE 3. Find the sum of $7y - x$, $5x - 3y$, $-2y + 8x$, $-2x - y$.

Solution.
$$
\begin{array}{r}
-\ x + 7y \\
5x - 3y \\
8x - 2y \\
-2x -\ y \\
\hline
10x +\ y
\end{array}
$$

The *difference* of two expressions is the sum of the first and the additive inverse of each term of the second.

EXAMPLE 4.   Find the difference of $3x^2y + 2y^2$ and $5x^2y - 3y^2 + 2z$.

Solution.   Arrange the expressions so that like terms are in the same column and find the difference of the terms in each column.

$$3x^2y + 2y^2$$
$$\underline{5x^2y - 3y^2 + 2z}$$
$$-2x^2y + 5y^2 - 2z$$

You may prefer to write the sign of the additive inverse of each term in the second row just above the original sign, putting each in parentheses, and then add.

$$3x^2y + 2y^2$$
$$(-) \quad (+) \quad (-)$$
$$\underline{+ 5x^2y - 3y^2 + 2z}$$
$$- 2x^2y + 5y^2 - 2z$$

If you follow the latter plan, be sure you do not mark over or alter in any way the original signs of the second expression.   The new signs in parentheses are not part of the problem.   They are crutches that aid in recalling what the additive inverse of a term really is.

EXERCISE SET 2.4

COMBINE similar terms:

1. $5x - 7 + 6x + 4 - 3x + 3 - 2x$
2. $a + b + c - 4a - 2b + 8c + 5a - 2b + 7c$
3. $3xy^2 - 4x^4 + 7x^2y^2 - 2x^2y + 6x^4 + x^2y^2 + 3x^2y - 2x^2y^2$
4. $x^2 + 2xy + y^2 + 6x^2 - 5xy - 2y^2 - 3x^2 + 4y^2$
5. $3ax + 3bxy - 4x^2 + 2bx - bx^2 + 2axy + 3x - 2y$
6. $5x^3 - 3x^2 + 4x - 11 - 8x^3 + 12x^2 - 15x - 7 + 6x^3 - 9x^2 + 10x + 14$

FIND the sum of the expressions:

7. $x^3 - 7x + 1, 1 - x^3, 5x - 2$ 8. $x^4 - x^3, 3x^4 - 5x + 2, x^4 - 3, -(2x^4 + 3)$

SUBTRACT the second expression from the first:

9. $2x^2 + 2x - 5, 2x^2 - 3x + 4$ 10. $3x^3 - 2x^2 + 7, 5x^3 - 2x^2 + 6$

ADD:

11. $14x + 5y$ 12. $6x^2 - 8x + 1$
        $-12x + 4y$ $-3x^2 + 4x$
        $\underline{-2x - 6y}$ $\underline{ - 2x - 1}$

13. $5x^4 - 3x^3 + 4x^2 - x - 4$
$ 3x^3 - 2x + 8$
$\underline{3x^4 + 6x^3 - 2x^2}$

14. $8x^2 - 3xy - 12y^2$
$x^2 + 2xy - 4y^2$
$-8x^2 + 3xy + 5y^2$
$\underline{- 2xy - 13y^2}$

SUBTRACT:

15. $3xy^2 - 4xy + 6z$
$\underline{5xy^2 - 3z}$

16. $6x^2 + 6xy + y^2$
$\underline{4x^2 - 8xy + 3y^2}$

17. $5h^3 - 4hk + 3k^2$
$\underline{-4h^3 + 5hk - k^2}$

18. $x^3 - 3x^2y + 3xy^2 - y^3$
$\underline{x^3 + 3xy^2 - 4}$

19. Subtract $-2x^3 + 3x^2 - 1$ from 0.   **20.** Subtract $3xy - 4zy - 1$ from 1.
21. From the sum of $21x^2 - 4y$ and $-18x^2 + 3y$ subtract $14x^2 - 7x + 2y - 1$.

2.13   REMOVING GROUPING SYMBOLS

In order to simplify expressions, it is often necessary to remove sets of parentheses from the expression.   This is done by use of the distributive law. Since

$$+a = +1 \cdot a$$

it follows that

$$+(a + b) = +1 \cdot (a + b) = +1 \cdot a + (+1) \cdot b = a + b$$

Likewise, since

$$-a = -1 \cdot a$$

we have

$$-(a + b) = -1 \cdot (a + b) = -1 \cdot a + (-1) \cdot b = -a - b$$

Thus,

$$+(2x - 3y) = +1 \cdot (2x - 3y) = 2x - 3y$$
$$-(2x - 3y) = -1 \cdot (2x - 3y) = -1 \cdot (2x) + (-1) \cdot (-3y)$$
$$= -2x + 3y$$

Expressions like $3(2x - 3y)$, $-5(7x + 4y - 2z)$ are handled the same way. For example, by the extended distributive law,

$$-3(5x - 2y + 3) = -3 \cdot (5x) + (-3) \cdot (-2y) + (-3) \cdot (+3)$$
$$= -15x + 6y - 9$$

Sequence of
removal
When groups occur within groups, we usually remove the innermost symbols of grouping first.

EXAMPLE 1. Simplify by removing parentheses, brackets, and combining like terms:

$$2x^2 - [2x^2 - (x - 1) + 2x] - 5x + 3$$

Solution. $2x^2 - [2x^2 - (x - 1) + 2x] - 5x + 3$

$$= 2x^2 - [2x^2 - x + 1 + 2x] - 5x + 3 \qquad \text{removing inner ( )}$$
$$= 2x^2 - 2x^2 + x - 1 - 2x - 5x + 3 \qquad \text{removing brackets}$$
$$= -6x + 2 \qquad \text{combining terms}$$

We may enclose terms within parentheses to be preceded by a plus sign without altering the sign of any term thus enclosed. If the parentheses are to be preceded by a minus sign, however, then we must change the sign of any term thus enclosed.

EXAMPLE 2. Enclose the last two terms of $3x + 7y - 6z - 3w$ within parentheses preceded by a minus sign. $3x + 7y - [-(6z - 3w)]$

Solution. $3x + 7y - 6z - 3w = 3x + 7y - (6z + 3w)$

EXAMPLE 3. Enclose the last three terms of $6 + x - y - z$ within parentheses preceded by a minus sign.

Solution. $6 + x - y - z = 6 - (-x + y + z)$

EXAMPLE 4. Enclose the last two terms of $3a - 2b + 5c$ within parentheses preceded by a plus sign.

Solution. $3a - 2b + 5c = 3a + (-2b + 5c)$

EXERCISE SET 2.5

REMOVE symbols of grouping and COMBINE like terms:

1. $x + y + z + (4x + 3y)$
2. $(2x - 3y + 8) - (4x - 2)$
3. $[5z - (3x - y)] + (2z - 3y)$
4. $6x^2 - 9x + 1 - (x^2 - 3x + 3)$
5. $(z - w) - (3z - 2w) - (2x + 3w)$
6. $\{[2x - (x - y)] - [(2y + 3) + 7]\} - 1$
7. $5x - (2y - 3x) - [(x + y) - 2]$
8. $[(x + y) - (4x + y)] - [2x - (7y + 3x)]$
9. $2x + y - [6x + 7y + (x - 4y) - (3x - y)]$
10. $x^2 - [(6x + 4) - (-3x^2 - 6x - 7)]$
11. $9 + \{z^2 - [5z^2 - 4 - (2z^2 - y)]\}$
12. $10x - \{3y + [4x - (6 - 6x)] - (x + 7y)\}$
13. $5n - \{8n - (3n + 6) - [-6n + (7n - 5)]\}$
14. $4a - (a - \{-7a - [8a - (5a + 3)] - [-6a - (2a - 9)]\})$

ENCLOSE the last two terms in parentheses preceded by a plus sign:

15. $2x + 7y + 16$ 16. $a - 3b - 8 - c$ 17. $2y - 5x - 10$
18. $4x - 5y + 6z + 7 - 8w$ 19. $3k - 6 + 7m$ 20. $2x - 4y - 4$

ENCLOSE the last two terms in parentheses preceded by a minus sign:

21. $-7x - 2y - 4$ 22. $13 - 2x - 7y$ 23. $8 - 2a + 3b$
24. $x - y + 2z$ 25. $x^2 + 2xy + y^2$ 26. $x^2 - y^2 - z^2$
27. $4a + 6b - 7c$ 28. $3x + 2y + z$ 29. $-a - 2b - 3c$
30. $-3x - 4y + 2$

31. Enclose the last three terms of $a + b - c + d - e$ in parentheses preceded by a minus sign; then in the result enclose the last two terms in the parentheses in brackets preceded by a minus sign.

2.14   MULTIPLICATION AND DIVISION

The product of two monomials is found by applying the commutative and associative axioms. For example,

$$(7x^2y^3)(3xy^6) = (7 \cdot 3)(x^2 \cdot x)(y^3 \cdot y^6)$$
$$= 21x^3y^9$$

As a consequence of the distributive axiom, the product of two multinomials is the sum of the products formed by multiplying each term of one by every term of the other. Like terms, if there are any, are then combined. For example,

$$(3x^2 + 7y)(2x^2 - 5y) = (3x^2 + 7y)(2x^2) + (3x^2 + 7y)(-5y)$$
$$= (3x^2)(2x^2) + (7y)(2x^2) + (3x^2)(-5y) + (7y)(-5y)$$

Each term of the right member is now the product of two monomials. Hence,

$$(3x^2 + 7y)(2x^2 - 5y) = 6x^4 + 14x^2y - 15x^2y - 35y^2$$
$$= 6x^4 - x^2y - 35y^2$$

It is often convenient to write one of the expressions below the other and arrange the "partial products" so that similar terms are in columns. To multiply $x + y$ and $x^2 - 2xy + y^2$ we might arrange the work as follows:

$$x^2 - 2xy + y^2$$
$$\underline{x \ + y}$$

| $x^3 - 2x^2y +$ | xy^2 | | multiply by x |

$$x^2 - 2xy + y^2$$
$$x + y$$
$$\overline{x^3 - 2x^2y + \ xy^2} \qquad \text{multiply by } x$$
$$\underline{\quad\quad x^2y - 2xy^2 + y^3} \qquad \text{multiply by } y$$
$$x^3 - \quad x^2y - \ xy^2 + y^3 \qquad \text{add}$$

The division of one monomial by another nonzero monomial can be accomplished by using the fundamental principle of fractions. Thus,

$$\frac{36x^3y^4}{9x^2y} = \frac{4xy^3(9x^2y)}{1(9x^2y)} = \frac{4xy^3}{1} = 4xy^3$$

and
$$\frac{36x^3y^4}{14xy^5} = \frac{18x^2(2xy^4)}{7y(2xy^4)} = \frac{18x^2}{7y}$$

With the aid of the distributive axiom we can divide a multinomial by a nonzero monomial:

$$\frac{9x^2y^2 - 21x^4y^3 + 36x^2y}{3x^2y} = \frac{(3y - 7x^2y^2 + 12)(3x^2y)}{1(3x^2y)}$$

$$= 3y - 7x^2y^2 + 12$$

Note that this is equivalent to dividing each term of the multinomial by the monomial.   Thus,

$$\frac{9x^2y^2 - 21x^4y^3 + 36x^2y}{3x^2y} = \frac{9x^2y^2}{3x^2y} - \frac{21x^4y^3}{3x^2y} + \frac{36x^2y}{3x^2y}$$

$$= 3y - 7x^2y^2 + 12$$

Polynomial
 A *polynomial* is an expression each of whose terms is a constant, or a variable that has a positive-integer exponent, or a product of constants and variables that have positive-integer exponents, and no term has a variable in a denominator.   Thus, 8, $5x$, 3^2x, $\frac{1}{2}x - 1$, $7x^2yz + 3z + 6$ are polynomials.   The expression $3x + (2/y^2)$ is not a polynomial, however, since $2/y^2$ cannot be written as the product of a constant and a positive-integer power of the variable y.   The expressions 3^2x and $\frac{1}{2}x - 1$ are polynomials because 3^2 and $\frac{1}{2}$ are considered as constants.

Degree of a term
 The *degree of a term* of a polynomial with respect to some selected set of the literal factors is the number of times these literal numbers occur as factors in the term.   For example, $7x^3y^2z^5$ is of degree 3 with respect to x, of degree 5 with respect to x and y, of degree 10 with respect to x, y, and z, of degree 8 with respect to x and z, and so on.

Degree of a polynomial
 The *degree of a polynomial* with respect to a selected set of literal factors is the degree of that term of the polynomial which is of highest degree with respect to those literal factors.   Thus, $5x^3y^2 + 2x^2y^4$ is of degree 3 with respect to x, of degree 4 with respect to y, and of degree 6 with respect to x and y.

Division of polynomials
 To divide one polynomial by another we use an algorithm (procedure) similar to that of long division in arithmetic:

   1.   Arrange the terms of the dividend and the divisor in descending powers of some literal factor common to both.

   2.   Divide the *first* term of the divisor into the *first* term of the dividend to obtain the first term of the quotient.

   3.   Multiply the *entire* divisor by the first term of the quotient found in step 2.   Subtract this product from the dividend.

4. Use the remainder thus obtained for a *new* dividend. Repeat steps 2 and 3 to obtain the second term of the quotient.

5. Continue the above procedure until a remainder has been obtained that is either zero or is of lower degree in the common literal number than the divisor.

If the remainder is 0, we express the result in the form

$$\frac{\text{dividend}}{\text{divisor}} = \text{quotient}$$

If the remainder $r \neq 0$, we express the result in the form

$$\frac{\text{dividend}}{\text{divisor}} = \text{quotient} + \frac{\text{remainder}}{\text{divisor}}$$

EXAMPLE 1. Divide $2a^4 - 3a^3 - 7a^2 - 1$ by $a^2 + 3a - 1$.

Solution.

$$
\begin{array}{r}
2a^2 - 9a + 22 \\
a^2 + 3a - 1 \overline{)\, 2a^4 - 3a^3 - 7a^2 \qquad - 1} \\
2a^4 + 6a^3 - 2a^2 \\
\hline
- 9a^3 - 5a^2 \qquad - 1 \\
- 9a^3 - 27a^2 + 9a \\
\hline
22a^2 - 9a - 1 \\
22a^2 + 66a - 22 \\
\hline
- 75a + 21
\end{array}
$$

Hence,

$$\frac{2a^4 - 3a^3 - 7a^2 - 1}{a^2 + 3a - 1} = 2a^2 - 9a + 22 + \frac{-75a + 21}{a^2 + 3a - 1}$$

EXERCISE SET 2.6

MULTIPLY:

1. $4x^5y^2$ by $-21x^2y^3$
2. $4x$ by $-2xy$
3. $-3x^4y^2z$ by $7xz$
4. $3xyz$ by $-x^2y$
5. $(5xy)(4xz)(-7x^3yz^2)$
6. $(-3x^2y)(-x)(5y^2)(6xy)$
7. $-4(2x - 7y + 3z)$
8. $-xy(5xy - x + 2)$
9. $6z(-3z^2 + 4wz + 1)$
10. $3a(2a^3 - 5ab - b^2)$
11. $(3a + 4)(2a + 1)$
12. $(3 + 2x)(x + 2)$
13. $(x^2 - 4x + 4)(x + 3)$
14. $(11 - 15x - 7x^2)(25 - 16x^2)$
15. $(x^2 - 5xy + 3y^2)(2x^2 + y)$
16. $(x^2 + y^2)(x^2 - y^2)$
17. $(x - 2y + 3z)(x - 2y + 3z)$
18. $(3m^3 + m^2 - 2m - 5)(m^2 - 5m - 6)$
19. $(x^2 - 8x - 1)(x^2 + 2x - 3)$
20. $(x^2 - xy + y^2)(x^2 + xy + y^2)$

PERFORM the indicated operations:

21. $(2x - y)^2$ **22.** $(3x + 4)^2$ **23.** $(x + y + z)^2$

24. $(a - b - c)^2$ **25.** $(x - 2y - 3z)^2$ **26.** $(x + y - 2z)^2$

FIND the quotient:

27. $\dfrac{16x^3}{8x}$ **28.** $\dfrac{-63x^3y^2}{-9x^2}$ **29.** $\dfrac{81xy^3}{-27x}$

30. $\dfrac{-25x^5y^2}{5x^5y^2}$ **31.** $(24x^4 - 16x^2) \div 4x$ **32.** $(x^4 - 5x^2y^2) \div x^2$

33. $(4x + 5y - 2z) \div (-1)$ **34.** $(9x^2y^2 - 27x^4y^3 + 21x^2y) \div 3x^2y$

35. $(12x^2y^2 + 48x^4y) \div (-12x^3y)$ **36.** $(-3xy^2 - 6xy + 18y) \div (-3y)$

37. $\dfrac{16x^2(a - b)}{4x(a - b)}$ **38.** $\dfrac{15(x + y - z)^2}{5(x + y - z)}$

39. $\dfrac{65(a + 2b - 3c)^2}{-9(a + 2b - 3c)}$ **40.** $\dfrac{-21x^2(4x - 9y)}{-2x(4x - 9y)}$

DIVIDE and check:

41. $x^3 + x^2 + 3x + 6$ by $x + 2$ **42.** $x^3 + 3x^2 + 6x - 9$ by $x - 3$

43. $x^3 - 3x + 4$ by $x - 4$ **44.** $2x^4 - 3x^3 - 7x^2 + 7x - 1$ by $x^2 + 3x - 1$

45. $(x^3 + 8y^3) \div (x^2 - 2xy + 4y^2)$ **46.** $(x^4 + x^2y^2 + y^4) \div (x^2 + xy + y^2)$

47. $(x + 8 - x^3) \div (x^2 + 4 + 2x)$ **48.** $(a^3 - 6a^2 + 12a - 8) \div (4 - 4a + a^2)$

49. $(x^4 - 4y^4) \div (x^2 + 2y^2)$ **50.** $(x^3 - b^3) \div (x - b)$

51. $(8y^3 + 64x^3) \div (8x + 2y)$ **52.** $(x^5 - y^5) \div (x - y)$

53. $(x^3 + y^3) \div (x + y)$ **54.** $(x^8 - y^8) \div (x^2 + y^2)$

55. $(9x^3 - 27x^2 + 14x + 8) \div (3x^2 - 5x - 2)$

56. $(x^2 + y^2 + z^2 + 2xy + 2xz + 2yz) \div (x + y - z)$

57. $\dfrac{6x^3 + 11x^2y - 25xy^2 + 9y^3}{2x - y}$ **58.** $\dfrac{x^3 - 8y^3}{x^2 + 2xy + 4y^2}$

59. $\dfrac{a^6 - b^6}{a + b}$ **60.** $\dfrac{125a^3 - 64}{5a - 4}$

3

EQUATIONS, FUNCTIONS, GRAPHS

An equation is a statement that two given expressions represent the same number. To write an equation, we set one of the given expressions equal to the other. The statement that the two expressions are equal may or may not be true. For example, $3 = 3$ and $9 = 13$ are equations. One of them is true, the other false.

The equation

$$ax + b = 0 \qquad a \text{ and } b \text{ constant, } a \neq 0 \qquad (3.1)$$

Solution set

is called an *equation of the first degree in the variable x.* The set of all elements which when substituted for the variable x make $ax + b = 0$ a true statement is called the *solution set* of the equation. When we recall that the domain of a variable is the set of eligible replacements for the variable, we see that the solution set is a subset of the domain of x. In determining the solution set of an equation, it is important to keep in mind the domain of the variable. For example, let us consider two first-degree equations in which the domain of x is restricted to the set $\{1,2,3,4,5,6\}$. Then the solution set of the equation $2x - 6 = 0$ is the set $\{3\}$. On the other hand, the solution set of the equation $2x - 5 = 0$ is the empty set $\varnothing$, because there is no element in the domain which makes the equation a true statement.

Let us now consider the equation

$$\frac{2x}{x-2} = 4 \qquad x \in R$$

which is not of the form of Eq. (3.1) but which illustrates a point frequently overlooked by students. The number 2 is not an eligible replacement for x, that is, 2 is not in the domain. The reason for this is that if x is replaced by 2, the left side of the equation becomes $\frac{4}{0}$ and is therefore meaningless. We say that the left side of the equation is undefined for $x = 2$.

We will always exclude from the domain of a variable any element for which an expression in the equation is undefined.

3.1 IDENTITIES AND CONDITIONAL EQUATIONS

An *identity* is an equation whose solution set is the domain of the variable. Hence, if the domain is the set of all real numbers, then the replacement of the variable by any numeral representing a real number always results in a true statement. For example, the equation

$$(x + 3)(x - 3) = x^2 - 9$$

is an identity because it is true for every real value of x. Thus, the solution set is the domain of x. The equation

$$\frac{x^3}{x} + 1 = x^2 + 1 \qquad x \neq 0$$

is also an identity because it is true for every real value of x except 0. The solution set is the domain of x.

Conditional equation   A *conditional equation* is an equation whose solution set is a proper subset of the domain of the variable. This means that a conditional equation is not true when the variable is replaced by some elements in the domain. If the domain of x is the set R of real numbers, the equation

$$x^2 - 5x = -6$$

is a conditional equation because it is true if and only if $x = 2$ or $x = 3$. The solution set $\{2,3\}$ is a proper subset of the domain.

It is easy to show that 2 is a solution. For if $x = 2$, the left member of the equation equals $(2)^2 - 5(2) = 4 - 10 = -6$. Since the right member is -6, we see that $x = 2$ satisfies the equation. The student may check the other solution by the same method.

3.2 EQUIVALENT EQUATIONS

If the solution set of one equation is exactly the same as the solution set of another, the equations are *equivalent*. For example, the equations

$$3x^2 - 15x + 18 = 0 \qquad \text{and} \qquad 2x^2 - 10x = -12$$

are equivalent. Each has the solution set $\{2,3\}$.

We recall that if a, b, $c \in R$ and if $a = b$, then $a + c = b + c$ and $ac = bc$. Since a polynomial represents a real number when the domains of the variables are real numbers, the following two theorems can be proved (we omit the proofs here):

THEOREM. If a polynomial is added to each member of an equation, the resulting equation is equivalent to the first. (3.2)

THEOREM. If each member of an equation is multiplied by a non-zero constant, the resulting equation is equivalent to the first. (3.3)

It is important that we understand the implications of these two theorems. Theorem (3.2) implies that if we add an expression which is not a polynomial to each side of an equation, the resulting equation may or may not be equivalent to the given equation. For example, the equation

$$x^2 - 7x = -12$$

has the solution set $\{3,4\}$. If we add $1/(x - 4)$ to each member, the transformed equation

$$x^2 - 7x + \frac{1}{x - 4} = -12 + \frac{1}{x - 4}$$

does not have 4 in its solution set because 4 is not in the domain of the variable. Hence, the two equations are not equivalent.

Theorem (3.3) implies that if we multiply each member of an equation by something other than a constant, we may not obtain an equivalent equation. For example, the equation

$$x - 3 = 0$$

has the solution set $\{3\}$. If we multiply each side by $(x - 5)$, the resulting equation

$$x^2 - 8x + 15 = 0$$

has the solution set $\{3,5\}$. Therefore, the transformed equation is not equivalent to the given equation.

As a further example of what can happen when we multiply each side of an equation by an expression involving the variable, consider the equation

$$x^2 - 4x - 21 = 0$$

which has the solution set $\{-3,7\}$. If we multiply each side by $1/(x - 7)$, the transformed equation

$$x + 3 = 0$$

has the solution set $\{-3\}$, and consequently is not equivalent to the given equation.

EXAMPLE 1. Find the solution set for $x + (3x - 7) = 4x - (x + 2)$.

Solution.

$$x + (3x - 7) = 4x - (x + 2) \qquad \text{given}$$
$$x + 3x - 7 = 4x - x - 2 \qquad \text{removing parentheses}$$
$$4x - 7 = 3x - 2 \qquad \text{combining terms}$$
$$4x = 3x + 5 \qquad \text{adding 7 to each side}$$
$$x = 5 \qquad \text{subtracting } 3x \text{ from each side}$$

Check.   If $x = 5$,

Left Member	Right Member
$5 + (15 - 7)$	$20 - (5 + 2)$
$= 13$	$= 13$

EXAMPLE 2.   Find the solution set for

$$\frac{x + 10}{6} + \frac{1}{3} - \frac{x}{4} = \frac{4 + 5x}{6} - x$$

Solution.

$$12\left(\frac{x + 10}{6}\right) + 12\left(\frac{1}{3}\right) - 12\left(\frac{x}{4}\right) = 12\left(\frac{4 + 5x}{6}\right) - 12(x) \qquad \text{why?}$$
$$2x + 20 + 4 - 3x = 8 + 10x - 12x \qquad \text{simplifying each term}$$
$$-x + 24 = -2x + 8 \qquad \text{combining terms}$$
$$x + 24 = 8 \qquad \text{adding } 2x \text{ to each side}$$
$$x = -16 \qquad \text{subtracting 24 from each side}$$

Check.   If $x = -16$,

Left Member	Right Member
$\dfrac{-16 + 10}{6} + \dfrac{1}{3} - \dfrac{-16}{4}$	$\dfrac{4 + 5(-16)}{6} - (-16)$
$= -1 + \dfrac{1}{3} + 4$	$= -\dfrac{76}{6} + 16$
$= \dfrac{10}{3}$	$= \dfrac{10}{3}$

Note.   In this example, we could have multiplied each side of the given equation by 24, 36, 48, or any other number into which each of the denomi-

nators 6, 3, 4 will divide. We chose the multiplier 12 because it is the least
of such numbers. It is called the *least common multiple* of the denominators.

EXERCISE SET 3.1

FIND the solution set of each of the following conditional equations:

1. $5x + 7 = 3(x + 1)$ **2.** $4(5x + 1) - 6(3x + 2) = 4x$

3. $12 \left(\dfrac{3x}{4} - 2 \right) = 12 \left(\dfrac{x}{6} + \dfrac{1}{3} \right)$ **4.** $(2x + 4)^2 + 5x = (x + 5)(4x + 3) + 9$

5. $\dfrac{2(x + 3)}{3} - \dfrac{4(2x + 5)}{5} + \dfrac{16}{15} = 0$ **6.** $\dfrac{y - 5}{4} = 2 - \dfrac{y + 5}{14}$

7. $\dfrac{y + 7}{3} - \dfrac{y - 2}{5} = \dfrac{7y - 3}{10} - 9$

SOLVE for the variable indicated:

8. $nE = I(R + nr)$ for r

9. $F = \dfrac{kmM}{d^2}$ for m

10. $V = \frac{1}{3} \pi h^2 (3r - h)$ for r

SOLVE for the whole parentheses. (The variables are enclosed by parentheses. For the present,
do not consider their meaning. They will be defined later.)

11. $4(\tan \theta) = (\tan \theta) + 9$ **12.** $9(\sec \phi) - 4 = 3(\sec \phi) - 16$
13. $3(\sin \alpha) - 3 = 2 - 7(\sin \alpha)$ **14.** $2[(\ln x) + 1] + 1 = (\ln x)$
15. $4[9 - 5(\ln x)] + 3[3(\ln x) - 1] = 0$ **16.** $6[4(\cot \theta) - 7] = 3 + 5[2(\cot \theta) + 5]$

17. $2(\cos x) + \frac{3}{2} = 3(\cos x) + 1$ **18.** $\dfrac{3(\sin \beta) + 2}{2} = 2 - \dfrac{3(\sin \beta) - 2}{3}$

19. $\dfrac{8(\log x) - 11}{9} - \dfrac{7(\log x) + 4}{12} - \dfrac{3(\log x) - 8}{8} = 0$

20. $\dfrac{2(\log x) - 1}{3} + \dfrac{3(\log x) + 1}{4} = \dfrac{3(\log x) - 1}{6} + 1$

3.3 SUBSETS OF $R \times R$

In Chap. 1 we defined the cross product of two sets A and B as the
set of all ordered pairs (x,y) such that x is an element of A and y is
an element of B. Thus,

$$A \times B = \{(x,y) \mid x \in A, y \in B\} \tag{3.4}$$

If R is the set of real numbers, then $R \times R = \{(x,y) \mid x, y \in R\}$. It follows
that if A and B are subsets of R, then $A \times B$ is a subset of $R \times R$. We turn
our attention to constructing graphs of certain subsets of $R \times R$ by setting
up a one-to-one correspondence between the set of all ordered pairs of real
numbers and the set of all points in a plane.

3.4   RECTANGULAR COORDINATE SYSTEM

A one-to-one correspondence between sets of points in a plane and sets of ordered pairs of real numbers can be made as follows.

Construction of coordinate plane

Let two perpendicular lines intersect at the point O. Let one of the lines be horizontal, and the other vertical. Set up on each line a number scale having O as the origin. Generally, the same unit of length is used on both number scales, but this is not necessary. Let each point on the horizontal scale to the *right* of the origin correspond to a positive real number; then each point to the *left* of the origin will correspond to a negative real number. On the vertical scale, let each point *above* the origin correspond to a positive real number; then each point *below* the origin corresponds to a negative real number. The origin corresponds to zero on each scale (see Fig. 3.1).

Let the symbols for numbers on the horizontal line be the elements which x stands for in the ordered pair (x,y) and the symbols for numbers on the vertical line be the elements which y stands for. Then the following statements can be made:

1.   The first coordinate of the ordered pair (x,y) represents the directed distance of a point from the vertical scale. The distance is measured to the right if the first coordinate of the ordered pair is positive, and to the left if the first coordinate is negative.

2.   The second coordinate of the ordered pair represents the distance of a point from the horizontal line. This distance is measured upward if the second coordinate is positive and downward if it is negative.

Hence, in the plane determined by the intersecting number scales each point corresponds to an ordered pair of real numbers, the first representing the directed distance of the point from the vertical scale and the second representing the directed distance of the point from the horizontal scale. Conversely, each ordered pair of real numbers corresponds to a point in the plane. Thus, there is a one-to-one correspondence between the points in the plane and the ordered pairs of real numbers.

Abscissa

Ordinate

We call the horizontal scale the *x axis* and the vertical scale the *y axis*. The first coordinate of an ordered pair of real numbers is called the *abscissa* of the point in the plane that corresponds to the ordered pair. The second coordinate of the ordered pair is called the *ordinate* of the point.

Quadrant

The x and y axes separate the plane into four parts, each of which is called a *quadrant*. The part of the plane that is above the x axis and to the right of the y axis is the first quadrant, or quadrant I. The portion of the plane above the x axis and to the left of the y axis is the second quadrant, or quadrant II. The third quadrant consists of the part of the plane below the x axis and to the left of the y axis, and quadrant IV consists of the portion of the plane lying below the x axis and to the right of the y axis. The location of the quadrants is shown in Fig. 3.1.

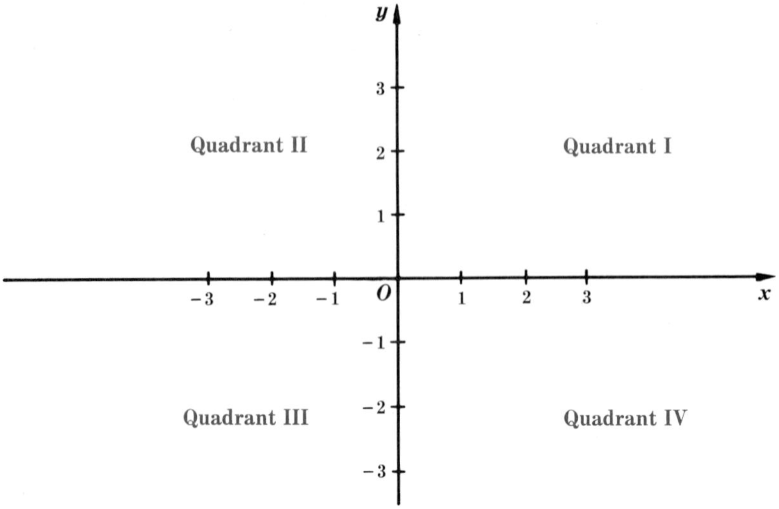

FIGURE 3.1 The coordinate axes

It is evident from the figure that the point corresponding to the ordered pair of real numbers (x,y) lies in

1. Quadrant I if x and y are both positive
2. Quadrant II if x is negative and y positive
3. Quadrant III if both x and y are negative
4. Quadrant IV if x is positive and y negative

The plane determined by the x and y axes is the *rectangular coordinate plane* (or the *cartesian plane*). The system just described for establishing a one-to-one correspondence between ordered pairs of real numbers and points in the coordinate plane is called the *rectangular coordinate system.*

Rectangular coordinate system

A point in the coordinate plane can be located when its coordinates are given. For example, the point $(-5,2)$ is located 5 units to the left of the y axis and 2 units above the x axis. Conversely, the coordinates of a point located at given distances from the axes can be found. For example, the point that is 3 units to the left of the y axis and 4 units below the x axis has the coordinates $(-3,-4)$.

Plotting points

Locating points in the coordinate plane and marking them in some manner is called *plotting* the points. The points $(-5,2)$, $(-3,-4)$, $(3,3)$, $(1,-4)$ are plotted in Fig. 3.2.

3.5 GRAPH OF A SUBSET OF $R \times R$

Let $A = \{0,1,2\}$ and $B = \{3,4\}$. Then $A \times B = \{(0,3),(0,4),(1,3),(1,4),(2,3),(2,4)\}$ is a subset of $R \times R$. Take each of the ordered pairs of $A \times B$ as the coordinates of a point in a rectangular coordinate system. The set consisting of these points, and no others, is called the *graph* of $A \times B$, shown in Fig. 3.3.

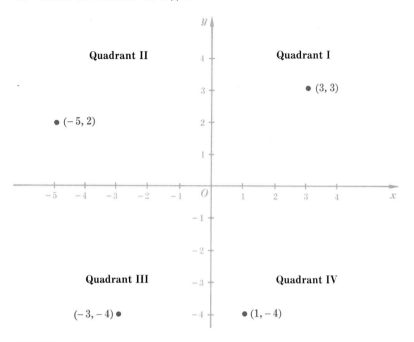

FIGURE 3.2 The rectangular coordinate system

The preceding discussion is summarized in the following definition.

DEFINITION. The graph of $A \times B$ is the set of all points, and only those points, whose coordinates are the ordered pairs of $A \times B$. **(3.5)**

Note that if $P = \{1,2,3, \ldots\}$, the graph of $P \times P$ is the set of all points in the first quadrant whose coordinates are positive integers. If $I = \{\ldots, -3, -2, -1, 0, 1, 2, 3, \ldots\}$, the graph of $I \times I$ is the set of all

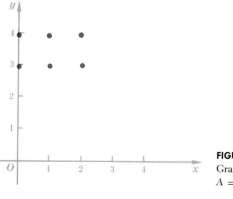

FIGURE 3.3
Graph of $A \times B$, where $A = \{0,1,2\}$, $B = \{3,4\}$

points in the coordinate plane whose coordinates are integers. If R is the set of all real numbers, the graph of $R \times R$ consists of all points in the plane.

3.6   SOLUTION SETS OF EQUATIONS IN TWO VARIABLES

Consider the statement

$$y = 3x + 5$$

If $x = 1$ and $y = 8$, the statement is true. If $x = 1$ and $y = 7$, the statement is false. Any ordered pair of numbers, such as $(2,11)$, for example, that makes the statement true is called a solution of the equation in two variables. The set of all such ordered pairs is the solution set of the equation. We may denote this solution set as

$$\{(x,y) \mid y = 3x + 5\}$$

which is read "the set of all ordered pairs (x,y) such that $y = 3x + 5$." Each ordered pair of the solution set corresponds to a point in the coordinate plane. The set of all such points is called the *graph of the equation*. There are an infinite number of ordered pairs of real numbers that satisfy the equation if the domain of each variable is the set of real numbers. It will be shown later that in the case where the domain of both x and y is the set R of real numbers, the graph of $y = 3x + 5$ is a straight line.

Graph of an equation

DEFINITION.   The solution set of an equation or an inequality in two variables x and y is the set of all ordered pairs (x,y) whose co-   **(3.6)** ordinates satisfy the equation or inequality.

For example, suppose the domain of both x and y is the set $\{0,1,2,3\}$. Then the solution set of the equation

$$y = x + 1$$

is the set $\{(0,1),(1,2),(2,3)\}$. The graph of the equation is the set of points shown in Fig. 3.4.

For the same domain of both x and y, the solution set of the inequality

$$y < x + 1$$

is the set $\{(0,0),(1,0),(1,1),(2,0),(2,1),(2,2),(3,0),(3,1),(3,2),(3,3)\}$. The graph of the inequality is shown in Fig. 3.5.

3.7   RELATIONS AND FUNCTIONS

In this section we shall define what we mean by a *relation*. We shall then use relations to introduce functions, one of the most important concepts of mathematics.

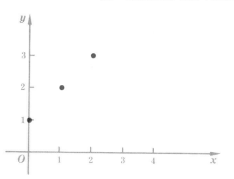

FIGURE 3.4 Graph of $y = x + 1$,
$x, y \in \{0,1,2,3\}$

FIGURE 3.5 Graph of $y < x + 1$,
$x, y \in \{0,1,2,3\}$

DEFINITION.   A relation is a set of ordered pairs.   (3.7)

Thus, if

$$S = \{(1,1),(1,2),(3,7)\}$$

Domain of a relation

then S is a relation.   The set of all first coordinates of the ordered pairs of a relation is called the *domain* of the relation.   We shall denote the domain of a relation S by $D(S)$.   Hence,

$$D(S) = \{x \mid (x,y) \in S\}$$

In the present example, $D(S) = \{1,3\}$.

The second coordinate of any ordered pair of a relation is called the *image* of the first coordinate.   The set of all images (second coordinates) of the ordered pairs is called the *range*, or *image set*, of the relation.   We shall denote the range, or image set, of a relation S by $R(S)$ or by $I(S)$.   Hence,

Range of a relation

$$R(S) = \{y \mid (x,y) \in S\} = I(S)$$

In the present example, $R(S) = \{1,2,7\}$.

As another example, consider the relation

$$T = \{(0,3),(2,5),(4,7)\}$$

Here $D(T) = \{0,2,4\}$ and $R(T) = \{3,5,7\}$

Note that the second coordinate of any ordered pair of T is 3 more than the corresponding first coordinate.   Hence, we can describe T as the set of all ordered pairs (x,y) satisfying the equation $y = x + 3$, where $x = 0, 2, 4$. We can write this relation as

$$T = \{(x,y) \mid y = x + 3, x = 0,2,4\}$$

To define the relation

$$S = \{(2,8),(3,27),(4,64)\}$$

by means of an equation, we observe that the coordinates of each ordered pair are related so that the second coordinate is the third power of the first. This relation can be written as

$$S = \{(x,y) \mid y = x^3, x = 2,3,4\}$$

A relation such as

$$Q = \{(1,1),(2,4),(5,7)\}$$

Description of a relation

may be more difficult to define by means of an equation. The coordinates of the ordered pairs are related in such a manner that when the first component is 1, the second is 1; when the first component is 2, the second is 4; and when the first is 5, the second is 7. It may not be obvious that the relation Q can be defined as

$$Q = \{(x,y) \mid 2y = -x^2 + 9x - 6, x = 1,2,5\}$$

When a relation is defined by the use of an equation, the domain is either specified, as in the preceding examples, or it is understood. If the domain is not specified, we understand that it consists of all numbers for which the equation can be satisfied. For example, the relation

$$R = \{(x,y) \mid y = \sqrt{4 - x}\}$$

has for its domain all real numbers less than or equal to 4.

Since $A \times b$ is the set of all ordered pairs (x,y) such that $x \in A$ and $y \in B$, it follows that any subset of $A \times B$ is a relation whose domain is in set A and whose range is in set B. A subset of $A \times B$ is called a relation *from A to B*. If $A = B$, we say that the relation is *in A* rather than that it is from A to A. Thus, a relation in A is a subset of $A \times A$. Consider, for example, a universe

$$U = \{0,1,2,3\}$$

and the relation

$$S = \{(x,y) \mid (x,y) \in U \times U \text{ and } y = x + 1\}$$

Here S is a relation in U defined by the equation $y = x + 1$. This equation states that the second coordinate of each ordered pair is related to the first coordinate so that the second is one more than the first. Hence,

$$S = \{(0,1),(1,2),(2,3)\}$$

The domain of a relation will be specified or understood, or obtainable from the conditions that define the relation.

EXAMPLE 1. If $U = \{0,1,2,3\}$ and if

$$T = \{(x,y) \mid (x,y) \in U \times U \text{ and } y = 2x + 1\}$$

list the elements of T in set notation.

Solution. $T = \{(0,1),(1,3)\}$

EXAMPLE 2. If $U = \{0,1,2,3\}$ and if

$$W = \{(x,y) \mid (x,y) \in U \times U \text{ and } y > x + 1\}$$

use set notation and list the elements in W.

Solution. $U \times U = \{(0,0),(0,1),\mathbf{(0,2)},\mathbf{(0,3)},(1,0),(1,1),(1,2),\mathbf{(1,3)},(2,0),(2,1),$
$(2,2),(2,3),(3,0),(3,1),(3,2),(3,3)\}$. Hence,

$$W = \{(0,2),(0,3),(1,3)\}$$

Note that we have set in boldface type the three ordered pairs having second coordinates greater than the sum of the first coordinate plus 1.

We are now ready to define a class of very important relations called *functions*. Many definitions of a function have been formulated over the course of centuries. The following one seems satisfactory at the present.

Function

DEFINITION. A function is a relation in which no two distinct ordered pairs have the same first coordinate. **(3.8)**

The relation $L = \{(1,1),(2,3),(4,6)\}$ is a relation that is also a function. The domain is $D(L) = \{1,2,4\}$. Note that no two distinct ordered pairs have the same first coordinate.

The relation $M = \{(1,1),(2,4),(1,6)\}$ is *not* a function, because the distinct ordered pairs $(1,1)$ and $(1,6)$ have the same first coordinate.

Notation

Functions will be denoted by a letter such as f, F, g, G, h, H, etc.

If f is a function and if $(x,y) \in f$, we usually denote the second coordinate y of the ordered pair by the symbol $f(x)$. Thus,

$$y = f(x)$$

means that y is the second coordinate of the ordered pair whose first coordinate is x. With this agreement, we can write the ordered pair (x,y) as

$$(x,f(x))$$

We read $f(x)$ as "f at x" or as "f of x."

The second coordinate y of the ordered pair (x,y) of a given function is usually represented by an expression that involves the first coordinate x. The function $f = \{(x,y) \mid y = x^2 - 5x + 6\}$ is such a function. Because y and $f(x)$ are symbols that may be used to represent the same thing, we could just as well write this function as $f = \{(x,f(x)) \mid f(x) = x^2 - 5x + 6\}$.

The important thing here is to understand that the ordered pairs of f are

$$(x,y) = (x, x^2 - 5x + 6)$$

This means, for example, that if $x = 1$ is the first coordinate of an ordered pair, then the second coordinate is $(1)^2 - 5(1) + 6 = 2$.   Therefore $(1,2)$ is one of the ordered pairs of f.   The equation

$$f(x) = x^2 - 5x + 6$$

thus provides us with a rule for determining the ordered pairs of f.   For example, to find the second coordinate when $x = 2$, we write

$$f(2) = (2)^2 - 5(2) + 6 = 0$$

Hence, $(x, f(x)) = (2,0)$ is another of the ordered pairs of f.
   If $x = -3$, then

$$f(-3) = (-3)^2 - 5(-3) + 6 = 30$$

and $(x,y) = (x,f(x)) = (-3,30)$ is an ordered pair of f.
   If $x = k$, then

$$f(k) = k^2 - 5k + 6$$

and $(k, k^2 - 5k + 6)$ is an ordered pair of the function.

EXAMPLE 3.   If $f(x) = 3x^3 - 2$, find (1) $f(3)$; (2) $f(-2)$; (3) $f(u + 3v)$; (4) $f(h) + f(2h + 1)$.

Solution.

(1) $$f(3) = 3(3)^3 - 2 = 79$$
(2) $$f(-2) = 3(-2)^3 - 2 = -26$$
(3) $$f(u + 3v) = 3(u + 3v)^3 - 2$$
(4) Since $\quad f(h) = 3h^3 - 2 \quad$ and $\quad f(2h + 1) = 3(2h + 1)^3 - 2$
$$f(h) + f(2h + 1) = 3h^3 - 2 + 3(2h + 1)^3 - 2$$
$$= 3h^3 + 3(2h + 1)^3 - 4$$

The multiplications indicated in (3) and (4) are left to the student.

   The graph of a function in the coordinate plane is the set of all points, and only those points, whose coordinates are the ordered pairs of the function. In Fig. 3.6, we have constructed the graph of the function

$$f = \{(-2,1),(-1,0),(0,-1),(1,2),(2,3)\}$$

   It is common practice to define functions by merely stating a method for finding the second coordinates of the ordered pairs.   For example, $f(x) = x^3$ means that f is a function whose ordered pairs are (x,x^3), and

$h(x) = 3x^2 + 5x$ means that h is a function whose ordered pairs are $(x, 3x^2 + 5x)$. If a function is a set of ordered pairs (x,y), we call x the *in-dependent* variable and y the *dependent* variable of the function.

EXAMPLE 4. Given that the domain of the function $f(x) = x^3$ is the set of all integers x such that $-2 \leq x \leq 2$, find the ordered pairs of the function f and sketch its graph.

Solution. $f(-2) = (-2)^3 = -8$; $f(-1) = -1$; $f(0) = 0$; $f(1) = 1$; $f(2) = 8$. Hence, the ordered pairs of f are $(-2,-8)$, $(-1,-1)$, $(0,0)$, $(1,1)$, $(2,8)$. The graph is shown in Fig. 3.7.

Most of the time it will be impossible to list all the ordered pairs of a given function and therefore impossible to plot all the points on the graph of the function. In this case, we find a few of the ordered pairs, plot the corresponding points, and connect them with a smooth curve.

EXAMPLE 5. Sketch the graph of the function defined by $f(x)=x^2-x-6$, $x \in R$.

FIGURE 3.7 Graph of $f(x) = x^3$, $x \in \{-2,-1,0,1,2\}$

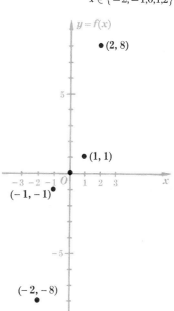

FIGURE 3.6 Graph of $f = \{(-2,1),(-1,0), (0,-1),(1,2),(2,3)\}$

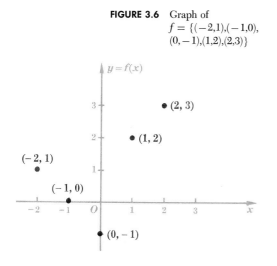

Solution. Assign a few convenient values to x and find the corresponding values of $y = f(x)$. Tabulate these ordered pairs as shown:

x	-3	-2	-1	0	1	2	3	4
$y = f(x)$	6	0	-4	-6	-6	-4	0	6

Plot the points whose coordinates are the ordered pairs of the table, and draw a smooth curve connecting these points in order from left to right. The graph is shown in Fig. 3.8.

Functions frequently require more than one equation in their definition. For example,

$$f(x) = x + 2 \qquad \text{if } x \geq 0$$
$$= -x + 2 \qquad \text{if } x < 0$$

defines a function. To sketch the graph of this function, we tabulate a few of the ordered pairs as shown:

x	-2	-1	0	1	2
$f(x)$	4	3	2	3	4

The graph is shown in Fig. 3.9.

FIGURE 3.8 Graph of
$$f(x) = x^2 - x - 6, \ x \in R$$

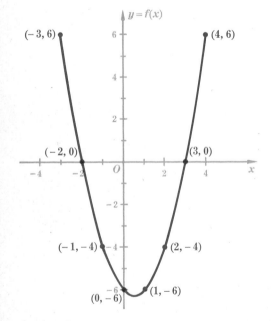

FIGURE 3.9 Graph of
$$f(x) = x + 2 \quad \text{if } x \geq 0$$
$$= -x + 2 \quad \text{if } x < 0$$

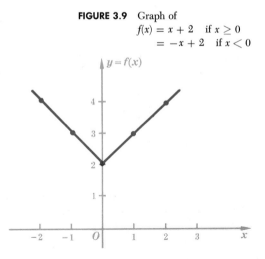

It is important to understand that in the preceding examples we selected convenient numbers from the domain of the function when we found some of the ordered pairs. Unless otherwise indicated, we shall take the domain of functions to be the set of all real numbers for which the function is defined.

EXERCISE SET 3.2

1. Which of the following relations are functions?

    (a) $\{(2,3),(3,1),(4,3),(5,7)\}$
    (b) $\{(-2,-1),(-3,-2),(-8,-6)\}$
    (c) $\{(1,4),(2,6),(1,5),(3,9)\}$
    (d) $\{(3,3),(2,1),(1,2),(2,3)\}$
    (e) $\{(-2,-3),(-4,-2),(-3,-2),(1,3)\}$

2. List the elements in the domain of each of the functions:

    (a) $\{(-3,2),(-2,3),(-1,4),(0,5),(1,6)\}$ $\{-3,-2,-1,0,1\}$
    (b) $\{(\frac{1}{2},0),(\frac{3}{2},-1),(2,-2),(\frac{5}{3},\frac{4}{3})\}$
    (c) $\{(5,-1),(4,-2),(0,1),(-2,3),(-3,-3)\}$
    (d) $\{(7,a),(5,b),(3,b),(2,c),(4,d)\}$
    (e) $\{(p,3),(r,4),(v,-2),(w,2)\}$
    (f) $\{(a,b),(d,f),(h,j),(n,p)\}$

3. List the elements in the range of each of the functions of Prob. 2 above.
4. If $S = \{(3,4),(2,-1),(1,0),(x,2)\}$ is to be a function, list the integers that x cannot represent.
5. If $T = \{(-2,-1),(0,3),(x,-1),(3,7)\}$ is to be a function, list the values of x that must be excluded.
6. Given the relation $R = \{(4,2),(3,3),(2,y),(1,5)\}$, is R a function for every real value of y?
7. When a real function is determined by an equation, we will always assume that the domain is the set of all numbers for which the equation has meaning. Let f be a function determined by the rule

$$f(x) = \frac{x}{x-1}$$

Determine the domain of the function.
8. Determine the domain of the function defined by $f(x) = \sqrt{1/x}$.
9. Determine the domain of the function defined by $f(x) = \sqrt{x^2 - 5x + 6}$.
10. Determine the domain of $f(x) = \sqrt{7x - x^2 - 12}$.
11. The domain of the function defined by $f(x) = 2x + 8$ is given to be $\{-3,-2,0\}$. Find the range of f. $2, 4, 8$
12. Let the domain of $f(x) = 1/(x-1)$ be the set $\{0,2,3\}$. Find the range of f.
13. If $g(x) = x^2 - 3x$, evaluate

    (a) $g(0)$ (b) $g(2)$ (c) $g(-1)$ (d) $g(\sqrt{3})$ (e) $g(a^2)$ (f) $g(x-3)$

14. If $h(x) = x + 4$, evaluate

    (a) $h(4) + h(1)$ (b) $h(6) - h(2)$ (c) $h(a+1) - h(a)$ (d) $[h(a+t) - h(a)]/t$

15. If $f(x) = 4^x$, evaluate

    (a) $f(0)$ (b) $f(2)$ (c) $f(-1)$ (d) $f(-3)$ (e) $f(\frac{1}{2})$ (f) $f(-2)$

16. If $f(x) = 9^{-x}$ evaluate

    (a) $f(-1)$ (b) $f(-2)$ (c) $f(0)$ (d) $f(1)$

17. Determine an equation of the form $y = f(x)$ which defines a function consisting of the number pairs $\{(1,1),(2,4),(3,9),(4,16)\}$.

18. Determine an equation of the form $y = f(x)$ which defines a function whose domain is the set of positive integers and which contains the ordered pairs $(7,0)$, $(8,1)$, $(10,3)$, $(12,5)$.

19. Determine an equation of the form $y = f(x)$ which defines a function whose domain is the set of positive integers and which contains the ordered pairs $(0,1)$, $(1,\frac{1}{3})$, $(2,\frac{1}{9})$, $(3,\frac{1}{27})$.

SKETCH the graph of the functions defined by the following:

20. $f(x) = x^2 + 2$ 21. $f(x) = x^2 - 4x + 5$ 22. $f(x) = x^3 - x$

23. $f(x) = -x^2 + x - 1$ 24. $f(x) = x^2 - 3x - 3$ 25. $f(x) = x^3 - 2x^2 + 3$

26. $f(x) = \begin{cases} x \text{ if } x \geq 0 \\ -x \text{ if } x < 0 \end{cases}$ 27. $f(x) = \begin{cases} x + 1 \text{ if } x \geq 0 \\ -x - 1 \text{ if } x < 0 \end{cases}$ 28. $g(x) = \begin{cases} x \text{ if } x < 0 \\ \frac{1}{2} \text{ if } x = 0 \\ x + 1 \text{ if } x > 0 \end{cases}$

3.8   INVERSE RELATIONS AND FUNCTIONS

A relation T on the set of real numbers R is a subset of $R \times R$. Consider the relation

$$T = \{(-1,-2),(0,0),(1,2),(2,4),(3,6)\}$$

whose domain is $D(T) = \{-1,0,1,2,3\}$ and whose range is

$$R(T) = \{-2,0,2,4,6\}$$

If we interchange the first and second coordinates of each ordered pair of T, we have the relation

$$S = \{(-2,-1),(0,0),(2,1),(4,2),(6,3)\}$$

Domain and range We observe that *the domain of T is the range of S, and that the range of T is the domain of S*. Thus, $D(T) = R(S)$, and $R(T) = D(S)$. We call S the *inverse relation* of T and we write

$$S = T^{-1}$$

DEFINITION. The inverse of the relation T is $\{(y,x) \mid (x,y) \in T\}$, denoted by T^{-1}.   (3.9)

EXAMPLE 1. Find the inverse relation of $f = \{(0,3),(2,7),(3,9),(-1,1)\}$.

Solution. Interchanging the first and second coordinates of each ordered pair of f, we have

$$f^{-1} = \{(3,0),(7,2),(9,3),(1,-1)\}$$

The inverse relation of a function may or may not be a function. For example, $f = \{(1,2),(2,3),(3,4),(6,4)\}$ is a function, but the inverse relation

$f^{-1} = \{(2,1),(3,2),(4,3),(4,6)\}$ is not a function because the ordered pairs $(4,3)$ and $(4,6)$ have the same first coordinate. On the other hand,

$$h = \{(-1,2),(0,3),(1,4),(2,5)\}$$

and

$$h^{-1} = \{(2,-1),(3,0),(4,1),(5,2)\}$$

are both functions.

Note that if f is a function having the property that no two distinct ordered pairs of f have the same *second* coordinate, then f^{-1} has the property that no two distinct ordered pairs of f^{-1} have the same *first* coordinate. Hence, f^{-1} is a function. In this case, when f^{-1} is also a function, it is called the *inverse function* of f.

Restriction for inverse function

When a relation is defined by an equation, the inverse relation is found by interchanging the variables in the equation. For example, if

$$f = \left\{ (x,y) \mid y = \frac{4x + 3}{2} \right\}$$

then

$$f^{-1} = \left\{ (x,y) \mid x = \frac{4y + 3}{2} \right\}$$

If we solve the equation $x = (4y + 3)/2$ for y, we get $y = (2x - 3)/4$. The inverse relation can then be written

$$f^{-1} = \left\{ (x,y) \mid y = \frac{2x - 3}{4} \right\}$$

The inverse relation written in this form is convenient for graphing.

We generally denote the function $f = \{(x,y) \mid y = f(x)\}$ by the equation $y = f(x)$. For example, the function

$$f = \left\{ (x,y) \mid y = \frac{1}{2} x + 5 \right\}$$

is abbreviated to $y = \frac{1}{2} x + 5$.

EXAMPLE 2. Let the function f be defined by the equation

$$y = f(x) = \frac{1}{3} x - 7$$

(1) Find $f^{-1}(x)$. (2) Show that $f(f^{-1}(x)) = x$. (3) Show that $f^{-1}(f(x)) = x$.

Solution. (1) Interchanging x and y in the equation $y = \frac{1}{3} x - 7$,

$$x = \frac{1}{3} y - 7$$

Then

$$y = 3x + 21$$

Hence, $f^{-1}(x) = 3x + 21$

(2) $f(f^{-1}(x)) = f(3x + 21)$

$$= \frac{1}{3}(3x + 21) - 7$$

$$= x$$

(3) $f^{-1}(f(x)) = f^{-1}\left(\frac{1}{3}x - 7\right)$

$$= 3\left(\frac{1}{3}x - 7\right) + 21$$

$$= x$$

It can be verified that when the inverse function of a function of x exists, then

$$f(f^{-1}(x)) = x$$

and $$f^{-1}(f(x)) = x$$

EXERCISE SET 3.3

1. Given the function $T = \{(0,-3),(1,3),(2,2),(3,5),(4,2)\}$. Write the ordered pairs of f^{-1}. Is f^{-1} the inverse function of f?
2. Given the function $f = \{(1,3),(2,2),(3,3),(4,0)\}$. Is f^{-1} the inverse function of f?
3. Is the inverse of the function $g = \{(-2,1),(-1,3),(0,5),(2,7)\}$ the inverse function of g?
4. If f is the function defined by $y = 4x + 3$, write the equation which defines f^{-1} in the abbreviated form $y = g(x)$. $4y + 3 = x$ $4y = x - 3$ $x - 3/4 = y$
5. Write the equation which defines f^{-1} if f is defined by the equation $y = 1/(2x + 3)$.
6. Write the equation which defines the inverse function of $y = 3/x$.
7. The function defined by $f(x) = |x|$ is called the *absolute-value* function. Its domain is the set of real numbers. The graph is shown in Fig. 3.10a. Write the ordered pairs of $f(x) = |x|$ having the following first coordinates:

   (a) $-2, -1, 0, 1, 2$ (b) $-5, -3, 2, 6$

   Is f^{-1} a function? $5, 3, 2, 6$

8. The *bracket function* defined by $f(x) = [x]$ of any real number x is the greatest integer which does not exceed x. Thus, $[x] = n$, provided $n \leq x < n + 1$, where n is an integer. Figure 3.10b is the graph.

   Write the ordered pairs of the bracket function $f(x) = [x]$ having the following first coordinates:

   (a) $5, 5.1, 5.4, 5.8, 5.99$ (b) $3, 3.2, 3.5, 3.9, 3.99$

   Is f^{-1} a function?
9. If $f(x) = x - [x]$, find $f(-5)$ and $f(3.6)$.
10. If $f(x) = [x] + [3 - x] - 2$, find $f(0)$ and $f(4)$.

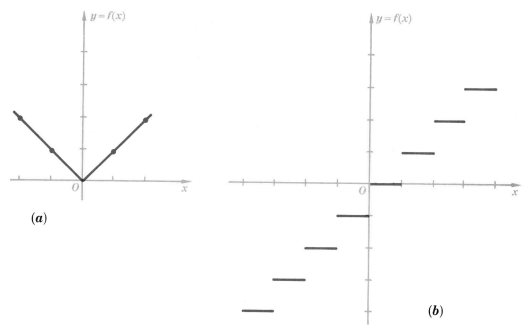

FIGURE 3.10 (*a*) Graph of $f(x) = |x|$, $x \in R$
(*b*) Graph of $f(x) = [x]$, $x \in R$

3.9 DISTANCE FORMULA

Consider any two distinct points $A(x_1,0)$ and $B(x_2,0)$ on the x axis. By convention, the distance d between A and B is always positive. Hence,

$$d = x_2 - x_1 \quad \text{or} \quad d = x_1 - x_2$$

This means that the distance between any two points on the x axis is the absolute value of the difference of the x coordinates of the points. Thus,

$$d = |x_2 - x_1|$$

Now, let $P_1(x_1,y_0)$ and $P_2(x_2,y_0)$ be any two points in the cartesian plane having the same y coordinate. The distance between P_1 and P_2 is the same as the distance between $A(x_1,0)$ and $B(x_2,0)$ (Fig. 3.11*a*). We denote this distance by $|P_1P_2|$. Then

$$|P_1P_2| = |x_2 - x_1|$$

By similar reasoning, we can show that the distance between two points $Q_1(x_0,y_1)$ and $Q_2(x_0,y_2)$ having the same x coordinate is

$$|Q_1Q_2| = |y_2 - y_1|$$

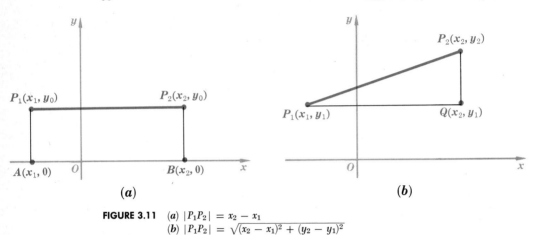

FIGURE 3.11 (a) $|P_1P_2| = x_2 - x_1$
(b) $|P_1P_2| = \sqrt{(x_2 - x_1)^2 + (y_2 - y_1)^2}$

A general formula for the distance between any two points $P_1(x_1,y_1)$ and $P_2(x_2,y_2)$ in the plane can be derived as follows, provided the same unit of length is used on both axes:

Let Q be the point whose coordinates are (x_2,y_1) in Fig. 3.11b. Then P_1QP_2 is a right triangle with hypotenuse P_1P_2.

From the theorem of Pythagoras,

$$|P_1P_2|^2 = |P_1Q|^2 + |QP_2|^2 = |x_2 - x_1|^2 + |y_2 - y_1|^2$$

and $\qquad |P_1P_2| = \sqrt{(x_2 - x_1)^2 + (y_2 - y_1)^2}$ $\qquad$ (3.10)

Distance formula

This formula is called the *distance formula* and will be referred to frequently in this book.

EXAMPLE 1. Find the distance between the points $(-5,2)$ and $(6,-1)$.

Solution. It makes no difference which point is designated as P_1. Call the first one listed P_1 and the second P_2. Then

$$|P_1P_2| = \sqrt{[6 - (-5)]^2 + (-1 - 2)^2} = \sqrt{130}$$

Equation of a circle

The distance formula can be used to obtain the standard form of the equation of a circle. Let the point $C(h,k)$ be the center of the circle whose radius is r. Let $P(x,y)$ be any point on the circle. Then

$$|CP| = r \qquad \text{and} \qquad \sqrt{(x - h)^2 + (y - k)^2} = r$$

Hence, $\qquad\qquad (x - h)^2 + (y - k)^2 = r^2$

If the center is at the origin, so that the coordinates of C are $(0,0)$, the equation becomes

$$x^2 + y^2 = r^2$$

EXAMPLE 2. Find the equation of a circle with its center at $(3,-5)$ and with radius 6.

Solution. Here $h = 3$, $k = -5$. Hence,

$$(x - 3)^2 + [y - (-5)]^2 = 6^2$$

and
$$(x - 3)^2 + (y + 5)^2 = 36$$

is the equation of the circle.

EXERCISE SET 3.4

1. Find the distance between the points

   (a) $(0,0)$ and $(3,4)$ (b) $(1,2)$ and $(4,6)$
   (c) $P(-1,3)$ and $Q(4,15)$ (d) $R(-5,-2)$ and $S(-2,6)$

2. Find the length of the line segment joining the points

   (a) $P(2,1)$ and $Q(14,6)$ (b) $S(-1,-7)$ and $T(4,5)$
   (c) $P(-9,-8)$ and $Q(-1,-2)$ (d) $P(5,-6)$ and $Q(-1,4)$

3. Show that the three points $P(-3,1)$, $Q(1,3)$, $R(9,7)$ lie on a straight line. HINT: Assume the points are in a straight line if $|PQ| + |QR| = |PR|$.
4. Show that the points $A(2,2)$, $B(6,4)$, $C(5,6)$ are the vertices of a right triangle.
5. Find the equation of the circle having the given center and radius.

   (a) $C(4,5)$, $r = 3$ (b) $C(-1,-3)$, $r = 2$

6. Find the equation of the circle whose center is $(1,4)$ and which is tangent to the x axis.

3.10   LINEAR FUNCTIONS

The function f defined by

$$f = \{(x,y) \mid y = mx + b\} \tag{3.11}$$

where m and b are constants, is called a *linear function in x*. We frequently denote this function by the equation

$$f(x) = mx + b$$

or by
$$y = mx + b$$

When the ordered pairs $(x,y) = (x, mx + b)$ of the function f are plotted on a rectangular coordinate system, the points lie on a straight line. Although a proof of this statement can be found in most textbooks on analytic geometry, the following approach leads to the conclusion. Let us select any three real values of x, say x_1, x_2, x_3 such that $x_1 < x_2 < x_3$. Then the points $P(x_1, mx_1 + b)$, $Q(x_2, mx_2 + b)$, and $R(x_3, mx_3 + b)$ are on the graph of f. By the distance formula,

$$|PQ| + |QR| = |PR|$$

Hence, we conclude that P, Q, and R lie on a straight line.

If $m = 0$, then f is the set of all ordered pairs (x, b), where x may have any value in the domain of f. The graph of $f(x) = b$ is, therefore, a straight line parallel to the x axis and b units from the origin. We call this graph the line $y = b$. For example, the line $y = 5$ is parallel to the x axis and 5 units above the origin; the line $y = -2$ is parallel to the x axis and 2 units below the origin. We call $f(x) = b$ a *constant function.*

Although the graph of a linear function is always a straight line, it is not true that any straight line is the graph of some linear function. For example, the set of ordered pairs (a, y), where y may have any value in a given set of real numbers, defines a relation that is not a function. For all values of y in this relation, $x = a$. Consequently, any two distinct ordered pairs have the same first coordinate, and the relation cannot be a function. The graph of the relation whose ordered pairs are (a, y) is readily seen to be a line parallel to the y axis at a distance of a units from the origin. We call this graph the line $x = a$. For example, the line $x = -3$ is parallel to the y axis and 3 units to the left of the origin.

Zeros of linear functions

The number x_0 is called a *zero* of the function $f(x) = mx + b$ if and only if $mx_0 + b = 0$. Thus, any x coordinate (or x value) which makes the second coordinate of an ordered pair of f equal to zero is a zero of the function. For example, consider the linear function $f(x) = 3x - 6$. To find the value of x that makes $f(x) = 0$, we set $3x - 6 = 0$ and solve the equation for x. The solution of this equation is $x = 2$. Hence, 2 is a zero of the function. Note that $f(2) = 0$.

In general, the zeros of any function are the elements in the solution set of the equation $f(x) = 0$. Geometrically, the real zeros of a function are the x coordinates of the points which the graph of the function has in common with the x axis. The graph of $f(x) = 3x - 6$ is shown in Fig. 3.12.

The linear function $f(x) = mx + b$ has only one zero; it is the element in the solution set of the linear equation $mx + b = 0$. Such equations were considered at the beginning of this chapter.

EXAMPLE 1. Find the zero of the function defined by $f(x) = 3x + 5$.

Solution. Set $3x + 5 = 0$. Then $x = -\frac{5}{3}$. Hence, $-\frac{5}{3}$ is the zero of f.

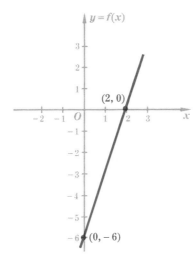

FIGURE 3.12
Graph of $f(x) = 3x - 6$. The
function has a zero at the point
$(2,0)$.

3.11   SLOPES AND EQUATIONS OF LINES

The equation

$$px + qy + r = 0 \qquad p \text{ and } q \text{ not both } 0$$

is called a *first-degree equation* or a *linear equation* in x and y.   This
equation defines a linear function whose graph is a straight line, for
if $q \neq 0$, we can write the equivalent equation

$$y = -\frac{p}{q}x - \frac{r}{q}$$

which is of the form $y = mx + b$.   If $q = 0$, the given equation reduces
to $px + r = 0$ so that

$$x = -\frac{r}{p}$$

In either case, the graph of the linear function is a straight line, and we
call it the graph of the linear equation.

If L denotes the straight-line graph of the linear equation $ax + by + c = 0$, we often write

$$L = \{(x,y) \mid ax + by + c = 0\}$$

or $L \mid ax + by + c = 0$

Any line not parallel to the y axis possesses an important property
which we now consider.   Suppose $P_1(x_1,y_1)$ and $P_2(x_2,y_2)$ are distinct
points on a nonvertical line L.   Draw a line through P_1 parallel to the

x axis and a line through P_2 parallel to the y axis. These lines meet in a point Q whose coordinates are (x_2, y_1) as in Fig. 3.13. By the distance formula

$$|P_1Q| = x_2 - x_1$$
$$|QP_2| = y_2 - y_1$$

we now define the *slope* of the line L to be the ratio of $|QP_2|$ to $|P_1Q|$ and denote this slope by the letter m. Thus,

$$m = \text{slope of } L = \frac{y_2 - y_1}{x_2 - x_1} \qquad x_2 \neq x_1 \qquad\qquad (3.12)$$

Since
$$m = \frac{y_2 - y_1}{x_2 - x_1} = \frac{-(y_2 - y_1)}{-(x_2 - x_1)} = \frac{y_1 - y_2}{x_1 - x_2}$$

it follows that the slope m is not changed if we change the order in which we select the points.

If L is parallel to the x axis, then $y_2 = y_1$ so that its slope is $0/(x_2 - x_1) = 0$. If L is perpendicular to the x axis, then $x_2 = x_1$ and $(y_2 - y_1)/0$ is not defined.

The slope of a line is also defined as the ratio of its vertical rise to its horizontal run. We chose the points P_1 and P_2 in Fig. 3.13 so that the vertical rise $y_2 - y_1 > 0$ and the horizontal run $x_2 - x_1 > 0$. In this case m is positive and the line rises from left to right. If m is negative, the line falls from left to right. If $m = 0$, the line neither rises nor falls from left to right.

EXAMPLE 1. The points $(1,7)$ and $6, -1$ are on a line. Find the slope of the line.

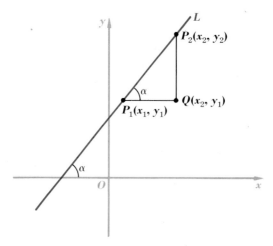

FIGURE 3.13

$$m = \text{slope} = \frac{y_2 - y_1}{x_2 - x_1}$$

Solution. $m = \dfrac{y_2 - y_1}{x_2 - x_1} = \dfrac{7 - (-1)}{1 - 6} = -\dfrac{8}{5}$

Since $m < 0$, the line falls from left to right.   To see that this is so, we need only draw the straight line through the points $(1,7)$ and $(6, -1)$.

We state two theorems concerning slopes of lines.

THEOREM.   Two lines are parallel if and only if they have the same slope.   (3.13)

THEOREM.   Two lines L_1 and L_2 having slopes m_1 and m_2, respectively, are perpendicular if and only if their slopes are negative   (3.14) reciprocals of each other.

The proofs of both these theorems are easy to construct.   We accept them here without proof.

EXAMPLE 2.   Show that if L_1 passes through $(2, -2)$ and $(4,2)$ and L_2 passes through $(1,1)$ and $(-3,3)$, then L_1 is perpendicular to L_2.

Solution.   Since $m_1 = 2$ and $m_2 = -\frac{1}{2}$, the slope of L_1 is the negative reciprocal of the slope of L_2.   Hence, the lines are perpendicular.

We turn our attention to deriving certain forms of the equation of a line. Let $P_1(x_1,y_1)$ be a fixed point on a line L having slope m.   Now suppose that $P(x,y)$ is an arbitrary point on L.   Then the slope of L is given by

$$m = \frac{y - y_1}{x - x_1}$$

Hence, $y - y_1 = m(x - x_1)$ (3.15)

Equation (3.15) is the equation of a line with slope m through the point (x_1,y_1).   It is called the *point-slope* form of the equation of the line.   This equation is satisfied by the coordinates of every point (x,y) on L.   It is not satisfied by the coordinates of a point not on L.

EXAMPLE 3.   Find the equation of the line through the point $(1, -2)$ with slope 2.

Solution.   From Eq. (3.15), we have

$$y - (-2) = 2(x - 1)$$

Hence, $y + 2 = 2x - 2$

or $2x - y - 4 = 0$

The x intercept of a line L is a if and only if L passes through the point $(a,0)$. The y intercept of L is b if and only if L passes through the point $(0,b)$.

Suppose a line L having slope m passes through the point $P_1(0,b)$. Then from Eq. (3.15)

$$y - b = m(x - 0)$$

Hence, $y = mx + b$ (3.16)

This form of the equation of L is called the *slope-intercept* form.

EXAMPLE 4. The equation of a line is $3x - 2y + 7 = 0$. Find the slope and the y intercept.

Solution. Since $3x - 2y + 7 = 0$,

$$y = \frac{3}{2}x + \frac{7}{2}$$

Hence, the slope of the line is $\frac{3}{2}$ and the y intercept is $\frac{7}{2}$.

EXAMPLE 5. Show that $L_1 \mid 3x - 2y + 4 = 0$ and $L_2 \mid 9x - 6y + 7 = 0$ are parallel.

Solution. Solving the equation of L_1 for y, we get

$$y = \frac{3}{2}x + 2$$

Hence, $m_1 = \frac{3}{2}$

Solving the equation of L_2 for y, we get

$$y = \frac{9}{6}x + \frac{7}{6}$$

Hence, $m_2 = \frac{9}{6} = \frac{3}{2}$

Since the slope of L_1 is the same as the slope of L_2, the lines are parallel.

EXAMPLE 6. Show that $7x - 5y + 8 = 0$ is perpendicular to

$$5x + 7y + 14 = 0$$

Solution. The slope of the first line is $\frac{7}{5}$, and the slope of the second is $-\frac{5}{7}$. Hence, the lines are perpendicular.

EXERCISE SET 3.5

1. The inclination of a line is $120°$. Find its slope.
2. The slope of a line is 1. Find its inclination.
3. Find the slope of the line determined by the points:

   (a) $(1,2)$ and $(2,5)$ (b) $(2,3)$ and $(8,2)$
   (c) $(-4,-2)$ and $(-5,-7)$ (d) $(0,5)$ and $(6,0)$

4. The slope of a line L passing through the point $(2,3)$ and $(8,k)$ is 2. Find the value of k.
5. Find the slope of a line perpendicular to the line determined by the points $(2,1)$ and $(8,7)$.
6. Find the slope of a line perpendicular to the line through $(-1,5)$ and $(6,-3)$.
7. Find the equation of the line through $(4,-3)$ that has slope -1.
8. Find the equation of line L if $m = \frac{2}{3}$ and $b = -2$. Recall that b is the y intercept.
9. Find the slope and the y intercept of each of the following lines:

   (a) $y = 2x + 3$ (b) $y = (-\frac{2}{3})x - 5$
   (c) $3x + 5y - 6 = 0$ (d) $x = 2y + 5$

10. Which of the following lines are parallel and which are perpendicular:

    (a) $2x + 3y + 1 = 0$ (b) $3x - 2y + 5 = 0$
    (c) $10x + 15y - 2 = 0$ (d) $2x + y - 4 = 0$

11. A line passes through the point $(2,3)$ and is parallel to the line $3x - 5y + 7 = 0$. Find the equation of the line.
12. Find the equation of the line through the point $(4,0)$ and perpendicular to the line $4x + 5y + 10 = 0$.
13. Find the equation of a line perpendicular to the line $2x - 3y + 4 = 0$ and passing through the origin.
14. Find the equation of the line which is parallel to the line $x + 3y - 7 = 0$ and which has x intercept 5.

3.12 VARIATION AND PROPORTION

There are many functional relationships in chemistry, physics, engineering, and other branches of science that are best stated in the language of variation. For example, if one end of a helical spring is attached to a support, and a weight is attached to the free end, the weight will cause the spring to stretch. The spring then exerts a force on the weight, and the farther the spring is stretched (within certain limits) the greater is the force exerted by it on the weight. We say that the force varies directly as the length of the stretch.

If the quotient $y/x^n = k$, so that

$$y = kx^n \qquad k \neq 0, n > 0$$

Direct variation

then y *varies directly* as the nth power of x, or y is *directly proportional* to the nth power of x. The constant k is called the *constant of proportionality*, or the *constant of variation*.

EXAMPLE 1. A body is dropped from a height near the surface of the earth, and the distance it falls in feet varies directly as the square of the time it falls, measured in seconds. Express this relation in the form of an equation.

Solution. Let y be the distance in feet and t be the number of seconds. Since y varies directly as the square of t, we have $y = kt^2$.

If the product $yx^n = k$, so that

$$y = kx^{-n} = \frac{k}{x^n} \qquad k \neq 0, n > 0$$

Inverse variation

then y *varies inversely* as the nth power of x, or y is *inversely proportional* to the nth power of x.

EXAMPLE 2. The attracting force of a certain electromagnet varies inversely as the square of the distance from it. Write an equation to express this relation.

Solution. Let f be the measure of the attracting force and d be the measure of the distance from the electromagnet. It follows that

$$f = \frac{k}{d^2}$$

Joint variation

When the variable z varies directly as the product $x^m y^n$, the relationship is called *joint variation*. The equation expressing this type of variation is given by

$$z = kx^m y^n \qquad k \neq 0, m \text{ and } n \text{ positive}$$

EXAMPLE 3. The volume of a right circular cone varies jointly as the square of the radius of its base and its height. Express this variation by an equation.

Solution. Let V be the volume, r the radius of the base, and h the height of the cone. Then

$$V = kr^2 h$$

Combined variation
When the variable z varies directly as x^m and inversely as y^n, the relationship is called *combined variation*. The equation which expresses this combined variation is

$$z = \frac{kx^m}{y^n} \qquad k \neq 0,\ m,\ n > 0$$

EXAMPLE 4. The intensity of sound varies directly as the strength of the source and inversely as the square of the distance from the source. Write the equation which describes the combined variation.

Solution. Let I be the intensity of the sound, S the strength of the source, and D the distance from the source. Then

$$I = \frac{kS}{D^2}$$

Solving problems that involve variation usually requires the determination of the constant of proportionality k. This constant can be found whenever the type of variation is known and one set of corresponding values of the variables is given.

EXAMPLE 5. If x varies directly as y and inversely as z, and if $x = 30$ when $y = 10$ and $z = 5$, find x when $y = 72$ and $z = 12$.

Solution. Since $x = ky/z$,

$$30 = \frac{10k}{5} \qquad \text{why?}$$

Therefore $\qquad\qquad\qquad\qquad k = 15$

The variation can now be expressed by the equation

$$x = \frac{15y}{z} \qquad \text{why?}$$

When $y = 72$ and $z = 12$, we have

$$x = \frac{15(72)}{12} = 90$$

Proportion

A *proportion* is a fractional equation of the form

$$\frac{a}{b} = \frac{c}{d}$$

Variation problems are often worked by use of a proportion. For example, consider the problem in which the resistance of a wire to the flow of electricity varies directly as its length and inversely as the square of its diameter, that is,

$$R = \frac{kL}{d^2}$$

For any set of values R_1, L_1, and d_1, we obtain

$$k = \frac{R_1 d_1^2}{L_1}$$

For a second set of values R_2, L_2, d_2

$$k = \frac{R_2 d_2^2}{L_2}$$

Hence,
$$\frac{R_1 d_1^2}{L_1} = \frac{R_2 d_2^2}{L_2}$$

If any five of these values are given, the sixth can be found.

EXERCISE SET 3.6

1. The volume of an enclosed gas varies inversely as the pressure. If a tank contains 10,000 cu ft of gas under a pressure of 20 lb/sq in., find the volume of the same gas under a pressure of 25 lb/sq in.

2. A body dropped from a height strikes the ground with a speed that varies directly as the square root of the height. If an object that falls from a height of 256 ft strikes the ground with a speed of 64 ft/sec, how far must an object fall in order to strike the ground with a speed of 32 ft/sec?

3. When a weight is attached to a spring, the spring is stretched a distance proportional to its weight. If a 3-lb weight stretches a certain spring half an inch, how much would an 18-lb weight stretch the spring, assuming the elastic limits of the spring are not exceeded?

4. If y varies as x and inversely as the cube of z, what change in y results from doubling x and halving z?

5. The area of a circle varies as the square of its radius. If the area of a circle is 154 sq in. when the radius is 7 in., find the area of a circle whose radius is 7 ft.

6. The resistance of a wire to the flow of electricity varies directly as its length and inversely as the square of its diameter. If a wire 350 ft long and 3 mm in diameter has a resistance of 1.08 ohms, find the length of a wire of the same material whose resistance is 0.72 ohm and whose diameter is 2 mm.

7. The amount of steam per second which will flow through a hole varies jointly as the steam pressure and the area of the cross section of the hole. If 40 lb of steam per second at a pressure of 150 lb/sq in. flows through a hole whose area is 12 sq in., how much steam at a pressure of 200 lb/sq in. will flow through a hole whose area is 18 sq in.?

8. If a body starts falling from rest, the total distance fallen varies as the square of the time during which it has fallen. If it falls 64 ft in 2 sec, how far will it fall in 8 sec?

9. The volume of a certain gas in a storage tank varies inversely as the pressure. If a certain amount of this gas occupies 48,000 cu ft when it is under a pressure of 3 lb/sq in., find the amount of space it would occupy under a pressure of 4 lb/sq in.

10. If z varies directly as x and inversely as y, and $z = 42$ when $x = 8$ and $y = 3$, find the value of z when $x = 3$ and $y = 5$.

11. At countdown an astronaut weighs 168.1 lb. Find his approximate weight 100 mi above the Earth's surface. The weight of a body above the Earth's surface varies inversely as the square of the distance from the center of the Earth. Take the radius of the Earth to be 4,000 mi, approximately.

12. Assume that the acceleration of gravity above the moon's surface varies inversely as the square of the distance from the center of the moon. If the acceleration of gravity on the surface of the moon is approximately 5.4 ft/sec^2, find the approximate acceleration of gravity of a lunar module 100 mi above the moon's surface. Take the radius of the moon to be approximately 2,200 mi

3.13 LITERAL EQUATIONS AND FORMULAS

An equation in which some of the known quantities (or constants) are represented by letters different from the letter representing the unknown quantity is called a *literal equation*. For example, if x is taken to be the unknown quantity, then $(k + x)^2 + 5k = (3 + x)^2 - 2k$ is a literal equation.

Formulas may be regarded as literal equations. In most cases, a formula can be solved for one particular letter in the formula by considering it to represent the unknown quantity. For example the formula for the simple interest I on a principal of P dollars at the rate of r percent per year for t years is given by the formula $I = Prt$. If we take P as the only unknown in this formula, we can solve for P by dividing each side of the equation by rt. Hence,

$$P = \frac{I}{rt}$$

If r is the unknown, then $r = I/Pt$, and if t is the unknown, $t = I/Pr$. We call this procedure *changing the subject of a formula*.

There are two kinds of formulas. *Mathematical formulas* are derived by analysis, and *empirical formulas* are derived from experimentation. Some of the familiar mathematical formulas are $A = \frac{1}{2}bh$, $A = \pi r^2$, $V = \pi r^2 h$, $V = e^3$, $A = \frac{1}{2}h(b_1 + b_2)$. In empirical formulas, the letters representing constants are chosen to agree with the results of experiments. Some examples of empirical formulas are the following:

1. The strength S of a rectangular beam is given by $S = kwd^2$, where w is the width and d is the depth of the beam, and k is a constant depending on the material.

2. A horizontal beam is supported at the ends. The width is w, the depth is d, the length is l. Experiments show that the safe load W of the beam is given by $W = kwd^2/l$, where k is a constant that depends on the material of which the beam is made.

3. The crushing load L of a square oak pillar is given by $L = kx^4/l$, where x is the thickness and l is the length of the pillar.

EXERCISE SET 3.7

SOLVE the following formulas as directed, in terms of the remaining letters.

1. $s = gt^2/2$ for g

2. $s = vt + s_0$ for v

3. $F = 9C/5 + 32$ for C

4. $V = \pi r^2 h/3$ for h

5. $S = n(a + l)/2$ for l

6. $S = (a - ar^n)/(1 - r)$ for a

7. $I = \dfrac{E}{r/n + R}$ for n

8. $s = 50t - g/t^2$ for g

9. The following table shows a linear (first-degree) relation between x and y. If y varies directly as x, construct a formula for y in terms of x.

x	1	2	5	8
y	-1	2	11	20

10. A circle whose radius is r in. has been cut from a square metal plate whose side is s in. If $r < s/2$, write a formula for the area A of the plate.

11. The geometric mean between two positive numbers a and b is the square root of their product. Construct a formula for the geometric mean g in terms of a and b.

4

EXPONENTS. EXPONENTIAL FUNCTIONS

In Sec. 2.6 we defined positive-integer exponents and then introduced certain basic theorems for operating with them. We now consider methods of defining the zero exponent, negative exponents, and rational exponents so that the basic theorems will also apply to them. The theorems of Sec. 2.6 are restated here as the *laws of exponents*. By the notation $n \in P$ we mean that n is an element of the set P of positive integers.

LAW 1. THE MULTIPLICATION LAW. If $m, n \in P$ and a is any number, then

$$a^m \cdot a^n = a^{m+n} \tag{4.1}$$

For a proof of this law, see Sec. 2.6.

LAW 2. THE DIVISION LAW. If $m, n \in P$ and $a \neq 0$, then

$$\frac{a^m}{a^n} = a^{m-n} \qquad \text{if } m > n$$

$$= 1 \qquad \text{if } m = n \tag{4.2}$$

$$= \frac{1}{a^{n-m}} \qquad \text{if } m < n$$

A proof was given in Sec. 2.6.

LAW 3. THE POWER OF A PRODUCT LAW. If $m \in P$ and a and b are any numbers, then

$$(ab)^m = a^m b^m \tag{4.3}$$

Proof. $(ab)^m = (ab)(ab)\cdots(ab)$ with m factors of ab

$\qquad\qquad = (aa\cdots a)(bb\cdots b)$ with m factors of a

followed by m factors of b, by the commutative and associative laws. Hence $(ab)^m = a^m b^m$ by definition.

LAW 4. THE POWER OF A QUOTIENT LAW.   If $m \in P$ and a and b are any numbers, $b \neq 0$, then

$$\left(\frac{a}{b}\right)^m = \frac{a^m}{b^m} \tag{4.4}$$

The proof of this law is left to the student.

LAW 5. THE POWER OF A POWER LAW.   If $m, n \in P$ and a is any number,

$$(a^m)^n = a^{mn} \tag{4.5}$$

The proof is left to the student.

4.1   ZERO EXPONENT

Thus far, we have defined a^n only when $n \in P$. Let us now extend our definition to the cases where n is zero, a negative integer, or a rational number. We are at liberty to define such exponents in many different ways. However, the logical thing to do is to define them so that the five laws of exponents for the positive integers will remain valid. Hence, to define the zero exponent, we reason as follows:

If Law 1 is to hold when one of the exponents, say n, is zero, then

$$a^m \cdot a^0 = a^{m+0} = a^m$$

Since $a^m \cdot 1 = a^m$ why?

it follows that $a^m \cdot a^0 = a^m \cdot 1$

and $a^0 = 1$ why?

Thus, it seems reasonable to define the zero exponent as follows:

DEFINITION.   If $a \in R$, $a \neq 0$, then

$$a^0 = 1 \tag{4.6}$$

Under this definition, all five of the laws for positive-integer exponents are preserved when any exponent is zero. The proof that such is the case is left as an exercise.

EXAMPLES.

$$3^0 = 1$$
$$2(3x - 7y)^0 = 2(1) = 2$$
$$[7(x + 2y - 1)^2]^0 = 1$$

4.2   NEGATIVE-INTEGER EXPONENTS

Let a be a real number and let m and n be positive integers. Suppose, also, that $m > n$. Now, if Law 1 is to be preserved, then

$$a^m \cdot a^{-n} = a^{m+(-n)} = a^{m-n}$$

If Law 2 is to hold also, then

$$a^m \cdot \frac{1}{a^n} = \frac{a^m}{a^n} = a^{m-n}$$

Hence, $$a^m \cdot a^{-n} = a^m \cdot \frac{1}{a^n}$$

and $$a^{-n} = \frac{1}{a^n}$$

This conclusion suggests the definition for a negative-integer exponent.

DEFINITION.   If $a \in R$, $a \neq 0$, and $n \in P$, then

$$a^{-n} = \frac{1}{a^n}$$

(4.7)

The five laws of exponents for the positive integers remain valid under this definition. The proof that the five laws do hold is left as an exercise.

EXAMPLES. $$7^{-2} = \frac{1}{7^2} = \frac{1}{49}$$

$$x^3 y^{-3} z^4 = x^3 \left(\frac{1}{y^3}\right) z^4 = \frac{x^3 z^4}{y^3}$$

$$\frac{x^{-2} z^4}{y^{-3}} = \frac{(1/x^2) \cdot z^4}{1/y^3} = \frac{y^3 z^4}{x^2}$$

$$\left(\frac{a}{b}\right)^{-1} = \frac{1}{a/b} = \frac{b}{a}$$

EXERCISE SET 4.1

PERFORM the indicated operations, using the laws of exponents. Simplify by removing all negative exponents, and reduce all complex fractions to simple fractions.

1. $x^{4m} \cdot x^{3m}$ 2. $a^{5x} \cdot a^x$
3. $20a^{x+y} \div 5a^y$ 4. $10^7 \div 10^6$
5. $(x + y)^3 (x + y)^2$ 6. $(2x + 3y)^m (2x + 3y)^{2m}$
7. $5^4 \cdot 2^4 \cdot 3^4$ 8. $[(x + y)^2 (x + y)^3] \div (x + y)^4$
9. $(x^{-2})^3$ 10. $(y^{-3})^{-4}$
11. $(x^{-4} y^{-2} z^0)^{-1}$ 12. $(\tfrac{2}{3})^{-3} (\tfrac{5}{2})^0$
13. $(3^{-2})(3^2)(7^0)(6)$ 14. $(x^0 y^{-3} z^{-2})^2$

15. $\dfrac{x^{-2}y^{-5}}{x^{-3}}$

16. $\dfrac{a}{b^{-1}} + \dfrac{b}{a^{-1}}$

17. $x^{-1} + y^{-1}$

18. $z^{-2} + w^{-2}$

19. $(x^n y^3)(x^{n+1} y^2)^{-2}$

20. $\dfrac{4^{-2} x^{-3}}{(2x)^{-4}}$

21. $\dfrac{[(2x + 3y)^2]^0}{(2x + 3y)^2}$

22. $\dfrac{(x^2 y^{-2})^3 (x^{-2} y)^2}{(xy^{-1})^6}$

23. $\dfrac{a^{-2}}{a^{-2} - b^{-2}}$

24. $(x^{-1} + y^{-1})^{-1}$

25. $(a^{-1} + b^{-1})(a^{-1} - b^{-1})$

26. $(a^y + b^{-y})(a^{-y} + b^y)$

27. $\dfrac{b^{-x}}{b^{-y}} + \dfrac{b^{-y}}{b^{-x}}$

28. $\dfrac{x^3}{y^{-2}} \div \dfrac{x^{-2}}{y^3}$

29. $\dfrac{1 - x^{-3}}{1 + x^{-3}}$

30. $\dfrac{x^{-1} + y^{-1}}{(x + y)^{-1}}$

31. $\dfrac{3 + 3x^2}{1 + x^{-2}}$

32. $\dfrac{9x^3 - x^{-1}}{3 - x^{-2}}$

4.3　RATIONAL EXPONENTS

Before we assign any meaning to powers with rational exponents, let us first consider the following definition.

DEFINITION.　If $a^n = b$, $n \in P$, then a is an *n*th root of b.　　　　(4.8)

For example, since $2^4 = 16$, it follows that 2 is a fourth root of 16. Since $(-2)^4 = 16$, it follows that -2 is also a fourth root of 16. Similarly, 2 is a fifth root of 32, and -2 is a fifth root of -32.

The following properties of real numbers can be established, but the proofs are not easy. At this point, we shall take them on faith.

1.　If $b \in R$, $b \neq 0$, then there exist exactly n distinct nth roots of b. Some (or all) of the n nth roots of b may not be real numbers. For example, there are two real fourth roots of 16 (2 and -2). There are two other fourth roots that are not real. The real number -16 has no real fourth root, but -16 has four fourth roots that are not real. We shall consider such nonreal roots in a later chapter.

2.　If b is a positive real number and n is an even positive integer, then there are two real nth roots of b. One of them is positive and the other is negative. The positive nth root is denoted by $\sqrt[n]{b}$ and is called the *principal nth root of b*. The negative nth root of b is denoted by $-\sqrt[n]{b}$. The two real roots are often taken together and denoted by $\pm\sqrt[n]{b}$. Thus, $\sqrt[4]{16} = 2$ (the principal fourth root) and $-\sqrt[4]{16} = -2$. The two roots that are real may be taken together and denoted by $\pm\sqrt[4]{16}$. For emphasis, we point out that $\sqrt[4]{16}$ is not ± 2.

Principal *n*th root

   3. If b is a negative number and n is an even positive integer, then b has no real nth root. This is evident from the fact that an even power of a real number is always positive, never negative. For example, no real number x exists such that $x^2 = -16$. Consequently, the real number -16 has no real second root.

   4. If b is a positive number and n is an odd positive integer, then there exists one and only one real nth root of b. This root is positive and is the principal nth root of b, denoted by $\sqrt[n]{b}$. For example, $\sqrt[3]{8} = 2$. The other two cube roots of 8 are not real numbers.

   5. If b is a negative number and n is an odd positive integer, there is again one and only one real nth root of b. This root is negative. For example, $\sqrt[3]{-27} = -3$. The two other third roots of -27 are nonreal.

   6. If $b = 0$, and $n \in P$, the nth root of b is zero.

   We are now ready to assign a value to the symbol $b^{1/n}$, where b is any number and n is a positive integer. If Law 5 is to be preserved in the case where $m = 1/n$, then we must have

$$(b^{1/n})^n = b^{(1/n) \cdot n} = b^1 = b$$

Hence, it seems reasonable to define $b^{1/n}$ to be the unique principal nth root of b.

DEFINITION.   If $b \in R$, $n \in P$, then

$$b^{1/n} = \sqrt[n]{b} \qquad\qquad (4.9)$$

except when b is negative and n is even.

   The symbol $\sqrt[n]{b}$ is called a *radical*. The positive integer n is the *index* and b is the *radicand* of the radical. If n is 2, we do not write the index of the radical. Thus, $\sqrt[2]{b}$ is written $\sqrt{b}$.

EXAMPLES.

(1) $\qquad\qquad\qquad\qquad 9^{1/2} = \sqrt{9} = 3$
(2) $\qquad\qquad\qquad\qquad 27^{1/3} = \sqrt[3]{27} = 3$
(3) $\qquad\qquad\qquad (-125)^{1/3} = \sqrt[3]{-125} = -5$

Note that such symbols as $(-9)^{1/2}$, $(-64)^{1/4}$ are not defined in the set of real numbers.

   From Definition (4.9) and Law 5, it follows that if m, $n \in P$, then we must have

$$b^{m/n} = (b^{1/n})^m = (\sqrt[n]{b})^m$$
$$b^{m/n} = (b^m)^{1/n} = \sqrt[n]{b^m}$$

This immediately suggests the following definition:

DEFINITION.   If the rational number m/n is in lowest terms and n is positive, then

$$b^{m/n} = (\sqrt[n]{b})^m = \sqrt[n]{b^m} \tag{4.10}$$

except when b is negative and n is even.

Note carefully the restrictions on b, m, and n stated in the definition. They are important. Without these restrictions, what would prevent us from making such statements as the following:

$$(-4)^{2/4} = \sqrt[4]{(-4)^2} = \sqrt[4]{16} = \sqrt[4]{(4)^2} = (4)^{2/4}$$

The definition applies only to rational exponents that are in lowest terms and have a positive denominator. It does not apply when $b < 0$ and n is even.

If $m = kp$ and $n = kq$, where p and q have no common divisors other than 1 or -1, and $k > 0$, it can be shown that no contradiction will result if we make the further definition that

$$b^{m/n} = b^{kp/kq} = b^{p/q} \qquad b > 0$$

Thus, $(16)^{3/12} = (16)^{1/4} = \sqrt[4]{16} = 2$. Also,

$$(27)^{8/6} = (27)^{4/3} = (\sqrt[3]{27})^4 = 81$$

If $b > 0$ and m, $n \in P$, we define $b^{-m/n}$ as follows:

$$b^{-m/n} = \frac{1}{b^{m/n}}$$

In order that the definitions of this section be acceptable, we need to show that all the laws of exponents hold under the definitions. The proofs that such is the case are left as exercises.

Simplifying roots   In simplifying roots of real numbers and in performing operations upon them, we may use either the exponential notation or the radical notation. Sometimes one is more convenient than the other. In the following examples, we assume that the indicated roots exist and that no denominator is zero.

EXAMPLE 1.   Find the value of $(^{27}\!/_{64})^{2/3}$.

Solution. $\left(\dfrac{27}{64}\right)^{2/3} = \left(\dfrac{3^3}{4^3}\right)^{2/3} = \left[\left(\dfrac{3}{4}\right)^3\right]^{2/3} = \left(\dfrac{3}{4}\right)^2 = \dfrac{9}{16}$

EXAMPLE 2.   Use a single radical to express the product $\sqrt{3} \cdot \sqrt[3]{4}$.

Solution. $\sqrt{3} \cdot \sqrt[3]{4} = 3^{1/2} \cdot 4^{1/3} = 3^{3/6} \cdot 4^{2/6} = (3^3 \cdot 4^2)^{1/6}$

$$= (432)^{1/6} = \sqrt[6]{432}$$

Simplifying radicals

We can often simplify radicals by changing them into equivalent forms which we consider more useful.

EXAMPLE 3. Simplify the radical $\sqrt{98x^3y^{-4}}$.

Solution. $\sqrt{98x^3y^{-4}} = (2 \cdot 7^2 x^3 y^{-4})^{1/2} = 2^{1/2} \cdot 7 \cdot x^{3/2} \cdot y^{-2}$
$$= 7xy^{-2}(2^{1/2}x^{1/2}) = 7xy^{-2}\sqrt{2x}$$

The radical of the preceding example can be simplified without using the exponential notation.   We recall that if $a > 0$ and $b > 0$, then

$$\sqrt[n]{ab} = (ab)^{1/n} \qquad \text{by definition}$$
$$= a^{1/n}b^{1/n} \qquad \text{by Law 5}$$

Hence $\sqrt[n]{ab} = \sqrt[n]{a} \cdot \sqrt[n]{b} \qquad$ by definition

If either $\sqrt[n]{a}$ or $\sqrt[n]{b}$ is rational, we say that $\sqrt[n]{ab}$ has been simplified. Thus,

$$\sqrt{98x^3y^{-4}} = \sqrt{49x^2y^{-4} \cdot 2x} = \sqrt{49x^2y^{-4}} \cdot \sqrt{2x} = 7xy^{-2}\sqrt{2x}$$
$$\sqrt[3]{250} = \sqrt[3]{125 \cdot 2} = \sqrt[3]{125} \cdot \sqrt[3]{2} = 5\sqrt[3]{2}$$

Certain radicals that have fractional radicands can also be simplified without using exponential notation.   If $a > 0$ and $b > 0$, then

$$\sqrt[n]{\frac{a}{b}} = \left(\frac{a}{b}\right)^{1/n} = \frac{a^{1/n}}{b^{1/n}} = \frac{\sqrt[n]{a}}{\sqrt[n]{b}}$$

If $\sqrt[n]{b}$ is rational, we say that the radical $\sqrt[n]{a/b}$ has been simplified.   For example,

$$\sqrt[4]{\frac{7}{16}} = \frac{\sqrt[4]{7}}{\sqrt[4]{16}} = \frac{\sqrt[4]{7}}{2} = \frac{1}{2}\sqrt[4]{7}$$

If an expression contains more than one term involving radicals, we use the distributive law to combine any radicals having the same index and radicand.

EXAMPLE 4. Combine like terms: $\sqrt{98} + 2\sqrt{2} - \sqrt{32}$.

Solution. $\sqrt{98} + 2\sqrt{2} - \sqrt{32} = \sqrt{49 \cdot 2} + 2\sqrt{2} - \sqrt{16 \cdot 2}$
$$= 7\sqrt{2} + 2\sqrt{2} - 4\sqrt{2}$$
$$= (7 + 2 - 4)\sqrt{2}$$
$$= 5\sqrt{2}$$

EXAMPLE 5. Combine: $\sqrt[3]{5x^2} + 3\sqrt[6]{25x^4}$.

Solution.

$$\sqrt[3]{5x^2} + 3\sqrt[6]{25x^4} = (5x^2)^{1/3} + 3(5^2x^4)^{1/6}$$
$$= 5^{2/6}x^{4/6} + 3(5^{2/6}x^{4/6})$$
$$= (1 + 3)5^{2/6}x^{4/6}$$
$$= 4(5^{1/3}x^{2/3})$$
$$= 4\sqrt[3]{5x^2}$$

If a fraction involves a radical in its denominator, it may be desirable to find an equivalent fraction which will not have a radical in the denominator. This process is called *rationalizing the denominator*. If the denominator consists of a single radical, we may proceed as follows.

Rationalizing
denominators

EXAMPLE 6. Rationalize the denominator of $\dfrac{3}{\sqrt[7]{2x^2}}$.

Solution.

$$\frac{3}{\sqrt[7]{2x^2}} = \frac{3}{2^{1/7}x^{2/7}} = \frac{3}{2^{1/7}x^{2/7}} \cdot \frac{2^{6/7}x^{5/7}}{2^{6/7}x^{5/7}}$$

$$= \frac{3(2^{6/7}x^{5/7})}{2x} = \frac{3\sqrt[7]{2^6x^5}}{2x} = \frac{3\sqrt[7]{64x^5}}{2x}$$

If the denominator of a fraction is the sum or difference of two numbers involving radicals, we proceed as in the following example:

EXAMPLE 7. Rationalize the denominator of $\dfrac{3}{6 + \sqrt{3}}$.

Solution. We know that $(a + b)(a - b) = a^2 - b^2$. Hence,

$$(6 + \sqrt{3})(6 - \sqrt{3}) = (6)^2 - (\sqrt{3})^2 = 36 - 3 = 33$$

Thus, if we multiply both numerator and denominator of the given fraction by $6 - \sqrt{3}$, the radical will be eliminated from the denominator. Therefore,

$$\frac{3}{6 + \sqrt{3}} = \frac{3}{6 + \sqrt{3}} \cdot \frac{6 - \sqrt{3}}{6 - \sqrt{3}} = \frac{3(6 - \sqrt{3})}{36 - 3} = \frac{3(6 - \sqrt{3})}{33}$$

EXAMPLE 8. Factor $(25 - x^2)^{3/2} + 2x^2(25 - x^2)^{1/2}$.

Solution. $(25 - x^2)^{1/2}$ is a common factor of each term of the given expression. Hence,

$$(25 - x^2)^{3/2} + 2x^2(25 - x^2)^{1/2} = (25 - x^2)^{1/2}[(25 - x^2) + 2x^2]$$
$$= (25 - x^2)^{1/2}(25 + x^2)$$

EXERCISE SET 4.2

FIND the rational values of each of the following:

1. $25^{1/2}$; $64^{1/3}$; $125^{1/3}$; $16^{3/2}$ 2. $4^{3/2}$; $4^{5/2}$; $8^{2/3}$; $9^{3/2}$
3. $27^{2/3}$; $8^{5/3}$; $32^{1/5}$; $4^{7/2}$; $(16\!/\!49)^{1/2}$; $(49\!/\!81)^{1/2}$ 4. $81^{1/4}$; $64^{2/3}$; $(1\!/\!27)^{2/3}$; $(1\!/\!32)^{3/5}$; $(4\!/\!49)^{3/2}$; $(27\!/\!8)^{5/3}$

REMOVE all possible fractions from the radical and rationalize denominators wherever necessary:

5. $\sqrt{x^6}$ 　　　　　　　　　6. $(\sqrt[3]{x})^6$ 　　　　　　　　7. $\sqrt{24x^3}$

8. $\sqrt{50x^4}$ 　　　　　　　　9. $\sqrt{147}$ 　　　　　　　　　10. $\sqrt{162x^4y^8}$

11. $\sqrt{45}\cdot\sqrt{5}$ 　　　　　　12. $\sqrt{7}\cdot\sqrt{28}$ 　　　　　　13. $\sqrt[3]{4}\cdot\sqrt[3]{14}$

14. $\sqrt[3]{108}\cdot\sqrt[3]{4}$ 　　　　15. $\sqrt[3]{3}\cdot\sqrt[3]{9}$ 　　　　　16. $\sqrt[4]{2}\cdot\sqrt[4]{8}$

17. $\dfrac{\sqrt{72}}{\sqrt[3]{18}}$ 　　　　　　18. $\dfrac{\sqrt[3]{5}}{\sqrt[3]{8}}$ 　　　　　　19. $\dfrac{\sqrt[3]{108}}{\sqrt[3]{4}}$

20. $\dfrac{\sqrt[3]{56}}{\sqrt[3]{-7}}$ 　　　　　21. $\sqrt{7\!/\!21}$ 　　　　　　　22. $\sqrt[3]{3\!/\!8}$

23. $\sqrt[3]{2\!/\!5}$ 　　　　　　　24. $\sqrt{\dfrac{3x}{2y^3}}$ 　　　　　25. $\sqrt[3]{\dfrac{3x^2}{2y^2}}$

26. $\sqrt[4]{\dfrac{4xy^5}{81z^4}}$ 　　　　　27. $\dfrac{6x\sqrt{2xy^3}}{x\sqrt{8x^3y}}$ 　　　　28. $\dfrac{\sqrt[4]{8}}{\sqrt[4]{3}}$

29. $(4 + x^{-1/2})(4 - x^{-1/2})$ 　　　　　30. $(x^{-1/2} + y^{-1/2})^2$

31. $(x^{1/2} - y^{1/2})(x^{1/2} + y^{1/2})$ 　　　32. $\dfrac{y^{1/2} + y^{-1/2}}{y^{1/2} - y^{-1/2}}$

33. $\dfrac{2}{2 - \sqrt{3}}$ 　　　　　　34. $\dfrac{1}{\sqrt{6} - 2}$

35. $\dfrac{3}{\sqrt{5} - 1}$ 　　　　　　36. $\dfrac{1}{\sqrt{2} - \sqrt{3}}$

37. $\dfrac{a}{a - \sqrt{a^2 - 16}}$ 　　　　38. $\dfrac{b + \sqrt{b^2 - 9}}{b - \sqrt{b^2 - 9}}$

39. $\dfrac{\sqrt{a} - \sqrt{a + 1}}{\sqrt{a} + \sqrt{a + 1}}$ 　　　　40. $\dfrac{\sqrt{a} + \sqrt{a^2 - b^2}}{\sqrt{a} - \sqrt{a^2 - b^2}}$

41. $\dfrac{x^{-1+1/(1-m)} - x^{1/(1-m)}}{x^{m/(1-m)}}$ 　　42. $\sqrt{4x - x^2} + \dfrac{x(1 - x)}{\sqrt{4x - x^2}}$

43. $\dfrac{(x^2/\sqrt{a - x^2}) + \sqrt{a - x^2}}{a - x^2}$ 　　44. $\dfrac{\sqrt{1 + x^2} - (x^2/\sqrt{1 + x^2}) + 1}{1 + \sqrt{1 + x^2}}$

45. $(a^2 + b^2)^{3/2} + a^2(a^2 + b^2)^{1/2}$ 　　46. $(4 - x^2)^{5/2} + x^2(4 - x^2)^{3/2} - (4 - x^2)$
47. $(x^2 - 25)^{1/2} - x^2(x^2 - 25)^{-3/2}$

4.4   RADICAL EQUATIONS

　　　An equation in which the variable has a fractional exponent or occurs in a radicand is called a *radical equation* or an *irrational equation*. For example,

$$x^{1/3} - 8 = 0 \qquad \text{and} \qquad \sqrt{x + 1} = 3$$

are both radical equations. The method of solving such equations is suggested by the following theorem:

THEOREM. If $a = b$ and if n is a rational number, then

$$a^n = b^n$$

(4.11)

The proof follows from the fact that $a = b$ means that a and b represent the same number. Hence, the expressions a^n and b^n represent the same number.

A frequent error results from assuming the converse of Theorem (4.11). The converse is not true, for $a^n = b^n$ does not imply that $a = b$. For example,

$$(-2)^2 = (+2)^2$$

but
$$-2 \neq +2$$

When each side of the equation $a = b$ is raised to the nth power, the equation $a^n = b^n$ results from the operation. In the process, extraneous solutions may be introduced. Hence, it is necessary to check each solution of the transformed equation $a^n = b^n$ to determine whether or not it satisfies the given equation.

EXAMPLE 1. Solve for x: $\sqrt{x - 6} + 3 = \sqrt{x + 9}$.

Solution. If we square each side of the equation, we get

$$(\sqrt{x - 6})^2 + 6\sqrt{x - 6} + 9 = (\sqrt{x + 9})^2$$

Then
$$x - 6 + 6\sqrt{x - 6} + 9 = x + 9$$
$$6\sqrt{x - 6} = 6$$

and
$$\sqrt{x - 6} = 1$$

If we square each side of this equation, we get

$$x - 6 = 1$$

Hence
$$x = 7$$

Check. If $x = 7$,

Left Member	Right Member
$\sqrt{7 - 6} + 3$	$\sqrt{7 + 9} = \sqrt{16}$
$= 4$	$= 4$

EXAMPLE 2. Solve for x: $\sqrt{30 - 3x} + \sqrt{23 - 2x} = 0$.

Solution. Write the equation so only one of the radicals will be on the left side of the equation:

$$\sqrt{30 - 3x} = -\sqrt{23 - 2x}$$

Then $$30 - 3x = 23 - 2x$$

and $$x = 7$$

Check. If $x = 7$, the left member is

$$\sqrt{30 - 21} + \sqrt{23 - 14} = \sqrt{9} + \sqrt{9} = 3 + 3 = 6$$

Since the right member is 0, we see that $x = 7$ is not a solution of the given equation. Since $x = 7$ is the only possible solution of the transformed equation, we must conclude that the original equation has no solution. Thus, the solution set is the empty set $\varnothing$.

In solving equations by raising each side to the same power, we assume the following:

If each side of an equation is raised to the same power, the solution set of the transformed equation contains all of the solutions of the original equation. For example, if $x = 5$, then the solution set of $x^2 = 25$ contains 5. If $x = 2$, then the solution set of $x^3 = 8$ contains 2.

EXERCISE SET 4.3

SOLVE and check. Recall that the solution set of any derived equation may or may not be the solution set of the given equation. Hence, the check is essential.

1. $(x + 4)^{1/2} - 3 = 0$ 2. $(3x - 6)^{1/2} - 6 = 0$
3. $(x + 5)^{1/2} + 4 = 0$ 4. $x - (x^2 - 1)^{1/2} = 1$
5. $\sqrt{4x^2 - 5} - 2x = 4$ 6. $2 + \sqrt{3x + 6} = 0$
7. $\sqrt{16y^2 + 25} - 4y = 2$ 8. $\sqrt{9x^2 + 4} = 3x - 2$
9. $(x + 2)^{1/3} = 2$ 10. $(2x - 1)^{1/3} = 3$
11. $(x^2 + 1)^{1/2} + x - 1 = 0$ 12. $4x + (16x^2 - 5)^{1/2} = 2$
13. $(9x - 4)^{1/3} = 3(x - 1)^{1/3}$ 14. $\sqrt{x} + \sqrt{x - 5} = 3$
15. $\sqrt{x - 4} = \sqrt{x + 11} - 3$ 16. $\sqrt{2x - 1} = \sqrt{2x + 11}$
17. $\sqrt{x - 5} + \sqrt{x + 10} + 3 = 0$ 18. $\sqrt{x + 4} = 2 - \sqrt{x + 16}$
19. $\sqrt{\sqrt{x - 4}} - 2 = 0$ 20. $\sqrt[3]{\sqrt{3x + 1}} = 2$
21. $\sqrt{\sqrt{400}} = \sqrt{5} + \sqrt{x}$ 22. $\sqrt{(x/2) - 1} - 3 = 0$
23. $\sqrt{(x/4) + 1} - 6 = 0$ 24. $\dfrac{5}{3} = \sqrt{(x/3) - 1}$
25. $\sqrt{3 + (x/2)} = \sqrt{(5x/2) - 13}$ 26. $\sqrt[3]{(2x/3) + 23} = \sqrt[3]{(3x/2) + 18}$
27. $\sqrt[3]{(3x^2 - 7x + 8)/4} = x^{2/3}$

4.5   EXPONENTIAL FUNCTIONS

If b is a real number, $b > 0$, we have defined b^n for all rational values of n. In the case where $b < 0$ and n is a rational fraction in lowest terms with an even denominator, b^n is not defined in the set of real numbers.

Irrational exponents
If b is positive, it is also possible to define b^n for n an irrational number. The definition requires the use of successive rational approximations to the irrational exponent and cannot be discussed fully in an elementary book. As an example, let us consider the meaning of $2^{\sqrt{2}}$. A very rough first approximation to $\sqrt{2}$ is the rational number 1.4. Thus, $2^{\sqrt{2}}$ is approximately equal to $2^{1.4}$, which is defined. A better approximation to $\sqrt{2}$ is 1.41. Hence, $2^{\sqrt{2}}$ is better approximated by $2^{1.41}$, which is defined. As we take better rational approximations to $\sqrt{2}$, we get better approximations to $2^{\sqrt{2}}$. In this way, we give meaning to $2^{\sqrt{2}}$.

In order to give meaning to the symbol b^n, $b > 0$ and n any real number, we accept the following statement:

b^x as unique
positive real
number

> If $b \in R$, $b > 0$, and $x \in R$, then we can assign to the symbol b^x one and only one positive real number such that all the laws of exponents are preserved.

The assignment of one and only one real number to b^x for any real x defines a function whose ordered pairs are (x, b^x).

DEFINITION.   If $b > 0$ and $x \in R$, then the function defined by

$$f(x) = b^x \qquad (4.12)$$

is called an exponential function.

Graph of the
exponential function
A study of the graph of the exponential function will aid us in understanding the behavior of the function. To construct the graph, construct a subset of the set of ordered pairs (x, b^x) of the function. Plot the points whose coordinates are the ordered pairs thus found and join them in order from left to right with a smooth curve. The figure then represents approximately the graph of the function.

EXAMPLE 1.   Sketch the graph of f if $f(x) = 2^x$.

Solution.   The following subset of the ordered pairs $(x, 2^x)$ of the function can be found by assigning arbitrary values to x:

$$\left\{ \left(-3, \frac{1}{8}\right), \left(-2, \frac{1}{4}\right), \left(-1, \frac{1}{2}\right), (0,1), (1,2), (2,4), (3,8) \right\}$$

These points are plotted and the graph sketched in Fig. 4.1a.

EXAMPLE 2. Sketch the graph of f if $f(x) = (\frac{1}{2})^x$.

Solution. A subset of the ordered pairs $(x,(\frac{1}{2})^x)$ is the set

$$\left\{(-3,8),(-2,4),(-1,2),(0,1),\left(1,\frac{1}{2}\right),\left(2,\frac{1}{4}\right),\left(3,\frac{1}{8}\right)\right\}$$

The graph is shown in Fig. 4.1b.

If $b > 1$, the graph of the exponential function b^x has the same general shape as the graph of the function 2^x and appears to have the following properties: (1) The value of the function is positive. Thus, the graph lies entirely above the x axis. (2) The value of the function increases as x increases. (3) The value of the function is 1 for $x = 0$. (4) If $x_2 > x_1$, then $b^{x_2} > b^{x_1}$. (5) If x is negative, $b^x < 1$.

If $b < 1$, the graph of the function b^x has the same general shape as the graph of the function $\frac{1}{2}^x$. The function then appears to have the following properties: (1) The value of the function is positive and the graph lies entirely above the x axis. (2) The value of the function decreases as x increases. (3) The value of the function is 1 when $x = 0$. (4) If $x_2 > x_1$, then $b^{x_2} < b^{x_1}$. (5) If $x < 0$, then $b^x > 1$.

How would you describe the function $f(x) = b^x$ and its graph when $b = 1$?

FIGURE 4.1 Graphs of (a) $f(x) = 2^x$; (b) $f(x) = (\frac{1}{2})^x$

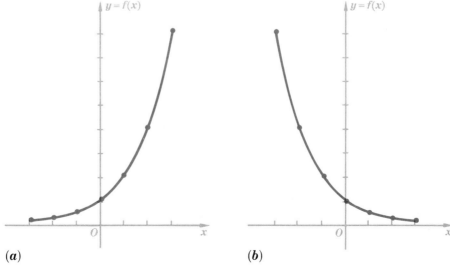

(a)   (b)

The preceding properties of the exponential function are established in more advanced courses. We accept them here without proof. We also accept two theorems from the calculus and state them as follows:

THEOREM. If $b > 0$, $b \neq 1$, and if $y > 0$, there exists one and only one real number x such that   (4.13)

$$b^x = y$$

THEOREM. If $b > 0$, $b \neq 1$, and if x_1 and x_2 are numbers such that

$$b^{x_1} = b^{x_2} \tag{4.14}$$

then $x_1 = x_2$

EXAMPLE 3. Show that the function 2^{-x} is the same as the function $(\frac{1}{2})^x$.

Solution. $(\frac{1}{2})^x = 1^x/2^x = 1/2^x = 2^{-x}$

EXAMPLE 4. If $9^x = 5$, find the value of 3^{4x}.

Solution. Since $9^x = 5$,

$$(3^2)^x = 5 \quad \text{and} \quad 3^{2x} = 5$$

Hence, $(3^{2x})^2 = 5^2$

and $3^{4x} = 25$

EXAMPLE 5. One of the ordered pairs of the exponential function b^x is $(3,216)$. Find a base b.

Solution. Since $f(x) = b^x$, we have

$$f(3) = b^3 = 216$$

Since $216 = 2^3 \cdot 3^3 = (2 \cdot 3)^3 = 6^3$

we see that $b^3 = 6^3$

and $b = 6$ meets the requirement.

4.6 EXPONENTIAL EQUATIONS

An equation of the form $b^x = a$, $b \neq 1$, is called an *exponential equation*. Certain equations of this type may be solved as in the following examples. After we have considered logarithmic functions in a later chapter, we shall be able to solve more difficult problems.

EXAMPLE 1. Find x if $3^{x+1} = 243$.

Solution. Since

$$3^{x+1} = 243 \qquad \text{and} \qquad 243 = 3(3)(3)(3)(3) = 3^5$$

we have $\qquad\qquad\qquad\qquad\qquad\qquad 3^{x+1} = 3^5$

Hence, $\qquad\qquad\qquad\qquad\qquad\qquad x + 1 = 5 \qquad$ by Def. (4.14)

and $\qquad\qquad\qquad\qquad\qquad\qquad\qquad x = 4$

Check. The check is left as an exercise.

EXAMPLE 2. Solve for x: $2^{4x+1} = 512$.

Solution. $\quad 2^{4x+1} = 512 = 2^9$
$$4x + 1 = 9$$
$$x = 2$$

Check. The check is left to the student.

EXERCISE SET 4.4

1. Sketch the graph of the function $f(x) = 3^x$ and find approximate values of the following:

   (a) $\sqrt{3}$ (b) $\sqrt[4]{3}$ (c) $\sqrt[3]{3}$
   (d) $3^{\pi/2}$ (e) $3^{\pi/3}$ (f) $3^{\pi/6}$
   (g) $3^{\sqrt{2}}$ (h) $3^{\sqrt{3}}$ (i) $3^{-\sqrt{3}}$

2. From the graph of Prob. 1, find approximate values of x when

   (a) $f(x) = 2$ (b) $f(x) = 4$ (c) $f(x) = 6$
   (d) $f(x) = \frac{1}{2}$ (e) $f(x) = \frac{1}{4}$ (f) $f(x) = 0.33$

3. Use the graph of Prob. 1 to show that the following inequalities are true:

   (a) $3^{\sqrt{2}} < 3^{\sqrt{3}}$ (b) $3^{-\sqrt{8}} > 0$

4. Use the laws of exponents to solve the following:

   (a) If $8^x = 3$, find the value of 2^{6x}.   (b) If $27^y = 4$, find the value of 3^{9y}.

5. Show that each of the following is true:

   (a) $3^{x-3} = \frac{1}{27}(3^x)$ (b) $4^{3-2x} = 64(4^{-2x})$

6. From the properties of the graphs of the exponential functions show that the following inequalities are true:

   (a) $4^{-\sqrt{8}} > 0$ (b) $2^5 > 4^{\sqrt{5}}$

7. If a certain radioactive substance decays at a rate such that at the end of 1 hr there is only one-half as much as there was at the beginning of the hour, how much of 50 g of the substance will be left after a period of t hr? HINT: Let $f(t)$ be the number of grams left at the end of t hr. Then $f(0) = 50$, $f(1) = 50(\frac{1}{2})$, $f(2) = 50(\frac{1}{2})(\frac{1}{2}) = 50(\frac{1}{2})^2$, etc.

8.   If there are 32 mg of the substance of Prob. 7 at 1:30 P.M., how much is left at 5:30 P.M.?
9.   When a certain capacitor is discharged through a resistance, the current (in amperes) flowing in the circuit at time t sec is given by

$$I = 0.5(2.7)^{-t/120}$$

Sketch the graph of this function and find the current flowing in the circuit after 60 sec.
10. If $f(x) = b^x$, prove that $f(u - v) = f(u)/f(v)$.
11. If $f(x) = b^x$, prove that $f(uv) = [f(u)]^v = [f(v)]^u$.
12. If $f(x) = b^x$, prove that $f(u/v) = [f(u)]^{1/v}$.

SOLVE the following exponential equations:

13.	$7^{2x} = 49$	14.	$3^{x+1} = 81$	15.	$4^x = \frac{1}{256}$
16.	$2^{3x} = \frac{1}{64}$	17.	$(\frac{2}{3})^x = 2\frac{7}{8}$	18.	$8^{-x} = \frac{1}{4}$
19.	$6^{-x} = \frac{1}{216}$	20.	$16^x = \frac{1}{8}$	21.	$3^{2x} = \frac{1}{81}$
22.	$5^{5x-2} = 125$	23.	$3^{x-7} = \frac{1}{27}$	24.	$2^{x-1} = \frac{1}{16}$
25.	$36^{x-3} = 216$	26.	$(49)^{x-3} = 343$	27.	$16^{2x+4} = 256$
28.	$25^{2x-4} = \frac{1}{625}$				

CHAPTER 5

TRIGONOMETRIC FUNCTIONS

The subject of trigonometry was originally developed to solve problems in astronomy and geometry. For example, it is well known that the acute angles of a right triangle are completely determined by the ratio of the lengths of any two sides of the triangle. Thus, if any two sides (or a side and an acute angle) are given, the remaining parts of the right triangle can be calculated.

In recent times, the subject matter has been extended considerably beyond the solution of triangles. The study of periodic phenomena, such as the current in an alternating current system or the vibration of a machine mounted on rubber supports, requires the use of certain periodic functions called *trigonometric functions* (or *circular functions*). These trigonometric functions are necessary to the study of engineering and the physical sciences and to mathematical analysis.

5.1 TRIGONOMETRIC POINTS

Unit circle

Trigonometric point

The circle whose radius is 1 unit and whose center is the origin of a rectangular coordinate system is called the *unit circle*. Let t be any real number. Start at the point $A(1,0)$ on the unit circle and measure along the circumference an arc of length $|t|$ units. If t is positive, measure the arc in the counterclockwise direction; if t is negative, measure the arc in the clockwise direction. This locates a unique point on the unit circle (Fig. 5.1). We call the point thus located the *trigonometric point* $P(t)$. The real number t and the point $P(t)$ constitute an ordered pair. Consequently, this procedure defines a function whose domain is the set of all real numbers and whose

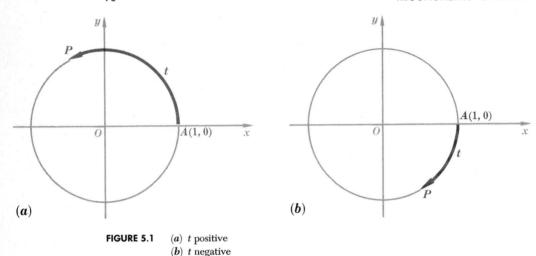

FIGURE 5.1   (*a*) *t* positive
                  (*b*) *t* negative

range is the set of points on the unit circle.   The ordered pairs of this function are $(t, P(t))$.

We assume here that every arc of a circle has length and that every real number can be paired with a unique point on the unit circle.   We justify the latter assumption intuitively by using a vertical (but flexible) number line as follows.   We fix the origin (zero point) of the number line on the point $A(1,0)$ of the unit circle and let the unit of length on the vertical number scale be the same as the length of the radius of the unit circle.   If the points on the upper part of the vertical number line represent the positive real numbers and we visualize winding the upper part of the line around the circle in a *counterclockwise* direction, we see that each point of the flexible scale (which represents a positive real number) falls on a unique point of the unit circle.   Since this winding can be continued indefinitely, each positive real number t can be paired with a unique point $P(t)$ on the unit circle.

Similarly, if we visualize winding the lower part of the flexible number line (whose points represent the negative real numbers) *clockwise* around the circle, each negative real number is also paired with a unique point on the unit circle.

Certain of these ordered pairs $(t, P(t))$ can be found easily.   When $t = 0$, $P(t) = P(0)$ is the trigonometric point $A(1,0)$.   Since the circumference of the unit circle is 2π units (approximately 6.28 units), one-half the circumference is π units, one-fourth the circumference is $\pi/2$ units, etc.   Hence, when $t = \pi/2$, $P(t) = P(\pi/2)$ is the point $(0,1)$.   Also, $P(\pi)$ is the point $(-1,0)$; $P(3\pi/2)$ is the point $(0,-1)$ (Fig. 5.2).

By similar reasoning, $P(-\pi/2)$ is the point $(0,-1)$; $P(-\pi)$ is the point $(-1,0)$; $P(-3\pi/2)$ is the point $(0,1)$.

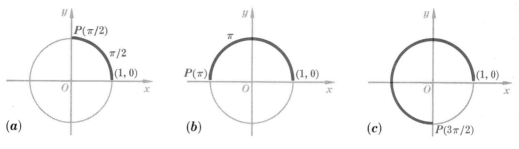

FIGURE 5.2   (a) $P(\pi/2)$
(b) $P(\pi)$
(c) $P(3\pi/2)$

To locate the trigonometric points $P(\pi/4)$, $P(3\pi/4)$, $P(5\pi/4)$, $P(7\pi/4)$, we reason as follows.   Let the coordinates of the trigonometric point $P(t)$ be the ordered pair (x,y).   Since $P(t)$ is 1 unit from the origin, it follows from the distance formula (Sec. 3.9)

$$x^2 + y^2 = 1 \tag{5.1}$$

If one of the coordinates of the point $P(t)$ is known, we can use Eq. (5.1) to find the other coordinate (except for sign).

From plane geometry, we see that the trigonometric point $P(\pi/4)$ is the midpoint of the arc on the unit circle from $(1,0)$ to $(0,1)$ and is equidistant from the x and y axes so that $x = y$ (see Fig. 5.3).   Substituting $y = x$ into the equation $x^2 + y^2 = 1$, we have

$$x^2 + x^2 = 1$$

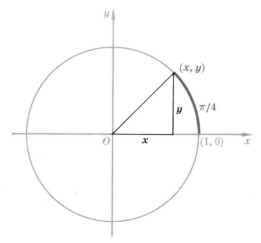

FIGURE 5.3 $P(\pi/4) = (\sqrt{2}/2, \sqrt{2}/2)$

Hence,
$$x = \frac{1}{\sqrt{2}} = \frac{\sqrt{2}}{2}$$

and
$$y = x = \frac{\sqrt{2}}{2}$$

Therefore, the coordinates of the trigonometric point $P(\pi/4)$ are

$$\left(\frac{\sqrt{2}}{2}, \frac{\sqrt{2}}{2} \right)$$

The problem of finding the coordinates of $P(3\pi/4)$, $P(5\pi/4)$, etc., will be left as exercises.

To find the coordinates of $P(\pi/3)$, consider the unit circle of Fig. 5.4. The arc AP = one-sixth of the circumference of the circle = $\pi/3$. Hence, from plane geometry, the triangle AOP is equilateral. Construct PM perpendicular to OA. Then, as you recall from your geometry, $OM = \frac{1}{2}OA = \frac{1}{2}$. Hence, $x = \frac{1}{2}$. Since $x^2 + y^2 = 1$, we have

$$\left(\frac{1}{2} \right)^2 + y^2 = 1$$

Therefore
$$y = \frac{\sqrt{3}}{2}$$

The coordinates of $P(2\pi/3)$, $P(4\pi/3)$, and so on, can be found in a similar manner.

The coordinates of $P(\pi/6)$ can also be found. In Fig. 5.5, let the arc $QP = \pi/3$. Then arc $AP = \pi/6$, and OM is perpendicular to the chord QP at its midpoint.

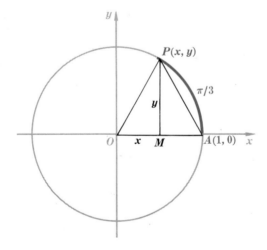

FIGURE 5.4 $P(\pi/3) = (\frac{1}{2}, \sqrt{3}/2)$

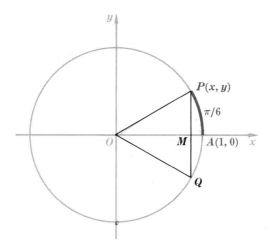

FIGURE 5.5

$P(\pi/6) = (\sqrt{3}/2, \frac{1}{2})$

Since triangle QOP is equilateral, it follows that

$$QP = OP = 1$$

and

$$MP = \frac{1}{2} QP = \frac{1}{2}$$

Hence,

$$y = \frac{1}{2}$$

On substituting this value for y into $x^2 + y^2 = 1$, we obtain

$$x = \frac{\sqrt{3}}{2}$$

Therefore, the point $P(\pi/6)$ is the point $(\sqrt{3}/2, \frac{1}{2})$. The coordinates of $P(5\pi/6)$, $P(7\pi/6)$, $P(-\pi/6)$, and so on, may now be obtained without too much difficulty. The following set of exercises will provide practice in finding the coordinates of $P(t)$ for many of the special values of t.

EXERCISE SET 5.1

1. Find the coordinates of each of the trigonometric points:

(a) $P(-2\pi)$ $(1,0)$ (b) $P(-3\pi)$ (c) $P(4\pi)$ $(1,0)$
(d) $P(-\pi)$ (e) $P(-3\pi/2)$ (f) $P(-4\pi)$ $(1,0)$
(g) $P(-5\pi/2)$ (h) $P(-7\pi/2)$
                     $(0,-1)$

2. Find the coordinates of each point:

(a) $P(-\pi/4)$ (b) $P(3\pi/4)$ (c) $P(-3\pi/4)$
(d) $P(5\pi/4)$ (e) $P(-5\pi/4)$ (f) $P(-7\pi/4)$

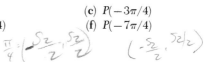

3. Determine the coordinates of each trigonometric point:

(a) $P(-\pi/3)$ (b) $P(2\pi/3)$ (c) $P(-2\pi/3)$
(d) $P(4\pi/3)$ (e) $P(5\pi/3)$ (f) $P(7\pi/3)$

4. Find the coordinates of $P(t)$ when t is

(a) $5\pi/6$ (b) $-\pi/6$ (c) $-5\pi/6$
(d) $7\pi/6$ (e) $11\pi/6$ (f) $-11\pi/6$

5. The trigonometric point $P(t)$ has the coordinates $(\sqrt{3}/2, \frac{1}{2})$ and $0 < t < \pi/2$. Find t.

6. Find the value of t if $P(t)$ is the point

(a) $(\sqrt{2}/2, -\sqrt{2}/2)$, and $3\pi/2 < t < 2\pi$
(b) $(-\frac{1}{2}, \sqrt{3}/2)$, and $\pi/2 < t < \pi$

5.2   TRIGONOMETRIC FUNCTIONS

Let t be any real number and let the coordinates of $P(t)$ be (x,y). We define the *cosine function* of t (written cos t) and the *sine function* of t (written sin t) in terms of the coordinates of the trigonometric point.

DEFINITION.   If the trigonometric point $P(t)$ has coordinates (x,y), then   **(5.2)**

$$\cos t = x \qquad \sin t = y$$

The point $P(x,y)$ on the unit circle can now be written as $P(\cos t, \sin t)$. Since $x^2 + y^2 = 1$, by Eq. (5.1), we have the very important identity

$$(\sin t)^2 + (\cos t)^2 = 1 \qquad \text{for all } t \tag{5.3}$$

Since $P(t)$ lies on the unit circle, neither of its coordinates can be greater than 1 in absolute value. It follows that $|x| \leq 1$ and $|y| \leq 1$. Hence,

$$-1 \leq \cos t \leq 1 \qquad \text{and} \qquad -1 \leq \sin t \leq 1$$

for all real numbers t.

Four additional functions are defined as follows. Note that the abbreviation for each is given.

$$\text{tangent } t = \tan t = \frac{\sin t}{\cos t} \qquad \cos t \neq 0$$

$$\text{cotangent } t = \cot t = \text{ctn } t = \frac{\cos t}{\sin t} \qquad \sin t \neq 0$$

$$\text{secant } t = \sec t = \frac{1}{\cos t} \qquad \cos t \neq 0 \tag{5.4}$$

$$\text{cosecant } t = \csc t = \frac{1}{\sin t} \qquad \sin t \neq 0$$

In terms of x and y, these functions are

$$\tan t = \frac{y}{x} \qquad x \neq 0$$

$$\cot t = \frac{x}{y} \qquad y \neq 0$$

$$\sec t = \frac{1}{x} \qquad x \neq 0 \tag{5.5}$$

$$\csc t = \frac{1}{y} \qquad y \neq 0$$

Circular functions of t

The six functions of t defined by Eqs. (5.2) and (5.4) are called *trigonometric functions* of t, or *circular functions* of t.

5.3 TRIGONOMETRIC FUNCTIONS OF SPECIAL REAL NUMBERS

The trigonometric functions of certain special real numbers can now be determined from the definitions of the functions. For example, we learned in Sec. 5.1 that if $t = \pi/4$, the coordinates of $P(\pi/4)$ are $(\sqrt{2}/2, \sqrt{2}/2)$. Since $\sin t = y$ and $\cos t = x$, we have

$$\sin \frac{\pi}{4} = \frac{\sqrt{2}}{2} \qquad \text{and} \qquad \cos \frac{\pi}{4} = \frac{\sqrt{2}}{2}$$

Since

$$\tan \frac{\pi}{4} = \frac{\sin \pi/4}{\cos \pi/4}$$

it follows that

$$\tan \frac{\pi}{4} = 1$$

Also,

$$\cot \frac{\pi}{4} = 1 \qquad \sec \frac{\pi}{4} = \sqrt{2} \qquad \csc \frac{\pi}{4} = \sqrt{2}$$

Functions of multiples of $\pi/2$ may be determined in the same manner. For example, if $t = 3\pi/2$, the coordinates of $P(3\pi/2)$ are $(0, -1)$. Hence,

$$\sin \frac{3\pi}{2} = -1 \qquad \qquad \cos \frac{3\pi}{2} = 0$$

$$\tan \frac{3\pi}{2} \text{ is undefined} \qquad \cot \frac{3\pi}{2} = 0$$

$$\sec \frac{3\pi}{2} \text{ is undefined} \qquad \csc \frac{3\pi}{2} = -1$$

If $t = 0$, the coordinates of $P(t) = P(0)$ are $(1,0)$. Hence,

$$\sin 0 = 0 \qquad \cos 0 = 1$$
$$\tan 0 = 0 \qquad \cot 0 \text{ is undefined}$$
$$\sec 0 = 1 \qquad \csc 0 \text{ is undefined}$$

If the point $Q(a,b)$ lies on the line joining the origin to the trigonometric point $P(t)$, we can find the trigonometric functions of t. This is very useful, and we hope the student will sketch figures and convince himself that the reasoning which follows is valid.

By the distance formula, $OQ = \sqrt{a^2 + b^2}$. If perpendiculars are constructed from P and Q to the x axis, two similar right triangles are formed. Then

$$\frac{x}{1} = \frac{a}{\sqrt{a^2 + b^2}} \quad \text{and} \quad \frac{y}{1} = \frac{b}{\sqrt{a^2 + b^2}}$$

Hence,

$$\sin t = y = \frac{b}{\sqrt{a^2 + b^2}}$$

and

$$\cos t = x = \frac{a}{\sqrt{a^2 + b^2}}$$

Then by Eq. (5.5), $\tan t = b/a$ if $a \neq 0$; $\cot t = a/b$ if $b \neq 0$; and so on.

EXAMPLE 1. If the trigonometric point $P(t)$ lies on the line joining the origin and the point $Q(12,5)$, find the values of the trigonometric functions of t.

Solution. The distance from the origin to the point $(12,5)$ is 13. If perpendiculars are constructed from $P(x,y) = P(t)$ and from $(12,5)$ to the x axis, two similar right triangles are formed. Then $y/1 = 5/13$ and $x/1 = 12/13$. Hence $\sin t = 5/13$, $\cos t = 12/13$, $\tan t = \sin t/\cos t = 5/12$, $\cot t = \cos t/\sin t = 12/5$, $\sec t = 1/\cos t = 13/12$, $\csc t = 1/\sin t = 13/5$.

EXERCISE SET 5.2

1. Find the value of each of the six trigonometric functions of

   (a) $\pi/2$ (b) $\pi/3$ (c) $\pi/6$
   (d) $-3\pi/2$ (e) $5\pi/6$ (f) $2\pi/3$

2. Find the value of

   (a) $\sin 3\pi/4$ (b) $\cos 11\pi/6$ (c) $\sec 7\pi/4$
   (d) $\tan (-5\pi/6)$ (e) $\sec (-2\pi/3)$ (f) $\csc (-\pi/3)$
   (g) $\sin (-\pi/2)$ (h) $\tan (-3\pi/2)$ (i) $\cos (-5\pi/6)$
   (j) $\sec (-\pi/6)$ (k) $\tan 5\pi/4$ (l) $\sin 11\pi/6$

3. If $\sin t = \frac{1}{2}$ and $\cos t < 0$, find

   (a) $\cos t$ (b) $\sec t$ (c) $\tan t$

4. If $\cos t = -\sqrt{3}/2$ and $\sin t < 0$, find

   (a) $\sin t$ (b) $\tan t$ (c) $\csc t$

5. If $\tan t = -1$ and $\sec t > 1$, find

   (a) $\cot t$ (b) $\sin t$ (c) $\cos t$

6. If $\cot t = -1/\sqrt{3}$ and $\sin t > 0$, find

   (a) $\cos t$ (b) $\sec t$ (c) $\csc t$

7. Complete the table below, showing the algebraic signs of the trigonometric functions in the indicated quadrants.

Quadrant in Which P(t) Lies	sin t	cos t	tan t	cot t	sec t	csc t
I	+	+	+			+
II		−		−	−	+
III	−	−				
IV		+	−			−

8. If t is a real number, determine the quadrant in which $P(t)$ lies when

   (a) $\sin t < 0$ (b) $\tan t < 0$ (c) $\cos t < 0$
   (d) $\cot t > 0$ (e) $\sec t > 0$ (f) $\csc t > 0$

9. Let $P(t)$ lie on the line joining the origin and the point $(5,12)$. Find the values of the trigonometric functions of t.   HINT: The distance from the origin to the point $(5,12)$ is 13. Hence the point $P(t)$ on the unit circle has coordinates $(5/13, 12/13)$ and $\sin t = 12/13$.

10. If $P(t)$ lies on the line segment joining the origin and the indicated point, find the values of the trigonometric functions of t:

   (a) $(3,4)$ (b) $(4,3)$ (c) $(24,7)$
   (d) $(12,-5)$ (e) $(8,-15)$ (f) $(-8,15)$
   (g) $(-10,-8)$ (h) $(0,-2)$ (i) $(15,-8)$

11. Find the values of $\sin t$, $\cos t$, $\tan t$ if

   (a) $\sin t = 12/13$, $P(t)$ in quadrant I   (b) $\cos t = -5/13$, $P(t)$ in quadrant III
   (c) $\tan t = \sqrt{3}$, $P(t)$ in quadrant III   (d) $\sin t = -1/5$, $P(t)$ in quadrant IV
   (e) $\sin t = 7/24$, $P(t)$ in quadrant II   (f) $\cos t = 3/4$, $P(t)$ in quadrant I
   (g) $\sin t = 5/6$, $P(t)$ in quadrant II

12. Show that $\sin(-\pi/4) = -\sin \pi/4$
13. Show that $\cos(-3\pi/4) = \cos 3\pi/4$
14. Show that $\tan(-5\pi/6) = -\tan 5\pi/6$
15. Determine which of the following statements are true.

   (a) $\sin(-t) = -\sin t$ (b) $\cos(-t) = -\cos t$ (c) $\tan(-t) = -\tan t$
   (d) $\cot(-t) = \cot t$ (e) $\sec(-t) = \sec t$ (f) $\csc(-t) = \csc t$

16. In which quadrants do $\sin t$ and $\cos t$ have the same sign?
17. In which quadrants do $\sin t$ and $\cos t$ have opposite signs?

5.4 TRIGONOMETRIC FUNCTIONS OF ANY REAL NUMBER

We shall accept without proof the statement that the real number π is irrational and cannot be expressed as the quotient of two integers. A rational *approximation* of the value of this number is 3.14. To indicate this, we write

$\pi = 3.14$ (approx.). Then $\pi/2 = 1.57$ (approx.), $\pi/4 = 0.78$ (approx.), $\pi/6 = 0.53$ (approx.), etc. We can then say that the trigonometric functions of 1.57, 0.78, 0.53 are approximately equal to the corresponding trigonometric functions of $\pi/2$, $\pi/4$, $\pi/6$. For example,

$$\sin 0.53 = \sin \frac{\pi}{6} = 0.5000 \text{ (approx.)}$$

Approximate values of the trigonometric functions have been computed and tabulated for values of t between 0 and 1.60 at intervals of 0.01 unit. Table I of the appendix lists these values correct to three or four decimal places. The table is easy to use. For example, to find tan 1.32, we look in the column headed t for 1.32 and then move horizontally across the page to the column headed tan t. There we find the entry 3.903, and we write

$$\tan 1.32 = 3.903$$

We can use Table I to find the values of the functions for $t > 1.60$. To show that this is so, let $P(t)$ be the trigonometric point corresponding to a real value of t, and let t_1 be the length of the *shorter* arc that joins $P(t)$ to the **Reference number** x axis (Fig. 5.6). The number t_1 is called the *reference number* for t. Locate the point $P(t_1)$ by measuring the arc length t_1 counterclockwise from the point $(1,0)$. Then $P(t_1)$ is in the first quadrant. Let (x,y) be the coordinates of $P(t)$ and (x_1,y_1) be the coordinates of $P(t_1)$. Then, as indicated by the figure,

$$|x| = x_1 \qquad |y| = y_1$$

It follows from the definitions of the trigonometric functions and these two equations that the values of the trigonometric functions of t and t_1 are numerically the same but may differ in sign. We sum up the discussion by the equation

$$|\text{any function of } t| = \text{same function of } t_1$$

The absolute values of the trigonometric functions of any real t can be determined when the reference number t_1 is known. We need only know the

FIGURE 5.6 (a) $t_1 = \pi - t$; (b) $t_1 = t - \pi$; (c) $t_1 = 2\pi - t$

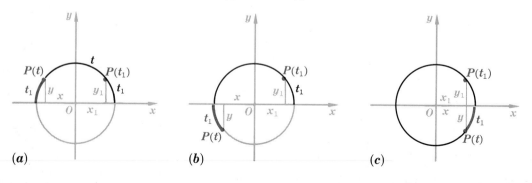

(a)   (b)   (c)

quadrant in which $P(t)$ lies (to determine the sign) and the value of t_1 (to determine the value of the function of t).

If $t > \pi/2 = 1.57$ (approx.), we determine the value of t_1 as in the following examples. Then from the values of the functions of t_1 we get the proper values of the functions of t. Thus, if $t = 3$, $t_1 = \pi - 3 = 0.14$ (approx.) (see Fig. 5.7a).

EXAMPLE 1. Find tan 4 and sec 4.

Solution. Since $t = 4$, and since $\pi < 4 < 3\pi/2$, we know that $P(4)$ lies in the third quadrant. Hence, tan 4 is positive and sec 4 is negative (Fig. 5.7b). The reference number $t_1 = 4 - \pi = 0.86$; therefore

$$|\tan 4| = \tan 0.86 = 1.162 \qquad \text{from Table I}$$

and

$$|\sec 4| = \sec 0.86 = 1.533$$

Hence,

$$\tan 4 = 1.162 \qquad \text{and} \qquad \sec 4 = -1.533$$

EXAMPLE 2. Find sin 5.84.

Solution. Since $3\pi/2 < 5.84 < 2\pi$, $P(5.84)$ lies in the fourth quadrant. Hence, sin 5.84 is negative. The reference number $t_1 = 2\pi - 5.84 = 0.44$ (Fig. 5.7c). Then

$$|\sin 5.84| = \sin 0.44 = 0.4259 \qquad \text{from Table I}$$

Hence,

$$\sin 5.84 = -0.4259$$

EXAMPLE 3. Find sin (-16).

Solution. Locate the point $P(-16)$ by moving 16 units clockwise around the unit circle from $(1,0)$. Since

$$-16 = -[2(6.28) + 3.44] \doteq -2(6.28) - 3.44$$

FIGURE 5.7 (*a*) $t_1 = \pi - 3$; (*b*) $t_1 = 4 - \pi$; (*c*) $t_1 = 2\pi - 5.84$

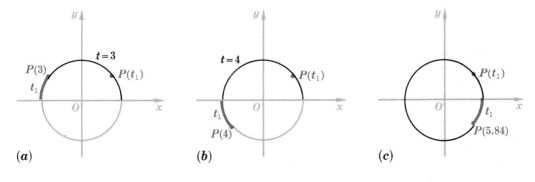

(*a*)　　　　　(*b*)　　　　　(*c*)

we proceed twice around the unit circle (clockwise) and then continue for 3.44 additional units. This means that $P(-16)$ is the same point as $P(-3.44)$. Hence, $P(-16)$ is in the second quadrant and $\sin(-16)$ is positive.

The reference number $t_1 = |-3.44 - (-\pi)| = 0.30$. Hence,

$$|\sin(-16)| = \sin 0.30 = 0.2955 \qquad \text{from Table I}$$

and
$$\sin(-16) = 0.2955$$

EXERCISE SET 5.3

1. Use Table I and evaluate (take $\pi = 3.14$):

   (a) $\sin 1$ (b) $\cos 1.3$ (c) $\tan 2.83$ (d) $\cos 4.87$
   (e) $\sin 2.12$ (f) $\sin(-3.48)$ (g) $\cos(-5.48)$ (h) $\tan(-2.32)$

2. Evaluate:

   (a) $\sin 7.2$ (b) $\sin(-8.28)$ (c) $\tan 9.42$ (d) $\tan(-9.42)$
   (e) $\cos 12.56$ (f) $\cos 6.66$ (g) $\cos(-8.38)$ (h) $\tan 15$
   (i) $\sin 24$ (j) $\cos 18$ (k) $\tan 35$ (l) $\cot 21$
   (m) $\sec(-30)$ (n) $\sec 48$ (o) $\cot 59$ (p) $\csc(-45)$

3. If $\pi/2 < t < \pi$ and $\cos t = -\sqrt{3}/2$, find t.
4. If $\pi/2 < t < \pi$ and $\tan t = -\sqrt{3}$, find t.
5. If $3\pi/2 < t < 2\pi$ and $\tan t = -0.5994$, find t.
6. Find t if $\sin t = -0.9044$ and $3\pi/2 < t < 2\pi$.
7. If $0 < t < \pi/2$ and $100 \sin t - 103 \cos t = 0$, find t.
8. Find t if $100 \sin t + 301 \cos t = 0$ and $\pi/2 < t < \pi$.

5.5   TRIGONOMETRIC IDENTITIES

Fundamental identities

There are certain relations, useful in later mathematical study, among the trigonometric functions of the real number t that we call the *fundamental identities*. These fundamental identities are statements that are true for every real value of t for which the functions are defined. The first of these relations is derived from the fact that $P(t)$ is a point on the unit circle. If (x,y) are the coordinates of $P(t)$, then $x^2 + y^2 = 1$. Since $x = \cos t$ and $y = \sin t$, it follows that $(\cos t)^2 + (\sin t)^2 = 1$. By agreement, we write this statement as follows:

$$\cos^2 t + \sin^2 t = 1 \qquad \text{for all real } t \qquad (5.6)$$

Positive powers of the trigonometric functions are usually written in this form. For example, $(\tan t)^3$ will be written $\tan^3 t$; $(\sec t)^2$ will be written $\sec^2 t$, etc. This convention, however, is not applied to the -1 powers of trigonometric functions. Thus, $(\cos t)^{-1}$ will not be written $\cos^{-1} t$. The symbol $\cos^{-1} t$ has a special meaning and will be defined in a later chapter.

If $\cos t \neq 0$, we can divide each side of Eq. (5.6) by $\cos^2 t$ and get

$$1 + \frac{\sin^2 t}{\cos^2 t} = \frac{1}{\cos^2 t}$$

Hence,

$$1 + \left(\frac{\sin t}{\cos t}\right)^2 = \left(\frac{1}{\cos t}\right)^2$$

Since

$$\frac{\sin t}{\cos t} = \tan t \qquad \text{and} \qquad \frac{1}{\cos t} = \sec t$$

we have the identity

$$1 + \tan^2 t = \sec^2 t \tag{5.7}$$

Similarly, if $\sin t \neq 0$, we can divide each side of Eq. (5.6) by $\sin^2 t$ and use other identities to obtain

$$1 + \cot^2 t = \csc^2 t \tag{5.8}$$

Since $\tan t = \sin t / \cos t$ and $\cot t = \cos t / \sin t$, it follows that

$$\tan t \cot t = 1 \tag{5.9}$$

We leave it to the student to show that

$$\sin t \csc t = 1 \tag{5.10}$$

and

$$\cos t \sec t = 1 \tag{5.11}$$

Proof of an identity

These six fundamental identities and the definitions of the functions enable us to prove many other identities. We say that an identity is proved when (1) the left member is reduced (or transformed) to the exact form of the right member, (2) the right member is reduced to the exact form of the left member, or (3) each member is reduced to the same form.

The reductions (or transformations) may be done by performing any valid reversible algebraic operations that may be indicated and by substitutions from the fundamental identities. Generally, we take what appears to be the more complicated member of the equation and attempt to reduce it to the exact form of the other member. When in doubt as to how to proceed, we often express the functions in terms of sines and cosines and then simplify. This method, however, may introduce radicals that carry the ambiguous $\pm$ sign.

EXAMPLE 1. Express each of the other five trigonometric functions of t in terms of $\cos t$.

Solution. (1) Since $\cos^2 t + \sin^2 t = 1$,

$$\sin t = \pm\sqrt{1 - \cos^2 t}$$

where the sign of the radical is determined by the quadrant in which $P(t)$ lies.

(2) $$\tan t = \frac{\sin t}{\cos t} = \frac{\pm\sqrt{1 - \cos^2 t}}{\cos t} \qquad \cos t \neq 0$$

(3) $$\cot t = \frac{1}{\tan t} = \frac{\cos t}{\pm\sqrt{1 - \cos^2 t}} \qquad \cos t \neq 1$$

(4) $$\sec t = \frac{1}{\cos t} \qquad \cos t \neq 0$$

(5) $$\csc t = \frac{1}{\sin t} = \frac{1}{\pm\sqrt{1 - \cos^2 t}} \qquad \cos t \neq 1$$

EXAMPLE 2. Transform the left member to the exact form of the right member and thus prove the identity:

$$\frac{1 - \cos t}{\sin t} = \frac{\sin t}{1 + \cos t}$$

Solution. (1) Multiply numerator and denominator of $(1 - \cos t)/\sin t$ by $1 + \cos t$. Then for $1 + \cos t \neq 0$,

$$\frac{1 - \cos t}{\sin t} = \frac{(1 - \cos t)(1 + \cos t)}{(\sin t)(1 + \cos t)}$$

$$= \frac{1 - \cos^2 t}{(\sin t)(1 + \cos t)}$$

(2) $$= \frac{\sin^2 t}{(\sin t)(1 + \cos t)} \qquad \text{by Eq. (5.6)}$$

(3) $$= \frac{\sin t}{1 + \cos t} \qquad \begin{array}{l}\text{dividing numerator and}\\ \text{denominator by } \sin t,\\ \sin t \neq 0\end{array}$$

EXAMPLE 3. Transform the right member to the exact form of the left member and thus prove that

$$2 \sin^2 t - 1 = \frac{\tan t - \cot t}{\tan t + \cot t}$$

is an identity.

Solution.

$$\frac{\tan t - \cot t}{\tan t + \cot t} = \frac{\dfrac{\sin t}{\cos t} - \dfrac{\cos t}{\sin t}}{\dfrac{\sin t}{\cos t} + \dfrac{\cos t}{\sin t}}$$

$$= \frac{\sin^2 t - \cos^2 t}{\sin^2 t + \cos^2 t}$$

$$= \frac{\sin^2 t - \cos^2 t}{1}$$

$$= \sin^2 t - (1 - \sin^2 t)$$

$$= 2 \sin^2 t - 1$$

EXAMPLE 4. Reduce each member to the same form and thus prove the identity (α is a real number):

$$2 \sec^2 \alpha = \frac{1}{1 + \sin \alpha} + \frac{1}{1 - \sin \alpha}$$

Solution.

Left Member	Right Member
	$\dfrac{1}{1 + \sin \alpha} + \dfrac{1}{1 - \sin \alpha}$
$2 \sec^2 \alpha$	$= \dfrac{(1 - \sin \alpha) + (1 + \sin \alpha)}{(1 + \sin \alpha)(1 - \sin \alpha)}$
$= 2\left(\dfrac{1}{\cos^2 \alpha}\right)$	$= \dfrac{2}{1 - \sin^2 \alpha}$
$= \dfrac{2}{\cos^2 \alpha}$	$= \dfrac{2}{\cos^2 \alpha}$

EXERCISE SET 5.4

1. Express each of the following functions of the real number t in terms of $\sin t$:

   (a) $\cos^2 t$ (b) $\cos t$ (c) $\tan t$
   (d) $\sec^2 t$ (e) $\sec t$ (f) $\cot t$
   (g) $\csc t$ (h) $\cot^2 t$ (i) $\tan^2 t$

2. Express each of the following in terms of $\cos \phi$:

   (a) $\sin \phi$ (b) $\tan \phi$ (c) $\cot \phi$
   (d) $\sec \phi$ (e) $\csc \phi$ (f) $\tan^2 \phi$

3. Express each of the following in terms of $\tan \beta$:

   (a) $\cot \beta$ (b) $\sec^2 \beta$ (c) $\sin \beta$
   (d) $\csc \beta$ (e) $\cos \beta$

PROVE that each of the following is an identity:

4. $\tan t \cdot \cos t = \sin t$

5. $\tan t + \cot t = \sec t \cdot \csc t$

6. $\dfrac{\sin t}{\csc t} = 1 - \dfrac{\cos t}{\sec t}$

7. $1 - 2 \sin^2 t = 2 \cos^2 t - 1$

8. $(\sin \alpha)(\cot \alpha) + (\sin \alpha)(\csc \alpha) = \cos \alpha + 1$

9. $\dfrac{1}{\tan \phi + \cot \phi} = (\sin \phi)(\cos \phi)$

10. $\dfrac{\csc^2 \beta}{\cot^2 \beta} - 1 = \tan^2 \beta$

11. $\tan^2 \beta - \sin^2 \beta = \tan^2 \beta \sin^2 \beta$

12. $\dfrac{\cot x - \tan x}{\tan x + \cot x} = 1 - 2 \sin^2 x$

13. $\dfrac{1 - \cos \alpha}{\sin \alpha} - \dfrac{\sin \alpha}{1 + \cos \alpha} = 0$

14. $\cot x + \tan x = \cot x \cdot \sec^2 x$

15. $\dfrac{1 - \cos t}{1 + \cos t} = (\csc t - \cot t)^2$

16. $(\sin x - \cos x)^2 + 2 \sin x \cos x = 1$

17. $\sec \phi - \dfrac{\cos \phi}{1 + \sin \phi} = \tan \phi$

18. $(\sec y - \tan y)^2 = \dfrac{1 - \sin y}{1 + \sin y}$

19. $\dfrac{2}{\sin t + 1} - \dfrac{2}{\sin t - 1} = 4 \sec^2 t$

20. $\dfrac{3 \cos t}{2 - 2 \csc t} + \dfrac{3 \cos t}{2 + 2 \csc t} = -3 \sin t \tan t$

21. $(\tan t + \cot t)^2 (\sin^2 t) - \tan^2 t = 1$

22. $\dfrac{(\sec t - \tan t)^2 + 1}{\sec t \csc t - \tan t \csc t} = 2 \dfrac{\sin t}{\cos t}$

5.6   PLANE ANGLES

In a geometry course we might define a plane angle as the geometric figure formed by two half-lines having in common their end points. However, for our study we extend this concept of an angle.

Let O be the common end point of the two half-lines OA and OB (Fig. 5.8a). The half-lines OA and OB are the *sides* of an angle, and the point O is the *vertex* of the angle.

We now assign to this angle a number which corresponds to "an amount of rotation" in the plane of OA and OB about the vertex O which is required to make one of the half-lines coincide with the other. This number is called the *measure* of the angle, or simply the angle. If OA is rotated to coincide with OB, we call OA the *initial side* of the angle and OB the *terminal side*. The direction of the rotation is usually indicated by a curved arrow from the initial side to the terminal side. In Fig. 5.8b OA is the initial side of the angle, and in Fig. 5.8c OA is the terminal side. This discussion leads to an important definition:

Measure of an angle

> **DEFINITION.**   An angle is an *amount of rotation of a half-line in a plane about its end point from an initial position to a terminal position.*   **(5.12)**

FIGURE 5.8   (a) The plane angle formed by two half-lines with common end point O.
(b) OA is the initial side, OB the terminal side.
(c) OB is the initial side, OA the terminal side.

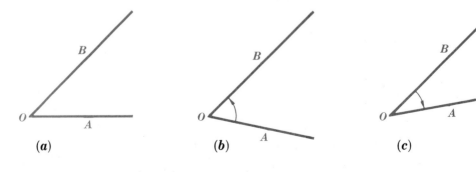

(a)   (b)   (c)

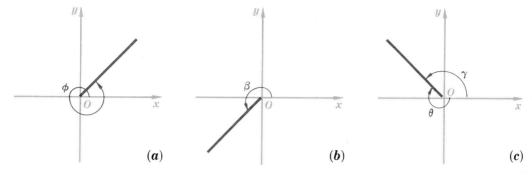

FIGURE 5.9 Angles in standard position

Angle in standard position

When the vertex of an angle is the origin of a rectangular coordinate system and the initial side coincides with the positive x axis, the angle is in *standard position.* An angle in standard position is said to be in the quadrant in which its terminal side lies. Thus, in Fig. 5.9, ϕ is in the first quadrant and β is in the third quadrant. An angle is called a *quadrantal* angle if it is in standard position and its terminal side coincides with one of the coordinate axes. Angles in standard position are called *coterminal* angles if their terminal sides coincide. In Fig. 5.9c, γ and θ are coterminal angles.

Coterminal angles

5.7 MEASURE OF ANGLES

The measure of an angle is positive if the rotation is in the counterclockwise direction and negative if the rotation is clockwise. To denote the measure of an angle we shall use two kinds of units, the *degree* unit and the *radian* unit.

Degree measure

A *degree* is defined to be an angle formed by a half-line rotated about its end point ⅟₃₆₀ of a complete rotation. To denote degree measure we use the symbol ° written just to the right of the measure number of the angle. Thus,

$$360° = 1 \text{ complete rotation}$$

$$180° = \frac{1}{2} \text{ of a complete rotation}$$

$$90° = \frac{1}{4} \text{ of a complete rotation}$$

and so on.

A *minute,* expressed by using ', is an angle formed by a rotation equal to ⅟₆₀ of a degree. Thus, $60' = 1°$. A *second* is an angle formed by a rotation equal to ⅟₆₀ of a minute and is designated by ". Thus, $60'' = 1'$.

Radian measure

The second kind of unit used in describing the measure of an angle, and by far the most important for future courses in mathematics, is the radian.

DEFINITION.   A radian is the angle formed by a half-line that has been rotated about its end point $1/2\pi$ of a complete rotation.   (5.13)

From this definition, we see that

$$1 \text{ radian} = 1 \text{ rad} = \frac{1}{2\pi} \text{ complete rotations}$$

$$2 \text{ rad} = 2\left(\frac{1}{2\pi}\right) \text{ complete rotations}$$

$$3 \text{ rad} = 3\left(\frac{1}{2\pi}\right) \text{ complete rotations}$$

$$2\pi \text{ rad} = 2\pi\left(\frac{1}{2\pi}\right) \text{ complete rotations}$$

$$= 1 \text{ complete rotation}$$

The definition of a radian can be described geometrically.   Let angle AOB (Fig. 5.10) be a central angle of the circle whose radius is r units of length.   Furthermore, let angle AOB be an angle of 1 rad and let the length of the intercepted arc be t units.   Since the ratio of the measure of two central angles of a circle equals the ratio of the lengths of their respective intercepted arcs, we have

$$\frac{1/2\pi \text{ complete rotations}}{1 \text{ complete rotation}} = \frac{t}{2\pi r}$$

from which $t = r$

Hence,

Angle of one radian

An angle of one radian is a central angle subtended by an arc equal in length to the radius of the circle.

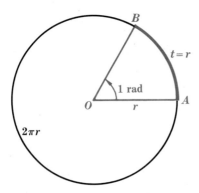

FIGURE 5.10
A central angle of 1 rad is subtended by an arc equal in length to the radius of the circle.

Radians to degrees   Since $360°$ = one complete rotation and also 2π rad = one complete rotation, we have the relation

$$2\pi \text{ rad} = 360°$$

Then

$$\pi \text{ rad} = 180°$$

$$\frac{\pi}{2} \text{ rad} = 90°$$

$$\frac{\pi}{3} \text{ rad} = 60°$$

and so on. Also,

$$1 \text{ rad} = \left(\frac{180}{\pi}\right)° = 57.295° \text{ (approx.)}$$

Then, from the fact that $180° = \pi$ rad, we have

$$1° = \frac{\pi}{180} \text{ rad} = 0.01745 \text{ rad (approx.)}$$

$$1' = 0.00029 \text{ rad (approx.)}$$

$$1'' = 0.00000485 \text{ rad (approx.)}$$

EXAMPLE 1.   Express $7\pi/4$ rad in degrees.

Solution.   Since π rad = $180°$, $7\pi/4$ rad = $(7\!\!/\!\!4)(180°) = 315°$.

EXAMPLE 2.   Express 2.6 rad in degrees, minutes, and seconds.

Solution.   (1) Since 1 rad = $57.295°$, 2.6 rad = $2.6(57.295°) = 148.967°$. Hence, 2.6 rad = $148° + 0.967°$.

(2) Now, $0.967° = 0.967(60') = 58.02' = 58' + 0.02'$.

(3) Also, $0.02' = 0.02(60'') = 1.2''$.

(4) Therefore, 2.6 rad = $148°58'1.2''$.

EXAMPLE 3.   Express $240°$ in radians.

Solution.   Since $1° = \pi/180$ rad, $240° = 240(\pi/180) = 4\pi/3$ rad.

EXAMPLE 4.   Express $31°$ in radians.

Solution. $31° = 31(\pi/180) = 31\pi/180$ rad

EXERCISE SET 5.5

EXPRESS each angle in degrees:

1.	$\pi/4$	**2.**	$\pi/5$	**3.**	$\pi/6$	**4.**	$3\pi/4$
5.	$2\pi/5$	**6.**	$5\pi/6$	**7.**	$-7\pi/6$	**8.**	$-13\pi/6$
9.	$-\pi/15$	**10.**	$-13\pi/15$	**11.**	$\pi/36$	**12.**	$5\pi/36$
13.	$7\pi/6$	**14.**	$-\pi/18$	**15.**	$-13\pi/18$	**16.**	$5\pi/12$
17.	$\frac{4}{9}\pi$	**18.**	$-\frac{3}{5}\pi$	**19.**	6π	**20.**	-7π
21.	36π	**22.**	4.3	**23.**	-21	**24.**	-7

EXPRESS each angle in terms of π rad:

25.	$30°$	**26.**	$45°$	**27.**	$120°$	**28.**	$-240°$
29.	$-270°$	**30.**	$-12°$	**31.**	$-330°$	**32.**	$150°$
33.	$72°$	**34.**	$225°$	**35.**	$108°$	**36.**	$9°$
37.	$-10°$	**38.**	$-315°$	**39.**	$300°$		

EXPRESS in radians (not in terms of π):

40.	$38°$	**41.**	$220°$	**42.**	$94°$	**43.**	$390°$
44.	$23°40'$	**45.**	$48°24'$	**46.**	$32°10'$	**47.**	$27°20'40''$
48.	$54°00'50''$						

5.8   ARC LENGTH.   AREA OF SECTOR

Consider a circle whose radius OA has length r (see Fig. 5.11a). Let arc AB be an arc of length r. Then angle AOB is an angle of 1 rad. Let arc AC be an arc of length s and let θ equal the number of radians in angle AOC. Since the ratio of the measures of two central angles equals the ratio of the measures of their respective intercepted arcs,

$$\frac{\text{measure of } \angle AOB}{\text{measure of } \angle AOC} = \frac{\text{length of arc } AB}{\text{length of arc } AC}$$

Hence,

$$\frac{1}{\theta} = \frac{r}{s}$$

and

$$s = r\theta \tag{5.14}$$

Thus, the length of any circular arc can be found by multiplying the number of units in the length of the radius by the number of radians in the angle subtended by the arc.

EXAMPLE 1.   The radius of a circle is 15 in.   Find the length of an arc of this circle which subtends a central angle of $60°$.

Solution.   The central angle $\theta = 60° = \pi/3$ rad.   Hence

$$s = r\theta = 15\left(\frac{\pi}{3}\right) = 5\pi = 15.7 \text{ in. (approx.)}$$

EXAMPLE 2.   The length of an arc of a circle is 25 ft, and it subtends a central angle of 2.5 rad.   Find the radius of the circle.

Solution.   Since $s = r\theta$, $25 = r(2.5)$; hence $r = 10$ ft.

In Fig. 5.11*b* let *OP* and *OQ* be radii of the circle and let angle *POQ* have measure θ rad.   Let *A* be the area of the sector *POQ*.   The ratio of the area of the sector to the area of the circle equals the ratio of the measure of θ to the measure of the angle subtended by the entire circle.   Since the whole circle subtends an angle of 2π rad, we have

$$\frac{A}{\pi r^2} = \frac{\theta}{2\pi}$$

and
$$A = \frac{1}{2}r^2\theta \qquad\qquad (5.15)$$

Note.   We must keep in mind that in Eqs. (5.14) and (5.15) the angle θ must be measured in radians.   Otherwise the equations are invalid.

EXAMPLE 3.   The radius of a circle is 8 cm.   Find the area of a sector whose central angle is 30°.

Solution.   The central angle $\theta = 30° = \pi/6$ rad.   Hence,

$$A = \frac{1}{2}(8^2)\frac{\pi}{6} = \frac{16}{3}\pi = 16.75 \text{ sq cm (approx.)}$$

FIGURE 5.11   (*a*) Length of arc $= r\theta$; (*b*) area of sector $= \frac{1}{2}r^2\theta$

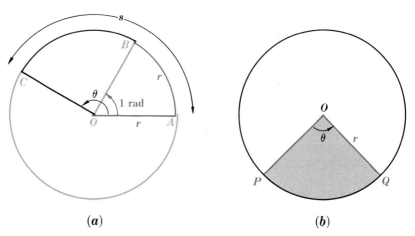

(*a*)   (*b*)

EXERCISE SET 5.6

1. A circle has a radius of 50 in. Find the length of the arc which subtends an angle whose radian measure is

   (a) $\pi/4$ (b) $3\pi/2$ (c) $\pi/10$ (d) $\pi/5$
   (e) 2   (f) 7   (g) 50   (h) 1.2

2. A circle has a radius of 30 ft. Find the number of radians in the central angle which intercepts an arc of

   (a) 30 ft   (b) 60 ft   (c) 15 ft   (d) 5 ft
   (e) 120 ft   (f) 2,400 ft   (g) 15π ft   (h) $105\pi/2$ ft

3. A central angle of $7\pi/4$ rad intercepts an arc s on a circle whose radius is r. Find r if the arc s is equal to

   (a) 7π ft   (b) 35π ft   (c) 105π ft   (d) 210π ft
   (e) π ft   (f) 7 ft   (g) 3 ft   (h) 28 ft

4. The radius of a circle is 10 in. Find the area of a circular sector whose central angle (measured in radians) is

   (a) 2   (b) 3   (c) 5   (d) 2.3
   (e) $\pi/4$ (f) $\pi/3$ (g) $\pi/2$ (h) $3\pi/5$

5. The area of a circular sector is 162 sq in. Find the radius if the central angle (measured in radians) is

   (a) 1   (b) 4   (c) 1,296   (d) 5,184

6. The area of a circular sector is 144 sq ft. Find the central angle if the radius is

   (a) 2 ft   (b) 3 ft   (c) 9 ft   (d) 18 ft

7. A wheel has a radius of 2 ft. If it rolls a distance of 6 ft without slipping, find the number of radians through which it turns. $s = \theta r$ $6 = 20$

8. A flywheel has a diameter of 20 in. If it turns at the rate of 60 rpm, through how many radians does it turn in 15 min? $(= 10$ $60 \cdot 15$

9. A pulley has a diameter of 20 in. It is driven by a belt moving 24 ft/sec. Through how many radians does the pulley turn per second?

10. A wheel having an outside diameter of 2 ft rotates at the rate of 200 rpm. Find the area swept over by a spoke of the wheel in 3 min. Assume the spoke reaches from the center of the axle to the outside rim of the wheel.

11. A satellite is launched into a circular orbit about the Earth. If the distance of the satellite from the center of the Earth is 5,000 mi, how far does it travel while its radius is sweeping an angle of 60°?

12. The electron in a hydrogen atom travels about the nucleus in a circle of radius r. How far does it travel while sweeping an angle of 150°?

13. The Earth orbits the sun in a path that is approximately a circle of radius 93,000,000 mi. If the radius sweeps about 1° per day, how far does the Earth travel in 1 day?

5.9   TRIGONOMETRIC FUNCTIONS OF ANGLES

Consider the unit circle (Fig. 5.12a) and the point $P(t)$. Let θ be the *radian measure of the central angle subtended by t*. Then, since $s = r\theta$,

$$t = 1(\theta) = \theta$$

Thus, the real number t is the same number as the radian measure of the angle. If $t = 1$, then $\theta = 1$ rad; if $t = 2$, $\theta = 2$ rad, and so on. Also, if $\theta = 1.52$ rad, then $t = 1.52$; if $\theta = 2.81$ rad, $t = 2.81$, and so on. Therefore, it seems logical to define any trigonometric function of an angle to be that same function of its radian measure.

> **DEFINITION.** Any trigonometric function of an angle whose radian measure is θ is equal to that same function of the real number θ.   (5.16)

Thus, $\sin (t \text{ rad}) = \sin t$; $\cos (x \text{ rad}) = \cos x$; $\tan (\theta \text{ rad}) = \tan \theta$, and so on. For example, if $\theta = 1.39$ rad, then $\sin \theta = \sin 1.39 = 0.9837$, from Table I. If $x = 0.55$ rad, then $\tan x = \tan 0.55 = 0.6131$.

An angle is frequently designated by its measure. Thus, we write $\theta = 30°$ to mean the "degree measure of the angle is 30." We write $\sin 30°$ to mean $\sin \theta$ where the measure of θ is $30°$. We write $\sin \pi/4$ to mean the trigonometric function $\sin$ (of an angle whose measure is $\pi/4$ rad).

Since the coordinates of $P(t)$ in terms of t are $(\cos t, \sin t)$, it follows that the coordinates of $P(t)$ in terms of θ are $(\cos \theta, \sin \theta)$. Now, let the point $R(x,y)$ be any point on the terminal side of the angle AOR (Fig. 5.12b), and

FIGURE 5.12   (a) Relation between θ (in radians) and t
(b) Relation between θ and its trigonometric functions

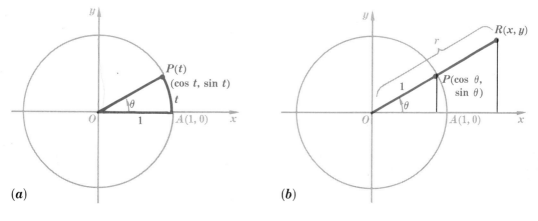

(a)   (b)

denote the length of OR by r. Then, from the properties of the similar triangles shown in the figure,

$$\frac{\sin \theta}{1} = \frac{y}{r} \quad \text{and} \quad \frac{\cos \theta}{1} = \frac{x}{r}$$

Therefore the following relations are valid unless they involve division by zero:

$$
\begin{aligned}
\sin \theta &= \frac{y}{r} & \cot \theta &= \frac{\cos \theta}{\sin \theta} = \frac{x}{y} \\
\cos \theta &= \frac{x}{r} & \sec \theta &= \frac{1}{\cos \theta} = \frac{r}{x} \\
\tan \theta &= \frac{\sin \theta}{\cos \theta} = \frac{y}{x} & \csc \theta &= \frac{1}{\sin \theta} = \frac{r}{y}
\end{aligned}
\tag{5.17}
$$

When $x = 0$, both $\tan \theta$ and $\sec \theta$ are undefined; when $y = 0$, $\cot \theta$ and $\csc \theta$ are undefined.

Equations (5.17) thus define the trigonometric functions of an angle in standard position whose terminal side passes through the point (x,y) of the coordinate plane. For any given angle, we should be able either to draw the axes so the angle will be in standard position or to move the angle to a convenient set of axes.

It is possible to represent the trigonometric functions of angles by directed line segments. Let $P(t)$ be a point on the unit circle. Construct a tangent to the circle at $A(1,0)$ and draw a line from the origin through $P(t)$ intersecting the tangent line at $T(x,y)$. Construct PM perpendicular to the x axis (Fig. 5.13). If θ is the radian measure of the angle AOP, then since $P(t)$ is on the unit circle,

$$\sin \theta = MP \quad \text{and} \quad \cos \theta = OM \qquad \theta \text{ in any quadrant}$$

If θ is in the first or fourth quadrant, we have from Eqs. (5.17),

$$\tan \theta = \frac{y}{x} = \frac{AT}{OA} = \frac{AT}{1} = AT$$

If θ is in the second or third quadrant, construct a second tangent line to the unit circle at $B(-1,0)$ and intersecting OP at $Q(x,y)$. Line OP extended through the origin again meets the first tangent at T. Then x and y, the coordinates of Q, are the negatives of the directed distances OA and AT, respectively. Again we find that $\tan \theta = AT$.

The only angles to which this discussion does not apply are the angles $\pi/2 + 2k\pi$ and $3\pi/2 + 2k\pi$, k an integer, for which $\tan \theta$ is undefined. Hence,

$$\tan \theta = AT \qquad \theta \text{ in any quadrant}$$

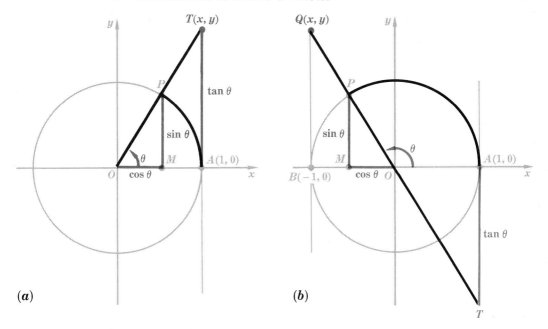

FIGURE 5.13   Trigonometric functions of angles in (*a*) first quadrant and (*b*) second quadrant

The magnitudes of the functions of θ are represented by the lengths of the segments, and the signs of the functions are given by the directions of the segments.   For example, if θ is in the second or fourth quadrant, then AT is directed downward and $\tan \theta$ is negative.

Since radians can be converted to degrees, and vice versa, the preceding discussion applies to angles measured in the degree system.

To estimate the numerical values of $\sin \theta$, $\cos \theta$, $\tan \theta$, estimate the lengths of the segments MP, OM, AT respectively.   Then $\cot \theta$, $\sec \theta$, $\csc \theta$ may be found by using appropriate identities.   For example, estimate the values of the functions of $122°$ (Fig. 5.13).   To make these estimates more accurate, draw the figure to a larger scale on a sheet of graph paper.   Since $P(t)$ is in the second quadrant, the sine of the angle is positive, the cosine and the tangent are both negative.   Hence,

$$\sin 122° = MP = 0.84 \qquad \csc 122° = \frac{1}{\sin 122°} = 1.19$$

$$\cos 122° = OM = -0.53 \qquad \sec 122° = \frac{1}{\cos 122°} = -1.9$$

$$\tan 122° = AT = -1.60 \qquad \cot 122° = \frac{1}{\tan 122°} = -0.63$$

In certain instances, the exact values of the trigonometric functions of angles can be found.

EXAMPLE 1. The angle θ is in standard position and the terminal side contains the point $P(-3,4)$. Find the values of the trigonometric functions of θ.

Solution. Since $x = -3$ and $y = 4$, $r = OP = 5$. Hence, $\sin \theta = y/r = \frac{4}{5}$, $\cos \theta = x/r = -\frac{3}{5}$, $\tan \theta = y/x = -\frac{4}{3}$, $\cot \theta = x/y = -\frac{3}{4}$, $\sec \theta = r/x = -\frac{5}{3}$, $\csc \theta = r/y = \frac{5}{4}$.

EXERCISE SET 5.7

In each of the following problems use graph paper and a protractor. Carefully construct the unit circle and the tangent at $A(1,0)$.

1. Construct the required angles and determine which of the following estimates is (are) not reasonable:

(a) $\sin 20° = 0.34$ (b) $\cos 20° = 0.94$
(c) $\tan 20° = 0.36$ (d) $\sin 170° = 0.49$
(e) $\cos 170° = -0.63$ (f) $\tan 170° = -0.56$
(g) $\sin 250° = -0.94$ (h) $\cos 250° = -0.84$
(i) $\tan 250° = -2.75$ (j) $\sin 88° = 1.20$

2. Construct the line segments of required length and verify the estimated values of θ:

(a) When $OM = 0.64$, then $\theta = 50°$.
(b) When $AT = 5.7$, then $\theta = 80°$.
(c) When $MP = 0.50$, then $\theta = 30°$.

3. Construct figures and estimate the values of the following:

(a) $\sin 50°$ (b) $\cos 50°$ (c) $\tan 50°$ (d) $\sin 150°$
(e) $\cos 150°$ (f) $\tan 150°$ (g) $\sin 230°$ (h) $\cos 230°$
(i) $\tan 230°$ (j) $\sin 340°$ (k) $\cos 340°$ (l) $\tan 340°$

FIND the trigonometric functions of the angle in each of the following, where the given point is on the terminal side of an angle in standard position:

4. $(12,5)$ 5. $(9,12)$ 6. $(3,5)$ 7. $(8,8)$
8. $(-15,-8)$ 9. $(6,4)$ 10. $(12,-16)$ 11. $(2\sqrt{6},5)$
12. $(-\sqrt{6},-\sqrt{10})$

DETERMINE for each of the following the quadrant in which the terminal side of the angle must lie:

13. The sine and tangent are both positive.   14. The sine and tangent are both negative.
15. The secant and cosecant are both positive.   16. The sine and cosine are both negative.
17. The sine and secant are both positive.

FIND the required quantities for each of the following, where the angle θ is in standard position and point $P(x,y)$ is on the terminal side:

18. If $y = 6$ and $\sin \theta = \frac{3}{5}$, find x and r.
19. If $x = 4$ and $\cos \theta = \frac{5}{6}$, find y and r.
20. If $y = 5$ and $\tan \theta = 1.5$, find x and r.

5.10 TABLES OF FUNCTIONS OF ANGLES

If an angle is measured in the degree system, we can convert degrees to radians. For example, if $\theta = 40°$, then $\theta = 40(\pi/180) = 0.70$ rad (approx.). Hence, $\sin 40° = \sin 0.70$ rad $= \sin 0.70 = 0.6442$ (approx.) from Table I. However, when an angle is measured in the degree system, as is often the case, it is convenient to have a table such as Table II of the appendix. This table is constructed to give approximate values of the trigonometric functions of angles from $0°$ to $90°$. Its application to angles greater than $90°$ will be considered later.

Finding trigonometric functions

To use Table II, proceed as follows:

1. If the angle is less than $45°$, locate the angle in the column at the left of the page under the heading θ. Move horizontally across the page to the column headed by the name of the function whose value is to be found.

For example, to find $\cos 33°20'$, look for $33°20'$ in the left-hand column and move across the page to the column headed *cos θ*. The entry at this point is 0.8355; hence $\cos 33°20' = 0.8355$.

2. If angle θ is greater than $45°$, locate the angle in the column at the extreme right of the page. This column is read from the bottom of the page toward the top and is headed θ in the lower right-hand corner. Move horizontally across the page to the left to the column with the name of the required function at the bottom. For instance, to find $\tan 67°50'$, read up the extreme right-hand column to $67°50'$. Move horizontally to the left to the column which has *tan θ* at the bottom. Here the entry is 2.455; hence, $\tan 67°50' = 2.455$.

EXAMPLE 1.

(1) $\sin 37°10' = 0.6041$
(2) $\sec 28°40' = 1.140$
(3) $\tan 15°20' = 0.2742$
(4) $\csc 71°30' = 1.054$
(5) $\sin 84°50' = 0.9959$

When the value of a trigonometric function of an angle is given, Table II enables us to find the corresponding angle. Let us assume at this point that the angle is acute. The following example then illustrates the method.

EXAMPLE 2. If $0 < \theta < 90°$ and tan $\theta = 0.7954$, find θ.

Solution. Find 0.7954 in the column headed *tan θ* at the top or *tan θ* at the foot. In this case, we find 0.7954 in the column headed *tan θ* at the top. This tells us to read the corresponding angle 38°30′ in the left-hand column. Hence, if tan $\theta = 0.7954$, then $\theta = 38°30′$.

EXAMPLE 3. Given sin $\theta = 0.9304$; find θ if $0 < \theta < 90°$.

Solution. Find 0.9304 in a column headed *sin θ* at the top or *sin θ* at the bottom. In this case, we find 0.9304 in the column with *sin θ* at the foot. This tells us to read the corresponding angle, 68°30′, in the right-hand column. Hence, if sin $\theta = 0.9304$, then $\theta = 68°30′$.

The two things to remember in using Table II are:
1. If the angle is less than 45°, read "down" and across the "top."
2. If the angle is greater than 45°, read "up" and across the "bottom."

EXERCISE SET 5.8

FIND the value of the given function, using the table of natural functions:

1. sin 28°	2. cos 37°	3. tan 21°
4. cot 20°	5. sec 17°	6. csc 32°
7. cos 20°10′	8. sin 43°30′	9. tan 41°40′
10. sec 25°50′	11. tan 18°20′	12. sin 44°40′
13. cos 10°30′	14. sin 15°50′	15. tan 22°10′
16. sin 47°	17. cos 62°	18. tan 75°
19. csc 82°	20. sec 65°	21. cos 89°
22. sin 48°10′	23. cos 52°50′	24. sin 74°30′
25. sin 68°50′	26. sin 89°50′	27. cos 89°10′
28. tan 1°10′	29. sin 2°40′	30. tan 89°50′
31. cot 72°20′	32. csc 45°10′	33. sec 64°30′

FIND the value of θ, assuming that $0 \le \theta \le 90°$, if:

34. sin $\theta = 0.6583$	35. cos $\theta = 0.7916$	36. tan $\theta = 0.7355$
37. sec $\theta = 1.241$	38. tan $\theta = 0.6088$	39. csc $\theta = 1.812$
40. sin $\theta = 0.3692$	41. cos $\theta = 0.8704$	42. cos $\theta = 0.9911$
43. tan $\theta = 0.0787$	44. csc $\theta = 2.323$	45. sec $\theta = 1.121$
46. cos $\theta = 0.7294$	47. sin $\theta = 0.7030$	48. tan $\theta = 0.9942$
49. tan $\theta = 1.124$	50. sin $\theta = 0.7470$	51. cos $\theta = 0.5688$

FIND the angle, assuming it to be acute, if:

52. sin $\phi = 0.9838$	53. cot $\alpha = 6.561$	54. tan $\beta = 4.989$
55. cot $\beta = 0.9490$	56. tan $\phi = 1.054$	57. cot $\phi = 1.303$
58. sec $\alpha = 1.722$	59. sin $\phi = 0.8936$	60. cos $\phi = 0.1305$

5.11 INTERPOLATION

We now turn to the problem of finding trigonometric functions of numbers, or angles, that are between those listed in the tables. For example, if $t = 1.485$, we cannot find sin t directly from Table I. However, since 1.485 lies halfway between 1.480 and 1.490, we suspect that sin 1.485 lies approximately halfway between sin 1.480 and sin 1.490. To determine approximate values of the functions of such numbers, we rely upon the following assumption:

Small changes in t

> The change in the value of a trigonometric function of t resulting from a small change in t is approximately proportional to the change in t.

Actual calculations show that the approximations obtained from this assumption are reasonably close. The approximations for the cotangent and cosecant functions become more incorrect for values of t near 0, and the approximations for the tangent and secant functions become more incorrect for values of t near $\pi/2$.

We shall consider a change of 0.01 in the number t to be a small change in t, and a change of $10'$ in the angle θ a small change in θ.

The process of discarding digits from the right end of a given number is called *rounding off* the number. The result of rounding off is an approximation of the given number. For example, 3.14159 rounds off to 3.1416 which is an approximation of 3.14159. We will use the following rules.

1. (a) If the right-hand digit to be discarded from a number is less than 5, do not change the digit to its immediate left. Thus, 78.34 rounds off to 78.3.
    (b) If the right-hand digit to be dropped is greater than 5, increase by 1 the digit to its immediate left. For example, 78.36 rounds off to 78.4.
    (c) If the right-hand digit to be discarded is 5, do not change the digit to the immediate left if that digit is even, but increase it by 1 if it is odd. Thus, 78.45 rounds off to 78.4, but 78.75 rounds off to 78.8.
2. (a) If more than one digit is to be discarded and if the left-hand digit of those to be dropped is less than 5, do not change the digit to its immediate left. For example, 78.4398 rounds off to 78.4.
    (b) If more than one digit is to be discarded and if the left-hand digit of those to be dropped is 5 or more, increase by one the digit to its immediate left. Thus, 78.3598 rounds off to 78.4.
3. When discarding digits from the right end of a whole number, replace each discarded digit with 0. For example, 634 rounds off

to 630, 6,356 rounds off to 6,400, 63.5 rounds off to 60, and 67.5 rounds off to 70.

The process of finding approximate values of trigonometric functions of numbers or angles that are between those listed in the tables is called *interpolation*. The process is illustrated in the following examples.

EXAMPLE 1. Find sin 1.483

Solution. The number 1.483 is greater than 1.480 and is less than 1.490. Thus, we assume that sin 1.483 is somewhere between sin 1.480 and sin 1.490; that is, sin 1.483 = sin 1.480 + some number to be determined. This number is approximated by using the assumption made at the beginning of this section. We reason as follows:

Since the difference between 1.480 and 1.490 is 0.10 and the difference between 1.480 and 1.483 is 0.003, we see that the number 1.483 is $0.003/0.010 = {}^3\!/_{10}$ of the way between 1.480 and 1.490. Hence, by the assumption, the approximate value of sin 1.483 is ${}^3\!/_{10}$ of the way between sin 1.480 and sin 1.490. From Table I,

$$\sin 1.480 = 0.9959 \qquad \text{and} \qquad \sin 1.490 = 0.9967$$

Therefore, the difference between 0.9959 and 0.9967 (called the *tabular difference*) is 0.0008, and ${}^3\!/_{10}$ of this tabular difference correct to four decimal places is 0.0002. Hence,

$$\sin 1.483 = \sin 1.480 + 0.0002$$
$$= 0.9959 + 0.0002 = 0.9961 \text{ (approx.)}$$

The following tabular arrangement is frequently used.

$$0.01 \left\{ 0.003 \begin{cases} \sin 1.480 = 0.9959 \\ \sin 1.483 = 0.9959 + d \end{cases} d \right\} 0.0008$$
$$\sin 1.490 = 0.9967$$

By the assumption, the change in sin *t* is approximately proportional to the change in *t*.

$$\frac{0.003}{0.01} = \frac{d}{0.0008}$$

and $d = 0.00024 = 0.0002 \text{ (approx.)}$

Therefore, $\sin 1.483 = 0.9959 + 0.0002$
$$= 0.9961 \text{ (approx.)}$$

EXAMPLE 2. Find cos 1.413.

Solution. Since cos 1.413 is between cos 1.410 and cos 1.420, we set up the following tabular arrangement. We must keep in mind that cos t *decreases* as t increases from 0 to $\pi/2 = 1.57$. Table I illustrates this important fact.

$$0.01 \left\{ 0.003 \left\{ \begin{array}{l} \cos 1.410 = 0.1601 \\ \cos 1.413 = 0.1601 - d \\ \cos 1.420 = 0.1502 \end{array} \right\} d \right\} 0.0099$$

Hence,
$$\frac{0.003}{0.01} = \frac{d}{0.0099}$$

and
$$d = 0.00297 = 0.0030 \text{ (approx.)}$$

Therefore,
$$\cos 1.413 = 0.1601 - 0.0030 = 0.1571 \text{ (approx.)}$$

When a trigonometric function of t is given, we may use a similar arrangement for finding the approximate value of t.

EXAMPLE 3. Find t if tan $t = 1.892$ and $0 < t < \pi/2$.

Solution. From Table I, tan $1.08 = 1.871$ and tan $1.09 = 1.917$. Hence, we conclude that t is between 1.08 and 1.09.

$$0.01 \left\{ d \left\{ \begin{array}{l} \tan 1.08 = 1.871 \\ \tan (1.08 + d) = 1.892 \\ \tan 1.09 = 1.917 \end{array} \right\} 0.021 \right\} 0.046$$

Hence,
$$\frac{d}{0.01} = \frac{0.021}{0.046}$$

and
$$d = 0.005 \text{ (approx.)}$$

Thus,
$$\tan (1.08 + 0.005) = 1.892 \text{ (approx.)}$$

Hence,
$$t = 1.08 + 0.005 = 1.085 \text{ (approx.)}$$

Table II is used in a similar manner.

EXAMPLE 4. Find sec $40°13'$.

Solution. The following arrangement can be used:

$$10' \left\{ 3' \left\{ \begin{array}{l} \sec 40°10' = 1.309 \\ \sec 40°13' = 1.309 + d \\ \sec 40°20' = 1.312 \end{array} \right\} d \right\} 0.003$$

Thus,
$$\frac{3}{10} = \frac{d}{0.003}$$

and $\qquad\qquad\qquad d = 0.0009 = 0.001$ (approx.)

Hence, $\qquad\quad$ sec $40°13' = 1.309 + 0.001 = 1.310$ (approx.)

EXERCISE SET 5.9

FIND the value of the given function (interpolate, using Table I):

1. sin 1.035	**2.** cos 0.685	**3.** tan 0.565
4. tan 1.326	**5.** cot 1.564	**6.** sec 1.172
7. csc 0.616	**8.** sin 0.928	**9.** tan 1.064
10. cos 0.352	**11.** tan 0.157	**12.** sin 0.108
13. sin 1.116	**14.** cos 1.024	**15.** tan 1.538

FIND the value of the given function (interpolate, using Table II):

16. sin 28°32'	**17.** cos 25°44'	**18.** tan 18°56'
19. sec 21°21'	**20.** sin 38°28'	**21.** cos 10°17'
22. sin 32°38'	**23.** cos 19°06'	**24.** tan 4°08'
25. tan 47°18'	**26.** sin 73°22'	**27.** cos 72°43'
28. cos 63°33'	**29.** tan 84°24'	**30.** sin 87°16'
31. sec 63°02'	**32.** cot 55°56'	**33.** csc 71°23'
34. cot 49°22'	**35.** csc 52°27'	**36.** sin 49°48'
37. cos 51°37'	**38.** sin 67°01'	**39.** cos 70°02'

FIND the real number, less than $\pi/2$, whose trigonometric function is given (interpolate, using Table I):

40. sin $t = 0.1320$	**41.** tan $t = 0.2200$	**42.** cos $t = 0.9371$
43. tan $x = 6.120$	**44.** cos $x = 0.2415$	**45.** cot $y = 0.2480$
46. sec $z = 2.938$	**47.** csc $z = 1.036$	**48.** sec $x = 1.688$
49. sin $x = 0.8698$	**50.** cos $x = 0.8746$	**51.** tan $z = 0.2592$

FIND the angle, less than 90°, whose trigonometric function is given (interpolate, using Table II):

52. tan $\phi = 0.4100$	**53.** cos $\phi = 0.9540$	**54.** sin $\phi = 0.5350$
55. sin $x = 0.6887$	**56.** sin $y = 0.7988$	**57.** cos $x = 0.3190$
58. tan $x = 1.3268$	**59.** cos $x = 0.6441$	**60.** sin $\phi = 0.8879$
61. cos $\phi = 0.2954$	**62.** sin $\beta = 0.9911$	**63.** tan $\beta = 2.7575$

5.12 TRIGONOMETRIC FUNCTIONS OF ANY ANGLE

In order to find trigonometric functions of a real number t greater than $\pi/2$ we found it convenient to define the reference number t_1 for t (see Sec. 5.4). Similarly, to find trigonometric functions of an angle θ when θ is greater than $\pi/2$ rad $= 90°$, we define the reference angle θ_1 for the angle θ.

Reference angle $\qquad$ The reference angle θ_1 for the angle θ is *the positive acute angle between the x axis and the terminal side of θ* (Fig. 5.14). Thus, if θ is in the second quadrant, then $\theta_1 = 180° - \theta$. If θ is in quadrant III, $\theta_1 = \theta - 180°$. If θ is in the fourth quadrant, then $\theta_1 = 360° - \theta$.

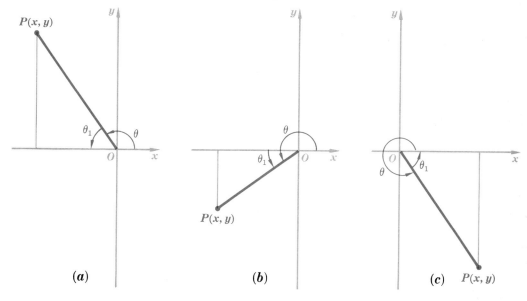

FIGURE 5.14   (a) $\theta_1 = 180° - \theta$
(b) $\theta_1 = \theta - 180°$
(c) $\theta_1 = 360° - \theta$

A discussion similar to that of Sec. 5.4 shows that any trigonometric function of the angle θ is numerically equal to the same function of the reference angle θ_1 but may differ in sign. Therefore, if $F(\theta)$ is any trigonometric function of θ and if θ_1 is the reference angle for θ, then

$$|F(\theta)| = F(\theta_1)$$

For example, if $\theta = 154°$, then θ is in the second quadrant and $\theta_1 = 26°$. Therefore, sin 154° is positive. Since

$$|\sin 154°| = \sin 26° = 0.4384$$

it follows that sin 154° = 0.4384.

Also, cos 154° is negative. Then since

$$|\cos 154°| = \cos 26° = 0.8988$$

it follows that cos 154° = −0.8988.

Similarly, tan 154° is negative. Consequently, we have

$$|\tan 154°| = \tan 26° = 0.4877$$

and
$$\tan 154° = -0.4877$$

EXAMPLE 1.   Find $\sin \theta$ and $\tan \theta$ if $\theta = 242°$.

Solution.   The terminal side of θ lies in the third quadrant. Hence, $\sin \theta$ is negative and $\tan \theta$ is positive. The reference angle $\theta_1 = 62°$. Since

$$|\sin 242°| = \sin 62° = 0.8829$$

and
$$|\tan 242°| = \tan 62° = 1.881$$

we have
$$\sin 242° = -0.8829$$

and
$$\tan 242° = 1.881$$

EXAMPLE 2.   Find $\cos \beta$ and $\cot \beta$ if $\beta = -147°$.

Solution.   Since β is in the third quadrant, $\cos \beta$ is negative and $\cot \beta$ is positive. The reference angle is $\beta_1 = 33°$. Hence,

$$\cos(-147°) = -\cos 33° = -0.8387$$

and
$$\cot(-147°) = \cot 33° = 1.540$$

Thus, to find the value of a trigonometric function of an angle, say θ, (1) find the quadrant in which θ lies; (2) determine the sign of the given function; (3) determine the reference angle θ_1; (4) from a table determine the value of the given function of θ_1; (5) affix the sign which was determined in (2).

EXERCISE SET 5.10

FIND the value of each function (interpolate whenever necessary, using Table II):

1.	$\sin 150°$	**2.**	$\cos 120°$	**3.**	$\tan 172°$
4.	$\sec 106°50'$	**5.**	$\csc 123°10'$	**6.**	$\cot 133°20'$
7.	$\cos 117°40'$	**8.**	$\sin 174°30'$	**9.**	$\cos 96°30'$
10.	$\tan 212°$	**11.**	$\cot 243°$	**12.**	$\sin 268°$
13.	$\cos 215°20'$	**14.**	$\sin 244°50'$	**15.**	$\tan 258°30'$
16.	$\cot 225°10'$	**17.**	$\sec 195°50'$	**18.**	$\cos 182°40'$
19.	$\sin 275°$	**20.**	$\cos 314°$	**21.**	$\tan 325°$
22.	$\tan 344°10'$	**23.**	$\sin 358°20'$	**24.**	$\tan 280°30'$
25.	$\sin(-115°)$	**26.**	$\cos(-225°)$	**27.**	$\tan(-320°)$
28.	$\tan(-278°)$	**29.**	$\cot(-123°)$	**30.**	$\sec(-154°)$
31.	$\cos 117°23'$	**32.**	$\sin 229°47'$	**33.**	$\cot 138°06'$
34.	$\sec 227°33'$	**35.**	$\cos 318°42'$	**36.**	$\sin 344°17'$

FIND all values of the angle between $0°$ and $360°$ which satisfy the equation. Express your answers to the nearest minute. Interpolate whenever necessary.

37.	$\tan \phi = 0.4100$	**38.**	$\cos \phi = 0.9540$	**39.**	$\sin x = 0.5350$
40.	$\sin x = -0.6887$	**41.**	$\sin y = -0.7988$	**42.**	$\cos z = -0.3190$
43.	$\tan \beta = -1.3268$	**44.**	$\cos \alpha = -0.6441$	**45.**	$\sin w = -0.8879$

46. $\cos x = -0.2954$ **47.** $\sin \gamma = -0.9911$ **48.** $\tan \gamma = -2.7575$
49. $\sec A = 1.130$ **50.** $\csc B = -2.156$ **51.** $\cot y = -2.088$
52. $\cot B = 1.247$ **53.** $\cos \gamma = -0.6616$ **54.** $\sin \alpha = -0.7478$

5.13   VARIATION OF TRIGONOMETRIC FUNCTIONS

Let us consider again the trigonometric point $P(t)$ on the unit circle (Fig. 5.15) and observe the changes in the length of MP as t varies continuously from 0 to 2π.   Since $\sin t = MP$, we see that for $t = 0$, $\sin t = 0$.

FIGURE 5.15   Variations of functions

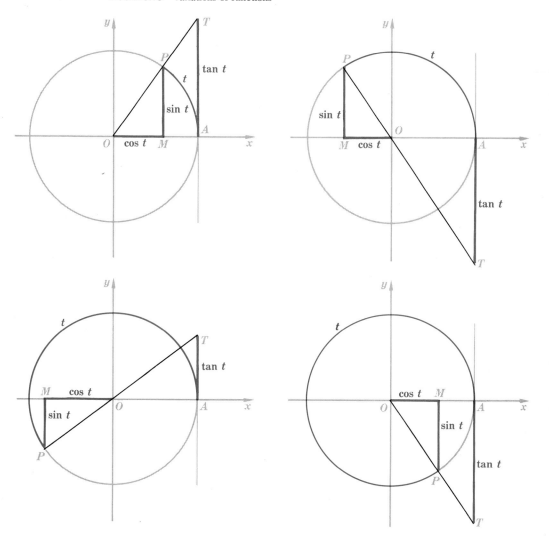

As t increases from 0 to $\pi/6$, sin t increases from 0 to sin $\pi/6 = 0.5$. As t continues to increase to $\pi/3$, sin t continues to increase to 0.87 (approx.). At $t = \pi/2$, sin t reaches a maximum value of 1.

As t increases from $\pi/2$ to π, sin t decreases from 1 to 0. In the third quadrant, sin t is negative and decreases from 0 to -1 as t increases from π to $3\pi/2$. As t then increases from $3\pi/2$ to 2π, sin t increases from -1 to 0.

By observing the changes in the length of OM, we find that as t increases from 0 to $\pi/2$, cos $t = OM$ decreases from 1 to 0. As t increases from $\pi/2$ to π, cos t continues to decrease from 0 to -1. Then as t increases from π to $3\pi/2$, cos t increases from -1 to 0. In the fourth quadrant, as t varies from $3\pi/2$ to 2π, we find that cos t increases from 0 to 1.

Since tan $t = AT$, we note that as t increases from 0, tan t increases from 0. The closer t gets to $\pi/2$ the greater tan t becomes. We describe this behavior by the statement: *When $0 < t < \pi/2$, tan t is positive and increases without bound as t approaches $\pi/2$.*

When t is in the second quadrant, AT is directed downward and tan t is negative. As t decreases toward $\pi/2$, AT increases in length in the downward direction. This means that tan t decreases algebraically. The nearer t gets to $\pi/2$, the less (more negative) tan t becomes. By choosing t near enough to $\pi/2$ we can make tan t as negatively large as we please. We describe this result by the statement: *When $\pi/2 < t < \pi$, tan t is negative, and tan t decreases without bound as t approaches $\pi/2$.*

From this discussion we see that as t increases from a value slightly greater than $\pi/2$ to the value π, tan t increases to 0. As t then increases from π to $3\pi/2$, tan t increases from 0 and repeats the behavior of tan t in the first quadrant. As t increases from $3\pi/2$ to 2π, tan t repeats the behavior of the function in the second quadrant.

The reciprocal relations enable us to determine the variation of the other three functions. For example, if $t = 0$, cot $t = 1/\tan t$ is undefined. Then as t varies from 0 to $\pi/2$, cot t decreases to 0. In the second quadrant,

Table 1.   Variation of Trigonometric Functions

As t varies from	sin t varies from	cos t varies from	tan t varies
0 to $\dfrac{\pi}{2}$	0 to 1	1 to 0	from 0 through all positive values
$\dfrac{\pi}{2}$ to π	1 to 0	0 to -1	through all negative values to 0
π to $\dfrac{3\pi}{2}$	0 to -1	-1 to 0	from 0 through all positive values
$\dfrac{3\pi}{2}$ to 2π	-1 to 0	0 to 1	through all negative values to 0

cot t is negative.   As t approaches π through values less than π, cot t decreases without bound.   As t increases from π to $3\pi/2$, and then from $3\pi/2$ to 2π, cot t repeats the behavior of the function in the first and second quadrants respectively.   Table 1 summarizes the variations in the sine, cosine, and tangent functions.

We leave to the student the task of constructing a similar table to exhibit the variation in cot t, sec t, and csc t as t varies from 0 to 2π.

EXERCISE SET 5.11

1.   Let $P(t)$ be a point on the unit circle in the first quadrant.   Construct the tangent to the circle at $B(0,1)$ and a line from the origin through $P(t)$ intersecting the tangent at Q.   Verify for this figure that cot t equals in magnitude the number of units in the length of the segment BQ.
2.   Use the figure of Prob. 1 to determine the variation in cot t as t varies from 0 to $\pi/2$.
3.   Use the figure of Prob. 1 to determine the variation in cot t as t varies from $\pi/2$ to π.
4.   Let $P(t)$ be a point on the unit circle in the first quadrant.   Construct the tangent to the circle at $A(1,0)$ and a line from the origin through $P(t)$ intersecting the tangent at the point T.   Show for this figure that sec t is equal in magnitude to the number of units in the length of OT.
5.   Use the figure of Prob. 4 to study the variation in sec t as t varies from 0 to 2π.
6.   Take $P(t)$ on the unit circle in the first quadrant.   Construct the tangent to the circle at $B(0,1)$ and the line from the origin through $P(t)$ intersecting the tangent at R.   Show that csc t is equal in magnitude to the number of units in the length of OR.
7.   Use the figure of Prob. 6 to study the variation of csc t as t varies from 0 to 2π.
8.   Make a table similar to the one in the preceding section showing the variation of cot t, sec t, and csc t as t varies from 0 to 2π.

5.14   GRAPH OF $y = \sin x$

Let x denote a real number.   Then the function defined by $y = \sin x$ is the function whose ordered pairs are $(x, \sin x)$.   It is evident from a consideration of the trigonometric point $P(x)$ as it moves around the unit circle in either direction that $\sin 2k\pi = 0$, where k is an integer.   Hence, the ordered pairs

$$\ldots, (-2\pi,0), (-\pi,0), (0,0), (\pi,0), (2\pi,0), \ldots$$

are ordered pairs of the function.   This means that the points whose coordinates are these ordered pairs are the points common to the graph of $y = \sin x$ and the x axis.

From the variation of the sine function, we know that $\sin (\pi/2) = 1$ and that $\sin x = 1$ only when x is coterminal with $\pi/2$.   Also, $\sin x$ is never greater than 1.   Now $\sin (3\pi/2) = -1$, and $\sin x = -1$ only when x is coterminal with $3\pi/2$.   The least value that $\sin x$ can have is -1.   With this information we could sketch a rough graph of the function.   However, a more accurate sketch can be obtained if a few additional points are plotted.   Some of the ordered pairs of $y = \sin x$ for values of x between -2π and 2π are tabulated:

x	$-\dfrac{7\pi}{4}$	$-\dfrac{3\pi}{2}$	$-\dfrac{5\pi}{4}$	$-\pi$	$-\dfrac{3\pi}{4}$	$-\dfrac{\pi}{2}$	$-\dfrac{\pi}{4}$
y	0.71	1	0.71	0	-0.71	-1	-0.71

x	0	$\dfrac{\pi}{4}$	$\dfrac{\pi}{2}$	$\dfrac{3\pi}{4}$	π	$\dfrac{5\pi}{4}$	$\dfrac{3\pi}{2}$	$\dfrac{7\pi}{4}$
y	0	0.71	1	0.71	0	-0.71	-1	-0.71

To get a well-proportioned graph, we use the same scale on both axes and mark the point on the x axis which represents π as nearly as possible at 3.14. We assume that the graphs of the trigonometric functions are smooth curves (have no corners or vertices).

When the ordered pairs $(x,\ \sin x)$ are plotted and joined by a smooth curve, that portion of the graph of $y = \sin x$ between $x = -2\pi$ and $x = 2\pi$ is obtained (see Fig. 5.16).

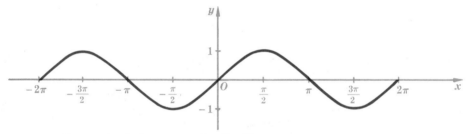

FIGURE 5.16   Graph of $y = \sin x$, $-2\pi \le x \le 2\pi$

5.15 GRAPH OF $y = a \sin x$

Consider the function defined by $y = a \sin x$, where a and x are real numbers. The ordered pairs of this function $(x,\ a \sin x)$ exhibit the role of the real number a as a multiplier. For example, let $a = 2$, so that $y = 2 \sin x$. When $x = \pi/6$, $y = 2 \sin (\pi/6) = 2(0.5) = 1$. When $x = \pi/2$, $y = 2 \sin (\pi/2) = 2(1) = 2$.

Since the greatest value $\sin x$ can have is 1, the greatest value that $a \sin x$ can have is $|a|$. The least value $a \sin x$ can have is $-|a|$. These maximum and minimum values occur when x is an odd multiple of $\pi/2$. We call the number $|a|$ the *amplitude* of the function.

DEFINITION. The amplitude of a function $y = f(x)$ is the greatest distance of any point on the graph of $f(x)$ from a horizontal line which passes halfway between the maximum and minimum values of the function.   (5.18)

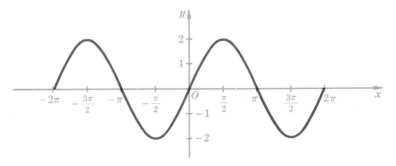

FIGURE 5.17 Graph of $y = 2 \sin x$, $-2\pi \leq x \leq 2\pi$

Thus the amplitude of $y = \sin x$ is 1, the amplitude of $y = 3 \sin x$ is 3, and the amplitude of $y = \frac{1}{3} \sin x$ is $\frac{1}{3}$.

The portion of the graph of $y = 2 \sin x$ for $-2\pi \leq x \leq 2\pi$ is shown in Fig. 5.17.

5.16 PERIODICITY OF TRIGONOMETRIC FUNCTIONS

The trigonometric functions belong to a large class of functions called *periodic* functions in which there is a regular repetition of the values of the function over a certain interval. For example, we note that as t increases from 0 to 2π, the trigonometric point $P(t)$ completes one circuit around the unit circle. Then as t continues to increase, $P(t)$ begins its second trip around the circle, and the trigonometric functions begin to repeat in the same order their earlier behavior. Thus, the variation of $\sin t$ as t increases from 0 to 2π is repeated as t increases from 2π to 4π.

DEFINITION. A function $f(x)$ is called periodic if there is a non-zero number p such that $f(x + p)$ is defined and $f(x + p) = f(x)$ for all values of x in the domain of $f(x)$. The number p is called a period of $f(x)$. (5.19)

If p is the smallest nonzero number for which $f(x + p) = f(x)$, then p is called the period of $f(x)$.

Since $P(t)$ and $P(t + 2\pi)$ represent the same point on the unit circle, we see that a function of $(t + 2\pi)$ must equal the function of t. We state this fact as a theorem.

THEOREM. Any trigonometric function of $(t + 2\pi)$ equals that same function of t. (5.20)

Hence, 2π is a period of each of the trigonometric functions. From a consideration of the variations of these functions, Sec. 5.13, we see that 2π is the period of the sine, cosine, secant, and cosecant functions. The period of the tangent and cotangent functions is π.

The importance of this conclusion is that if the values of a trigonometric function are known for one period of the variable, then the values of the function are known for all values of the variable. Hence, to sketch the graph of any trigonometric function of x, it is sufficient to sketch the graph for values of x in an interval of length 2π at most. The complete graph will consist of duplications of this portion.

5.17 GRAPH OF $y = a \sin bx$

The maximum value of $\sin bx$ is $+1$. Hence, the maximum value of $a \sin bx$ is $|a|$. The period can be determined as follows. For the sine function, the smallest nonzero number p for which $\sin (bx + p) = \sin bx$ is $p = 2\pi$. This means that $\sin (bx + 2\pi) = \sin bx$ for every x. Therefore

$$\sin b \left(x + \frac{2\pi}{b} \right) = \sin bx \qquad \text{for every } x$$

Let s be the smallest number for which

$$\sin b(x + s) = \sin bx \qquad \text{for every } x$$

Then
$$s = \frac{2\pi}{b}$$

Thus, $2\pi/b$ is the period of $\sin bx$. This means that as bx varies over an interval of length 2π, x varies over the length $2\pi/b$. For example, the amplitude of $y = 3 \sin 2x$ is 3 and the period is $2\pi/2 = \pi$. The amplitude of $y = \sin (\frac{1}{2}x)$ is 1, and the period is $2\pi/\frac{1}{2} = 4\pi$.

EXAMPLE 1. Sketch the graph of $y = \sin 2x$.

Solution. The amplitude is $a = 1$ and the period is $s = 2\pi/2 = \pi$. The amplitude is reached when $\sin 2x = 1$, that is, when

$$2x = \pm \frac{\pi}{2}, \pm \frac{3\pi}{2}, \pm \frac{5\pi}{2}, \cdots$$

and
$$x = \pm \frac{\pi}{4}, \pm \frac{3\pi}{4}, \pm \frac{5\pi}{4}, \cdots$$

A portion of the graph is shown in Fig. 5.18.

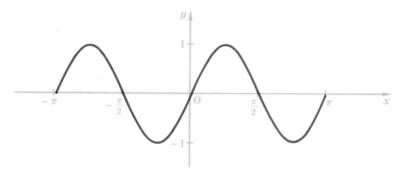

FIGURE 5.18 Graph of $y = \sin 2x$, $-\pi \leq x \leq \pi$

5.18 GRAPH OF $y = a \cos bx$

The graph of $y = a \cos bx$, $a > 0$, $b > 0$ can be constructed by the method employed in graphing the sine function. The following examples illustrate the procedure.

EXAMPLE 1. Sketch the graph of the function $y = \cos x$.

Solution. The period of $y = \cos x$ is 2π. The value of y is never greater than 1 nor less than -1. Hence, the amplitude of the function is 1. The amplitude is reached when $\cos x = 1$, and therefore when

$$x = 0, \pm 2\pi, \pm 4\pi, \ldots$$

The minimum value occurs at $x = \pm \pi, \pm 3\pi, \pm 5\pi, \ldots$.
The graph crosses the x axis when $\cos x = 0$, that is, when

$$x = \pm \frac{\pi}{2}, \pm \frac{3\pi}{2}, \pm \frac{5\pi}{2}, \ldots$$

Part of the graph is shown in Fig. 5.19.

FIGURE 5.19 Graph of $y = \cos x$, $-3\pi/2 \leq x \leq 5\pi/2$

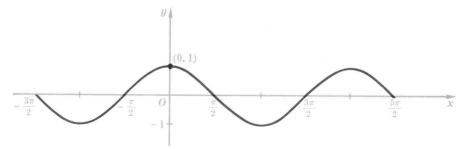

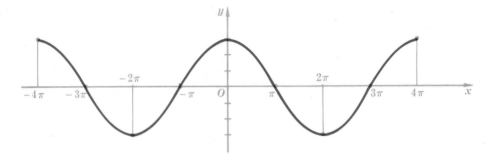

FIGURE 5.20   Graph of $y = 3 \cos (\tfrac{1}{2} x)$, $-4\pi \leq x \leq 4\pi$

EXAMPLE 2.   Draw the graph of $y = 3 \cos (\tfrac{1}{2} x)$.

Solution.   The amplitude is 3 and the period is 4π.   The amplitude is reached when $\cos (\tfrac{1}{2} x) = 1$, that is, when

$$\frac{1}{2} x = 0, \ \pm 2\pi, \ \pm 4\pi, \ \ldots$$

and $$x = 0, \ \pm 4\pi, \ \pm 8\pi, \ \ldots$$

The minimum (lowest) points on the graph occur when

$$\frac{1}{2} x = \pm \pi, \ \pm 3\pi, \ \pm 5\pi, \ \ldots$$

and $$x = \pm 2\pi, \ \pm 6\pi, \ \pm 10\pi, \ \ldots$$

The graph crosses the x axis whenever $y = 3 \cos (\tfrac{1}{2} x) = 0$.   These points occur when

$$\frac{1}{2} x = \pm \frac{\pi}{2}, \ \pm \frac{3\pi}{2}, \ \pm \frac{5\pi}{2}, \ \ldots$$

and $$x = \pm \pi, \ \pm 3\pi, \ \pm 5\pi, \ \ldots$$

Part of the graph is shown in Fig. 5.20.

5.19   GRAPHS OF OTHER TRIGONOMETRIC FUNCTIONS

To construct the graph of $y = \tan x$, we recall that the period of this function is π.   When $x = 0$, $\tan x = 0$.   As x increases from 0 to $\pi/2$, $\tan x$ is positive and increases without bound as x nears $\pi/2$.   As x increases from $\pi/2$ to π, $\tan x$ is negative and increases to 0.

If we construct the graph for $0 \leq x < \pi/2$ and for $\pi/2 < x \leq \pi$, we can obtain additional portions of the graph by duplicating the part we have constructed, as in Fig. 5.21.

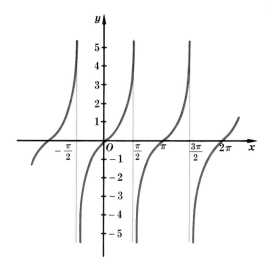

FIGURE 5.21 Graph of $y = \tan x$,
$-\pi < x < 2\pi$

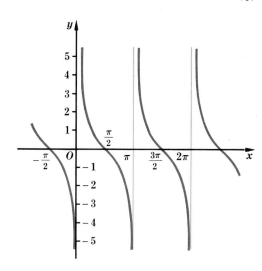

FIGURE 5.22 Graph of $y = \cot x$,
$-\pi/2 < x < 5\pi/2$

To draw the graph of $y = \cot x$, recall that $\cot x = 1/\tan x$. Hence, when $\tan x = 0$, $\cot x$ is undefined and when $\tan x$ is undefined, $\cot x = 0$. The graph of $y = \cot x$ crosses the x axis at $\pm(\pi/2)$, $\pm(3\pi/2)$, ... A portion of the graph is shown in Fig. 5.22.

EXERCISE SET 5.12

1. Sketch the graph of $y = \sec x$ from $x = -(\pi/2)$ to $x = 2\pi$. HINT: The period of sec x is 2π. Since $\sec x = 1/\cos x$, we know that $\sec x$ is undefined when $x = \pi/2$ and when $x = 3\pi/2$.
2. Sketch the graph of $y = \csc x$.

SKETCH the graph of the function on the interval from $x = -(\pi/2)$ to $x = 2\pi$:

3. $y = \sin(-x)$ 4. $y = \cos(-x)$ 5. $y = \tan(-x)$
6. $y = 3 \sin x$ 7. $y = \frac{1}{2} \cos x$ 8. $y = \frac{1}{3} \tan x$
9. $y = \sin 4x$ 10. $y = \cos 3x$ 11. $y = \tan 2x$
12. $y = 3 \sin(\frac{1}{2}x)$ 13. $y = 2 \cos 4x$ 14. $y = 3 \tan 3x$
15. $y = \cos[x - (\pi/2)]$ 16. $y = \tan(x - \pi)$

17. Use the graphs of the trigonometric functions to estimate the following numbers:

(a) $\sin 2$ (b) $\sin \frac{1}{2}$ (c) $\sin(-1.5)$
(d) $\tan(-1)$ (e) $\cos 4$ (f) $\sin 5$

18. Which of the following statements is (are) true?

(a) If $0 < x_1 < x_2 < \pi$, then $\cos x_1 < \cos x_2$.
(b) If $-\pi/2 < x_1 < x_2 < \pi/2$, then $\sin x_1 < \sin x_2$.
(c) If $0 < x_1 < x_2 < \pi/2$, then $\tan x_1 < \tan x_2$.

6

FACTORING POLYNOMIALS. OPERATIONS WITH FRACTIONS

In this chapter we present a short review of the fundamental methods for factoring certain polynomials and for operating with fractions. To factor a polynomial, we express the polynomial as a product of other polynomials, called factors of the given polynomial.

6.1 COMMON FACTORS

By the distributive law, we know that

$$ab + ac + ad + \cdots + am = a(b + c + d + \cdots + m)$$

This law can be applied to any multinomial if each of its terms has a factor common to all the terms.

EXAMPLE 1. Factor $5x^3 - 15x^2y + 20xy^2$.

Solution. Each term of the expression has $5x$ as a factor. Thus,

$$5x^3 - 15x^2y + 20xy^2 = 5x(x^2) - 5x(3xy) + 5x(4y^2) = 5x(x^2 - 3xy + 4y^2)$$

by the distributive law.

EXAMPLE 2. Factor $a(3x - 2y) + b(3x - 2y) - 2c(3x - 2y)$.

Solution. The common factor is $(3x - 2y)$. Hence, by the distributive law,

$$a(3x - 2y) + b(3x - 2y) - 2c(3x - 2y) = (3x - 2y)(a + b - 2c)$$

Grouping to factor

We often find it convenient to group the terms of an expression in order to find its factors. For example,

$$ad + bc + ac + bd = (ac + ad) + (bc + bd) \qquad \text{by commutative and associative laws}$$

$$= a(c + d) + b(c + d) \qquad \text{why?}$$

$$= (c + d)(a + b) \qquad \text{why?}$$

EXAMPLE 3. Factor $ax + 9by - 3ay - 3bx$.

Solution. $\quad ax + 9by - 3ay - 3bx = ax - 3ay - 3bx + 9by \qquad$ why?

$$= a(x - 3y) - 3b(x - 3y) \qquad \text{why?}$$

$$= (x - 3y)(a - 3b) \qquad \text{why?}$$

Check of factoring

We check a factoring problem by multiplying the factors we have found. If their product is the given expression, we are certain we have found at least one valid set of factors.

EXERCISE SET 6.1

FACTOR, if possible:

1. $5x^2 - 10x$
2. $18a^2 - 54a^2b$
3. $54x^2y^2 - 9xy$
4. $49xy - 98y^2$
5. $3a(2x - y) + 4(2x - y)$
6. $5a(x - 2y) - 3b(x - 2y)$
7. $2x(2a - b) + 3y(2a - b)$
8. $5x(x - y) - 2y(x - y)$
9. $a^2(x + y)^3 - 3b^2(x + y)^3$
10. $2x^2(x - 2) + 3x(x - 2) + x - 2$
11. $xy(x^2 + y^2) + yz(x^2 + y^2) + zw(x^2 + y^2)$
12. $x^2(2x - y) + x(2x - y) + x - y$
13. $m^2(x + 2y - 1) + 2m(x + 2y - 1) + 3(x + 2y - 1)$
14. $3(\tan \phi)^2 + 4 \tan \phi$
15. $2(\sin \beta)^3 - 3(\sin \beta)^2 + \sin \beta$
16. $3x^2(\ln x - 1) + 2x(\ln x - 1) + \ln x - 1$
17. $4xe^x + e^x$
18. $5a\sqrt{x} - 7\sqrt{x}$
19. $2xy\sqrt{3} + 2y\sqrt{3} + \sqrt{3}$
20. $4a(x - \sqrt{y}) + 2b(x - \sqrt{y})$
21. $7x(2x - \sqrt{7}) - 3y(2x - \sqrt{7})$
22. $4a^2(x - \sqrt{5}) + 3a(x - \sqrt{5}) + x - \sqrt{5}$

GROUP the terms and factor:

23. $x^2 + xy + px + py$
24. $rs + vs + rt + vt$
25. $8x^2 + 4xy - 6x - 3y$
26. $6x^3 - 2 - 4x^2 + 3x$
27. $9 + 12x^2 - 8x^3 - 6x$
28. $x^2 + 2xy + y^2 + 4x + 4y + 4$
29. $8xy + 12ay + 10bx + 15ab$
30. $6 - 10a + 27a^2 - 45a^3$
31. $m^4 + 6m^3 - 7m - 42$
32. $20ab - 28ad - 5bc + 7cd$
33. $ax - ay + az - bx + by - bz$
34. $3am - 6an + 4bm - 8bn + cm - 2cn$
35. $ax + ay - az - bx - by + bz + cx + cy - cz$

6.2   FACTORING POLYNOMIALS

We recall that an expression of the form

$$ax^n + bx^{n-1} + cx^{n-2} + \cdots + lx + m$$

Polynomial in x

where $a \neq 0$, n is a nonnegative integer, and $b, c, \ldots, m$ are any constants is a *polynomial in x*. If the coefficients $a, b, c, \ldots, m$ are rational numbers, the polynomial is called a *rational polynomial*.

A polynomial in x and y is an expression whose terms are of the form

$$ax^m y^n$$

where $a \neq 0$ and m, n are nonnegative integers, and where one of the factors x^m or y^n may not appear. For example, $3x^4 - 2y^2 + 4y^3$ and $x - 7 + 4xy$ are polynomials in x and y.

The process of factoring in elementary algebra is usually restricted to factoring polynomials having rational coefficients into a product of polynomials that also have rational coefficients. For example, we can verify by multiplication that

$$x^2 - 16 = (x - 4)(x + 4)$$

We can also verify that

$$x - 16 = (\sqrt{x} - 4)(\sqrt{x} + 4)$$

but in this case, $(\sqrt{x} - 4)$ and $(\sqrt{x} + 4)$ are not polynomials in x.

It is possible to express $x^2 - 2$ in the form

$$x^2 - 2 = (x - \sqrt{2})(x + \sqrt{2})$$

Here we see that the polynomials $(x - \sqrt{2})$ and $(x + \sqrt{2})$ do not have rational coefficients. Hence, we say that $x^2 - 2$ is not factorable in the system of rational numbers.

Prime polynomials

If a rational polynomial has no other rational polynomial factor than itself, or a constant, it is said to be *prime* or *irreducible* over the field of rational numbers. If it has other factors, it is called *composite*. A composite polynomial is *completely factored* when it is expressed as a product of all its prime factors.

6.3   SPECIAL TYPES OF FACTORING

TYPE 1. PERFECT-SQUARE TRINOMIALS. Since $(a \pm b)^2 = a^2 \pm 2ab + b^2$, we call the expression $a^2 \pm 2ab + b^2$ a *perfect-square trinomial* and use the identity as a guide for factoring such expressions.

EXAMPLE 1. $25a^2 + 40ab + 16b^2 = (5a + 4b)^2$

EXAMPLE 2.

$$x^2 + 2xy + y^2 + x + y = (x + y)^2 + (x + y)$$
$$= (x + y)(x + y + 1)$$

TYPE 2. THE DIFFERENCE OF TWO SQUARES. Recall that

$$(a + b)(a - b) = a^2 - b^2$$

Thus, the difference of the squares of any two numbers taken in a given order is the sum of the two numbers multiplied by their difference where the difference is taken in the given order.

EXAMPLE 3.

$$25x^2 - 64y^2 = (5x)^2 - (8y)^2$$
$$= (5x + 8y)(5x - 8y)$$

EXAMPLE 4.

$$(x + y)^2 - (z - w)^2 = [(x + y) + (z - w)][(x + y) - (z - w)]$$
$$= (x + y + z - w)(x + y - z + w)$$

TYPE 3. THE GENERAL TRINOMIAL. It is not possible to factor every trinomial. However, when a trinomial has rational polynomial factors, the factors may be found by trial and error. Since

$$(ax + b)(cx + d) = acx^2 + (ad + bc)x + bd$$

(a general trinomial), we search for binomial factors such that the following conditions are satisfied:

1. The first term of each is a factor of the first term of the trinomial.

2. The second term of each is a factor of the last term of the trinomial.

3. The product of the "outer" terms plus the product of the "inner" terms is the middle term of the trinomial.

Thus, to factor $3x^2 + 7x - 6$, we may try $(3x + 6)(x - 1)$, $(3x + 1) \times (x - 6)$, and $(3x - 6)(x + 1)$. None of these sets of factors will yield $3x^2 + 7x - 6$. However, when we try $(3x - 2)(x + 3)$, we find that the product is the given trinomial. Thus, $3x^2 + 7x - 6 = (x + 3)(3x - 2)$.

TYPE 4. THE SUM OR DIFFERENCE OF TWO CUBES. Actual multiplication provides the identities

$$a^3 + b^3 = (a + b)(a^2 - ab + b^2)$$
$$a^3 - b^3 = (a - b)(a^2 + ab + b^2)$$

which enable us to factor the sum or the difference of two cubes.

EXAMPLE 5.

$$27a^3 + 8b^3 = (3a)^3 + (2b)^3$$
$$= (3a + 2b)[(3a)^2 - (3a)(2b) + (2b)^2]$$
$$= (3a + 2b)(9a^2 - 6ab + 4b^2)$$

EXAMPLE 6. $125x^6 - 64y^3 = (5x^2)^3 - (4y)^3$
$$= (5x^2 - 4y)(25x^4 + 20x^2y + 16y^2)$$

Certain polynomials that are not factorable by any of the preceding methods may sometimes be factored as in the following example.

EXAMPLE 7.   Factor $9x^4 - 37x^2 + 4$.

Solution.   We note that if the middle term were $-12x^2$, the polynomial would be a perfect-square trinomial.   Hence, we add $25x^2$ to the middle term and then subtract $25x^2$ from the resulting expression.   This changes the form of the given polynomial but does not change its value.

$$
\begin{aligned}
9x^4 - 37x^2 + 4 &= 9x^4 - 37x^2 + 25x^2 + 4 - 25x^2 \\
&= (9x^4 - 12x^2 + 4) - 25x^2 \\
&= (3x^2 - 2)^2 - (5x)^2 \\
&= (3x^2 - 2 + 5x)(3x^2 - 2 - 5x) \\
&= (x + 2)(3x - 1)(3x + 1)(x - 2)
\end{aligned}
$$

It is interesting to note that this polynomial can be factored just as well by adding and subtracting $49x^2$.   We leave it to the student to show that the same set of binomial factors is obtained.

EXERCISE SET 6.2

FACTOR, if possible:

1. $81x^2 + 144x + 64$
2. $49x^6 - 168x^3y^4 + 144y^8$
3. $9x^2 - 6x(y + z) + (y + z)^2$
4. $-25x^2 - 60xy - 36y^2$
5. $25(a - b)^2 + 40(a - b)c + 16c^2$
6. $(a + b)^2 + 4(a + b)(a - b) + 4(a - b)^2$
7. $49a^2 - 36b^2$
8. $169x^4 - 81y^8$
9. $25 - x^6$
10. $a^2b^4 - 16c^6$

11. $\dfrac{x^2}{100} - \dfrac{1}{49}$
12. $\dfrac{1}{4}a^4 - \dfrac{1}{16}b^2$

13. $(a + b)^2 - 16$
14. $(x - y)^2 - 25$
15. $x^2 - (2y - 1)^2$
16. $a^2 - (b + 2)^2$
17. $(3x - 2y)^2 - 4$
18. $(2x + 3y)^2 - 16z^2$
19. $x^2 - 2xy + y^2 - 16$
20. $x^2 + 2xy + y^2 - 25$
21. $x^2 - y^2 - 2yz - z^2$
22. $6np + 16m^2 - 9p^2 - n^2$
23. $16a^2 - 8ab + b^2 - c^2 - 10cd - 25d^2$
24. $28xy - 36z^2 + 49y^2 + 60z - 25 + 4x^2$
25. $x^2 - 10x - 75$
26. $105 + 8a^3 - a^6$
27. $y^4 - 21y^2 + 104$
28. $68x^2y^4 + 36xy^2 + 1$
29. $77 - 4x - x^2$
30. $84 + 5n - n^2$
31. $a^2 - 6ab - 91b^2$
32. $70x^2 + 17x + 1$
33. $1 + 5ab - 14a^2b^2$
34. $3x^2 - 11x - 34$
35. $x^2 - (2m + 3n)x + 6mn$
36. $x^2 - (a - b)x - ab$

37. $4x^2 + 28x + 45$

38. $6x^2 + x - 2$

39. $25x^2 - 25ax + 6a^2$

40. $12x^2 + 11x + 2$

41. $36x^2 + 12x - 35$

42. $6 - x - 15x^2$

43. $7(x - y)^2 - 30(x - y) + 8$

44. $12(x + y)^2 + 17(x + y) - 7$

45. $x^6 - 27y^9z^3$

46. $x^6 + y^6$

47. $8a^3 - b^9$

48. $a^6 + 64$

49. $x^3y^3 + 125z^3$

50. $216x^3y^6 - 343z^9$

51. $x^3 - (x + y)^3$

52. $(x + y)^3 + (x - y)^3$

53. $(2x + y)^3 - (x + 2y)^3$

54. $(x - y)^3 - (x + y)^3$

FACTOR, if possible, by adding and subtracting the same expression:

55. $a^4 + 5a^2 + 9$

56. $x^4 - 21x^2y^2 + 36y^4$

57. $4x^4 - 33x^2 + 4$

58. $25a^4 + 6a^2b^2 + b^4$

59. $16x^4 - 81x^2 + 16$

60. $9x^4 + 6x^2y^2 + 49y^4$

FACTOR each of the following, if possible, in two different ways:

61. $16 - 17x^2 + x^4$

62. $64x^4 - 148x^2 + 9$

63. $16a^4 - 104a^2b^2 + 25b^4$

64. $36x^4 - 97x^2y^2 + 36y^4$

FACTOR the following trigonometric expressions, using the identities as indicated:

65. $3 \sin^2 \theta - 5 \sin \theta + \cos^2 \theta - 4$, using the identity $\sin^2 \theta + \cos^2 \theta = 1$

66. $6 \sec^2 \beta - 2 \tan \beta - 26$, using $1 + \tan^2 \beta = \sec^2 \beta$

67. $\cos \alpha - 15 \sin^2 \alpha + 13$

68. $6 \csc^2 t - 17 \cot t + 6$

6.4 SOLUTION OF EQUATIONS BY FACTORING

The solution set of some equations can be found by the use of factoring, as illustrated in the following examples. We recall that the solution set of an equation in x is the set of all numbers in the domain of x that make the equation a true statement.

EXAMPLE 1. Solve the equation $2x^2 = 12 - 5x$. This means we are to find $\{x \mid 2x^2 = 12 - 5x\}$.

Solution. Let us first write the equation in the equivalent form:

$$2x^2 + 5x - 12 = 0$$

Then

$$(x + 4)(2x - 3) = 0$$

We now see that if $x = -4$ or if $x = \frac{3}{2}$, then one of these factors is zero and therefore their product is zero. For these two values of x, the equation $(x + 4)(2x - 3) = 0$ is satisfied. On the other hand, if $x \neq -4$ and $x \neq \frac{3}{2}$ then neither factor is zero and therefore the product is not zero. Thus, the only two values of x that satisfy the factored form of the equation are $x = -4$

and $x = \frac{3}{2}$. Therefore, the solution set of the equation $(x + 4)(2x - 3) = 0$ is the set

$$\left\{-4, \; \frac{3}{2}\right\}$$

We strongly suspect that this set is the solution set of the original equation. It is possible, however, that we made an error somewhere along the line. Consequently, we must check each number in the set to see if it satisfies the given equation.

Check.   If $x = -4$,

Left Member	Right Member
$2(-4)^2$	$12 - 5(-4) = 12 + 20$
$= 32$	$= 32$

We leave it to the student to verify that $\frac{3}{2}$ is also a solution of the given equation.

EXAMPLE 2.   Solve $y^2 - 2y + 1 = 0$.

Solution.   Since $(y - 1)(y - 1)$

$$y^2 - 2y + 1 = 0 \qquad\qquad\qquad \text{given}$$
$$(y - 1)(y - 1) = 0 \qquad\qquad\qquad \text{factoring}$$

Then, $y - 1 = 0 \qquad y - 1 = 0 \qquad$ setting each factor equal to 0
Hence, $y = 1 \qquad\qquad y = 1$

The solution set is

$$\{1,1\}$$

We leave the check to the student.

Roots of an
equation   Note in these examples that there are as many roots (solutions) of the equation as there are factors of the left member when the equation is written so the right member is zero.   Also note that in Example 2, the two roots are equal.   We call $y = 1$ a *double root* of the given equation.   The solution set $\{1,1\}$ is usually written $\{1\}$.

EXAMPLE 3.   Find θ if $0 \le \theta \le \pi/2$ and $2 \sin^2 \theta - 3 \sin \theta + 1 = 0$.

Solution.   Since

$$2 \sin^2 \theta - 3 \sin \theta + 1 = 0 \qquad\qquad \text{given}$$
$$(2 \sin \theta - 1)(\sin \theta - 1) = 0 \qquad\qquad \text{factoring}$$

Hence, $\sin \theta = \dfrac{1}{2}$ $\sin \theta = 1$ why?

and $\theta = \dfrac{\pi}{6}$ $\theta = \dfrac{\pi}{2}$

Hence $\{\pi/6, \pi/2\}$ is the solution set. The check is left to the student.

EXERCISE SET 6.3

FIND the solution set of each of the following equations by use of factoring if possible. Check each solution.

1. $4x^2 - 20x = 0$

2. $3x^3 - 108x = 0$

3. $(2x - 4)(x^2 - 25) = 0$

4. $(7x + 3)(x^2 - 9) = 0$

5. $x^2 + 23x + 102 = 0$

6. $y^2 + 4y - 96 = 0$

7. $y^2 - 17y - 110 = 0$

8. $(x - 1)(x^2 + 22x + 121) = 0$

9. $y^4 - 18y^3 + 32y^2 = 0$

10. $6y^2 + 7y + 2 = 0$

11. $-7x + 10x^2 = 12$

12. $x + 2 = 15x^2$

13. Find $\sin \phi$ if $20 \sin^2 \phi - 3 \sin \phi - 2 = 0$.

14. Find $\cos \phi$ if $16 \cos^2 \phi + 2 \cos \phi - 3 = 0$.

15. Solve for $\tan t \mid 2 \tan^2 t - 15 \tan t + 7 = 0$.

In certain types of equations we can use factoring to find some (but not all) of the solutions. For example, we may factor the left side of

$$x^3 - 8 = 0$$

so that $(x - 2)(x^2 + 2x + 4) = 0$

Then $x = 2$ is a solution. The two remaining solutions are not real numbers and cannot be determined at this point.

FIND as many solutions as possible, using factoring:

16. $x^3 - 1 = 0$

17. $x^3 + 8 = 0$

18. $8y^3 - 64 = 0$

19. $27y^3 - 8 = 0$

SOLVE the following by the method of factoring:

20. Find two consecutive positive integers whose product is 272. HINT: Let $x =$ smaller integer, then $x + 1 =$ larger.

21. Find two consecutive positive integers whose product is 552.

22. Find two consecutive positive even integers whose product is 224.

23. Find two consecutive positive odd integers whose product is 195.

24. One leg of a right triangle is 12 in. long. The length of the hypotenuse is 3 in. more than twice the length of the other leg. Find the length of the hypotenuse.

25. One leg of a right triangle is 24 ft, and the length of the hypotenuse is 3 ft less than 4 times the length of the other leg. Find the hypotenuse.

26. The sum of the digits of a certain two-digit number is 13. The number itself is 25 more than the product of its digits. Find the number.

27. The fourth power of a real rational number is 96 more than 10 times the square of the number. Find the number.

28. Find the values of t such that $0 \leq t < 2\pi$, $t \in R$, and such that $2 \sin^2 t - 3 \sin t + 1 = 0$.

29. Find all real values of t such that $\sin^2 t + \sin t = 0$, $0 \leq t < 2\pi$.

30. If $\cos \phi - \cos^2 \phi = 0, 0 \leq \phi < 2\pi$, find ϕ.

31. If $\tan^2 x - 1 = 0, 0 \leq x < 2\pi$, find x.

32. Find ϕ if $\sec^2 \phi - 3 \sec \phi + 2 = 0, 0 \leq \phi < 2\pi$.

33. Find the smallest nonnegative value of t which satisfies $2 \sin^2 t + 9 \sin t - 5 = 0$.

34. Find, correct to two decimal places, the two real values of t, $0 \leq t < \pi$, which satisfy $3 \sin^2 t + 2 \sin t - 1 = 0$.

35. Find correct to two decimal places the values of x, $0 < x < 2\pi$, which satisfy $5 \cos^2 x + 4 \cos x - 1 = 0$.

6.5 SIMPLIFYING FRACTIONS

In Chap. 2, we proved that $ac/bc = a/b$, $b \neq 0$, $c \neq 0$. We use this theorem to simplify rational numbers. For example,

$$\frac{6}{8} = \frac{3 \cdot 2}{4 \cdot 2} = \frac{3}{4}$$

We may also simplify quotients of polynomials. To do this, we factor the numerator and denominator completely into products of irreducible polynomials and then divide the numerator and denominator by the product of all the factors common to both. The values of the variable must be chosen so that the denominator of a fraction is not zero.

EXAMPLE 1. Simplify the fraction

$$\frac{8x^2 - 2x - 15}{12x^2 + 23x + 10}$$

Solution.
$$\frac{8x^2 - 2x - 15}{12x^2 + 23x + 10} = \frac{(2x - 3)(4x + 5)}{(3x + 2)(4x + 5)}$$
$$= \frac{2x - 3}{3x + 2}$$

EXAMPLE 2.
$$\frac{2x^3 + 2x^2 - 3x - 3}{2x^3 - 2x^2 - 3x + 3} = \frac{2x^2(x + 1) - 3(x + 1)}{2x^2(x - 1) - 3(x - 1)}$$
$$= \frac{(x + 1)(2x^2 - 3)}{(x - 1)(2x^2 - 3)}$$
$$= \frac{x + 1}{x - 1}$$

By the distributive and commutative laws, $-(y - x) = (-1)(y - x) = -y + x = x - y$. For example, $4 - x = (-1)(x - 4)$. As a further example, $2y - 3x = (-1)(3x - 2y)$. We often use this idea in simplifying fractions as illustrated in the following example.

EXAMPLE 3. Simplify $\dfrac{2(x - 4)}{4 - x}$.

Solution. The denominator $4 - x$ can be expressed as $(-1)(x - 4)$. Hence,

$$\frac{2(x - 4)}{4 - x} = \frac{2(x - 4)}{(-1)(x - 4)} = \frac{2}{-1} = -2$$

EXERCISE SET 6.4

SIMPLIFY each fraction:

1. $\dfrac{2x^2 - 5x + 3}{2x^2 - x - 3}$

2. $\dfrac{7xy + 7y}{14xy^2 - 14y^2}$

3. $\dfrac{x^3 - 3x^2 + x - 3}{9 - x^2}$

4. $\dfrac{x - y - x^2 + 2xy - y^2}{y^2 - xy}$

5. $\dfrac{12y^2 + y - 1}{6y^2 - y - 1}$

6. $\dfrac{2x + 5}{8x^3 + 125}$

7. $\dfrac{4y^2 + 4y - 9x^2 + 1}{2y - 3x + 1}$

8. $\dfrac{8x^3 - 27}{4x^2 - 14x + 12}$

9. $\dfrac{16a^2 + 64ab + 64b^2 - c^2}{4a + 8b - c}$

10. $\dfrac{x^2 - 9y^2 - z^2 + 6yz}{x^2 - 9y^2 + z^2 - 2xz}$

11. $\dfrac{8x^3 - 125}{2x^3 + x^2 - 15x}$

12. $\dfrac{21 - x - 10x^2}{15xy - 20x - 21y + 28}$

13. $\dfrac{4x^2 + 4x - 24}{2x^3 + 6x^2 - 8x - 24}$

14. $\dfrac{6a - 11a^2 - 10a^3}{12a^2 - 38a^3 + 20a^4}$

15. $\dfrac{(x^2 - 49)(x^2 - 16x + 63)}{(x^2 - 14x + 49)(x^2 - 2x - 63)}$

16. $\dfrac{a^6 + 28a^3b^3 + 27b^6}{a^4 + 9a^2b^2 + 81b^4}$

17. $\dfrac{x^3 + 64}{(x^2 - 4x + 16)(x^2 + 4)}$

18. $\dfrac{x^4 + 2x^2y^2 + 9y^4}{x^2 + 2xy + 3y^2}$

19. $\dfrac{x^4 + 2x^2y^2 + 9y^4}{x^2 - 2xy + 3y^2}$

20. $\dfrac{4x^3 - 12x^2 - x + 3}{1 - 4x^2}$

6.6 ADDITION OF FRACTIONS

By Theorem (2.25),

$$\frac{a}{p} + \frac{b}{p} + \frac{c}{p} + \cdots + \frac{m}{p} = \frac{a + b + c + \cdots + m}{p}$$

provided $p \neq 0$. We use this theorem to add fractions having the same denominator.

EXAMPLE 1. $\dfrac{4x}{3y + 2} + \dfrac{2}{3y + 2} = \dfrac{4x + 2}{3y + 2}$

EXAMPLE 2. $\dfrac{3x}{x^2 + 1} - \dfrac{2y}{x^2 + 1} + \dfrac{3}{x^2 + 1} = \dfrac{3x - 2y + 3}{x^2 + 1}$

To add fractions having different denominators, we express the given fractions as equivalent fractions having a common denominator and then proceed as in Examples 1 and 2. It is desirable, but not essential, to use the *least common denominator* (abbreviated LCD) of all the fractions involved. The LCD is obtained as follows:

> Factor each denominator completely.
>
> Take the product of all the prime factors which occur in the given denominators each to the highest power in which it appears in any one of the denominators.

When a common denominator has been found (preferably the LCD) for the fractions to be added, we express each fraction as an equivalent fraction (having this common denominator). This is often done by inspection. We observe which factors are in the LCD but are not in the denominator of the given fraction. We then multiply the numerator and denominator by these factors.

EXAMPLE 3.   Express as a single fraction:

$$\frac{1}{4x^2 - 12x + 9} + \frac{2}{2x + 3} - \frac{1 + 6x}{4x^2 - 9}$$

Solution.   If the denominators are completely factored, the given expression may be written as follows:

$$\frac{1}{(2x - 3)^2} + \frac{2}{2x + 3} - \frac{1 + 6x}{(2x - 3)(2x + 3)}$$

Since the LCD is a product of all the prime factors of the denominators each to the highest power in which it occurs in any denominator, we see that the LCD is

$$(2x - 3)^2(2x + 3)$$

Expressing each fraction as an equivalent fraction whose denominator is the LCD, we have

$$\frac{1(2x + 3)}{(2x - 3)^2(2x + 3)} + \frac{2(2x - 3)^2}{(2x + 3)(2x - 3)^2} - \frac{(1 + 6x)(2x - 3)}{(2x - 3)(2x + 3)(2x - 3)}$$

$$= \frac{(2x + 3) + (8x^2 - 24x + 18) - (12x^2 - 16x - 3)}{(2x - 3)^2(2x + 3)}$$

$$= \frac{-4x^2 - 6x + 24}{(2x - 3)^2(2x + 3)}$$

We consider any algebraic expression which is not already a fraction as a fraction with denominator 1.   For example,

$$3x^2 + 4 - \frac{x^3}{2x - 1}$$

is taken to be $$\frac{3x^2 + 4}{1} - \frac{x^3}{2x - 1}$$

EXERCISE SET 6.5

PERFORM the indicated operations and simplify the result:

1. $\dfrac{x + 2y}{x^2 - 9y^2} + \dfrac{4x - y}{(x + 3y)(x - 3y)} - \dfrac{2x + y}{x^2 - 9y^2}$

2. $\dfrac{x - 2}{x^2 - 16x + 48} - \dfrac{x - 9}{(x - 4)(x - 12)} + \dfrac{2x + 12}{x^2 - 16x + 48}$

3. $\dfrac{1}{4x - 6} - \dfrac{2}{6 - 4x}$

4. $\dfrac{2}{x^2 - y^2} + \dfrac{x}{y^2 - x^2} + \dfrac{y}{x^2 - y^2}$

5. $\dfrac{1}{1 - 2x} + \dfrac{2}{2x - 1} - \dfrac{x}{2x - 1}$

6. $\dfrac{7x^2}{6 + x - x^2} - \dfrac{5}{(x - 3)(2 + x)}$

7. $\dfrac{4x^2}{x^4 - y^4} - \dfrac{1}{x^2 + y^2} - \dfrac{2}{x^2 - y^2}$

8. $\dfrac{x + 4}{x^2 + 4x} - \dfrac{x + 3}{x^2 + 7x + 12}$

9. $\dfrac{36}{2 + x - 6x^2} - \dfrac{1}{1 + x - 2x^2}$

10. $\dfrac{2}{a^2 - 2a + 1} - \dfrac{4}{a^2 + 2a + 1}$

11. $\dfrac{b}{6 + 5b + b^2} + \dfrac{3}{8 + 6b + b^2} - \dfrac{1}{12 + 7b + b^2}$

12. $\dfrac{2}{x^2 - 5x + 6} - \dfrac{2x}{x^2 - 4} + \dfrac{6}{x^2 - x - 6}$

13. $\dfrac{x + y}{2x - 3y} + \dfrac{x - y}{4(2x + 3y)}$

14. $\dfrac{2 - y}{y^2 - y - 6} + \dfrac{4 - y}{y^2 - 7y + 12}$

15. $\dfrac{x}{x^3 - y^3} - \dfrac{y}{x^2 + xy + y^2}$

16. $2x + \dfrac{2x + 4}{x - 1} + 3$

17. $\dfrac{14}{4 - x^2} - \dfrac{x - 6}{x + 2}$

18. $\dfrac{x + 2}{5(x + 3)} - \dfrac{3x + 9}{9 - x^2}$

19. $3x + 4 - \dfrac{5x + 7}{2x - 3}$

20. $\dfrac{2}{x + 3} - \dfrac{3}{x - 3} - \dfrac{4}{x + 4} + \dfrac{5}{x - 4}$

21. $\dfrac{3}{2n + 1} + \dfrac{3}{1 - 2n} - \dfrac{5n^2}{8n^3 + 1} - \dfrac{5n^2}{1 - 8n^3}$

22. $\dfrac{2a - 1}{a - 2} - \dfrac{2a + 1}{a + 2} + \dfrac{6a - 1}{a(2 - a)} - \dfrac{11}{4 - a^2} - \dfrac{8}{a(a^4 - 16)}$

23. $\dfrac{3}{x + 3} + \dfrac{x}{-3 - x} - \dfrac{2x^2}{x + 3}$

24. $2x - 3 + \dfrac{4x}{x - 1}$

25. $7y - \dfrac{y + 2}{y - 1} + 3$

26. $\dfrac{\sin\theta}{1+\sin\theta} + \dfrac{1}{\cos^2\theta}$ (HINT: Change $\cos^2\theta$ to $1-\sin^2\theta$ and factor.)

27. $\dfrac{1}{\sec\theta - 1} + \dfrac{3}{2\sec^2\theta + \sec\theta - 3}$

6.7   MULTIPLICATION AND DIVISION OF FRACTIONS

We have proved that

$$\frac{a}{b}\cdot\frac{c}{d} = \frac{ac}{bd} \qquad b\neq 0, d\neq 0$$

[Theorem (2.23)].   This implies that the product of two or more fractions is that fraction whose numerator is the product of the numerators of the original fractions and whose denominator is the product of the denominators. It is generally advisable to simplify each fraction as much as possible before multiplying.

EXAMPLE 1.

$$\frac{x^2 - 6x - 16}{x^2 - 2x - 8}\cdot\frac{x^2 - 8x + 15}{x^2 - x - 6} = \frac{(x-8)(x+2)}{(x+2)(x-4)}\cdot\frac{(x-3)(x-5)}{(x+2)(x-3)}$$

$$= \frac{x-8}{x-4}\cdot\frac{x-5}{x+2} = \frac{(x-8)(x-5)}{(x-4)(x+2)}$$

From Theorem (2.29), we have the identity

$$\frac{a}{b}\div\frac{c}{d} = \frac{a}{b}\cdot\frac{d}{c} \qquad b, c, d\neq 0$$

Thus, to divide a/b by c/d we multiply a/b by the reciprocal of c/d.

EXAMPLE 2.

$$\frac{a^2 - 7a + 12}{a^2 + 9a + 14}\div\frac{a^2 + 4a - 21}{a^2 - 6a - 16} = \frac{a^2 - 7a + 12}{a^2 + 9a + 14}\cdot\frac{a^2 - 6a - 16}{a^2 + 4a - 21}$$

$$= \frac{(a-3)(a-4)}{(a+2)(a+7)}\cdot\frac{(a-8)(a+2)}{(a+7)(a-3)}$$

$$= \frac{(a-4)(a-8)}{(a+7)^2}$$

EXERCISE SET 6.6

FIND the products and simplify:

1. $\dfrac{x^2 - 64}{x^2 - 4}\cdot\dfrac{x - 2}{x + 8}$

2. $\dfrac{x - 4}{x^2 - 4}\cdot\dfrac{x + 2}{x^2 - 16}$

3. $\dfrac{x^2 - 3x - 10}{x^2 + 2x - 35} \cdot \dfrac{x^2 + 4x - 21}{x^2 + 9x + 14}$

4. $\dfrac{2y^2 - y - 3}{4y^2 - 9} \cdot \dfrac{3 - 3y}{1 - y^2}$

5. $\dfrac{6x - 3y}{4x + 2y} \cdot \dfrac{4x^2 - y^2}{4x^2 - 4xy + y^2}$

6. $\dfrac{4y^2 - 4y - 3}{3y^2 + 7y - 6} \cdot \dfrac{2y^2 + 9y - 5}{2y^2 + 3y + 1} \cdot \dfrac{3y^2 + y - 2}{4y^2 - 8y + 3}$

7. $\dfrac{x^2y^2 - 3xy}{x^2y + xy^2} \cdot \dfrac{x^2y + xy^2 + x + y}{x^2y^2 - 2xy - 3}$

8. $\left(\dfrac{x^3 - 8y^3}{x^2 - 64y^2}\right)\left(\dfrac{x^2 - 11xy + 24y^2}{x^2 + 2xy + 4y^2}\right)\left(\dfrac{x^3 + 512y^3}{x^2 - xy - 6y^2}\right)$

9. $\left(\dfrac{a^3 + a^2 + a + 1}{2a^2 - 5a + 3}\right)\left(\dfrac{a - 1}{a^2 + 1}\right)\left(\dfrac{2a^2 - 3a}{a^2 - 1}\right)$

10. $\left(\dfrac{27a^3 + 1}{25a^2 - 4}\right)\left(\dfrac{25a^2 - 20a + 4}{15a^2 - a - 2}\right)$

11. $\dfrac{x^2 - y^2 + x - y}{y^2 - 2yz + z^2} \cdot \dfrac{y - z}{x - y}$

12. $\dfrac{a^2 - b^2}{a^2 + b^2} \cdot \dfrac{a^4 - b^4}{a^2 + 2ab + b^2}$

PERFORM the indicated operations and simplify:

13. $\dfrac{3x^2 + 9x}{16y^2} \div \dfrac{x + 3}{4xy}$

14. $\dfrac{b^2 + 4b}{2b^2 + 9b + 4} \div \dfrac{b^2 - 4}{2b^2 + 5b + 2}$

15. $\left(\dfrac{a^2 - 2a}{a^2 + a} \cdot \dfrac{a^2 + 2a}{a^2 - a}\right) \div \dfrac{a^2 - 4}{a^2 - 1}$

16. $\left(\dfrac{2b^2 + 11b + 5}{2b^2 + 3b + 1} \div \dfrac{3b^2 + b - 2}{4b^2 - 8b + 3}\right) \cdot \dfrac{3b^2 + 7b - 6}{4b^2 - 4b - 3}$

17. $\left(\dfrac{x^2y^2}{3x^2y - 3xy^2} \cdot \dfrac{x^2 + xy}{x^4 - y^4}\right) \div \dfrac{y}{x^2 - 2xy + y^2}$

18. $\dfrac{4y^2 - 9}{4y - 4} \div \left(\dfrac{2y^2 - y - 3}{4y - 7} \cdot \dfrac{2y + 3}{4y^2 - 4}\right)$

19. $\dfrac{x^2 - x - 20}{x^2 - 25} \cdot \left(\dfrac{x^2 - x - 2}{x^2 + 2x - 8} \div \dfrac{x + 1}{x^2 + 5x}\right)$

20. $\left(\dfrac{x^2 + x - 12}{4x^2 - 1} \div \dfrac{x^2 - 2x - 3}{6x^2 + x - 2}\right)\left(\dfrac{2x + 1}{3x + 2}\right)$

6.8 FRACTIONAL EQUATIONS

Clearing fractions

An equation in which a variable with a positive-integer exponent occurs in the denominator of a fraction is called a *fractional equation* in that variable. A fractional equation can be transformed into an equation that is not fractional by multiplying each side of the equation by a denominator common to all the fractions involved (preferably the LCD). We call this process "clearing the equation of fractions."

It is important to recall that multiplying each side of an equation in a variable x by a polynomial in x does not in general lead to an equivalent equation. This was pointed out in our discussion of Theorems (3.2) and (3.3). Hence, it is necessary to check each solution of the transformed equation by substituting it into the given equation. This check is a *must*; it is not an option. We shall consider the matter in more detail in the next section.

EXAMPLE 1.   Solve for x and check:

$$\frac{3x + 4}{6x - 5} - \frac{2x + 5}{4x - 1} = 0$$

Solution.   The LCD of the fractions is $(6x - 5)(4x - 1)$. Multiplying each side of the equation by the LCD,

$$\frac{3x + 4}{6x - 5}(6x - 5)(4x - 1) - \frac{2x + 5}{4x - 1}(6x - 5)(4x - 1) = 0$$

Hence, $(3x + 4)(4x - 1) - (2x + 5)(6x - 5) = 0$

Then $12x^2 + 13x - 4 - (12x^2 + 20x - 25) = 0$

and $-7x + 21 = 0$

Therefore $x = 3$

Check.   If $x = 3$,

Left Member	Right Member
$\dfrac{3(3) + 4}{6(3) - 5} - \dfrac{2(3) + 5}{4(3) - 1}$	
$= \dfrac{13}{13} - \dfrac{11}{11}$	
$= 0$	$= 0$

EXAMPLE 2.   Solve and check:

$$\frac{4x + 2}{x^2 - 1} - \frac{1}{x + 1} - \frac{1}{x - 1} = 0$$

Solution.   The left member of the equation is undefined if $x = -1$ or $x = 1$. At each step in the solution, we will indicate that these values of x do not satisfy the equation.   The LCD is $(x - 1)(x + 1) = x^2 - 1$. Multiplying by the LCD gives

$$4x + 2 = (x - 1) + (x + 1) \qquad x \neq -1, x \neq 1$$
$$4x + 2 = 2x \qquad\qquad\qquad x \neq -1, x \neq 1$$
$$2x = -2 \qquad\qquad\qquad\quad x \neq -1, x \neq 1$$

Hence, $x = -1$

Check.   Since $x = -1$ does not satisfy the given equation, we conclude that the equation has no solution.

EXERCISE SET 6.7

1. Find $\left\{ x \mid x \in R \text{ and } \dfrac{x - 1}{5x} + \dfrac{4}{x} = \dfrac{3}{x} + 1 \right\}$.

2. Find $\left\{y \mid y \in R \text{ and } \dfrac{3}{y+1} = \dfrac{2}{y+2} + \dfrac{1}{y-2}\right\}$.

FIND the solution set for each of the following equations and check by substituting into the original equation each purported root:

3. $\dfrac{1}{x^2 - 16} = \dfrac{2}{2x^2 - 7x + 3}$

4. $\dfrac{1}{x+1} + \dfrac{1}{2x+2} = \dfrac{1}{x+2}$

5. $\dfrac{-2}{z-3} + \dfrac{7}{z-2} = \dfrac{5}{z-1}$

6. $\dfrac{5}{y} = \dfrac{3}{y-1} + \dfrac{2}{y+1}$

7. $\dfrac{1}{x^2 - 4} - \dfrac{2}{2x^2 + 5x + 2} = 0$

8. $\dfrac{3}{y-2} = \dfrac{7}{y-1} + \dfrac{5}{2-y}$

9. $\dfrac{x-3}{2x} = \dfrac{x-2}{2x+1}$

10. $\dfrac{3}{z+1} + \dfrac{1}{1-z} = \dfrac{2}{z^2 - 1}$

11. $\dfrac{3}{\cos t + 1} + \dfrac{1}{1 - \cos t} = \dfrac{2}{\cos^2 t - 1}$. Find $\cos t$.

12. Find $\tan t$ if $\dfrac{3}{\tan^2 t - \tan t - 6} = \dfrac{\tan t}{\tan t + 2} - 1$.

13. $\dfrac{x+1}{x-1} + \dfrac{x+4}{x-4} = \dfrac{x+2}{x-2} + \dfrac{x+3}{x-3}$

14. $\dfrac{x+2}{x-3} + \dfrac{x-3}{x+4} + \dfrac{x+4}{x+2} = 3$

15. Mark can paint his bedroom walls in 8 hr. Matthew requires 10 hr to do the same job. How many hours are required if both work together? HINT: If $x =$ the number of hours required for both, then $1/x =$ the part of the job both can do in 1 hr.

16. The second digit of a two-digit number exceeds the first by 2. If the number is divided by the sum of its digits, the result is $3\frac{4}{7}$. Find the number.

17. The denominator of a fraction is 7 more than the numerator. If 6 is added to the numerator, the fraction becomes $\frac{18}{19}$. Find the original fraction.

18. How long will it take two machines to do a job together if the first can do it alone in 6 hr and the second can do it alone in 9 hr?

19. The cost of manufacturing a certain transistor radio is $72. What should be the sale price so the manufacturer can give a discount of 10 percent of the sale price and still make a profit of 10 percent of the sale price?

20. An investor has four equal amounts of money invested at 6, 5, 4, and 3 percent respectively. The annual return on the four investments is $144. Find the amount of each investment.

6.9 EXTRANEOUS SOLUTIONS

When solving certain equations in the variable x, it is often convenient to multiply each side of the equation by a polynomial in the variable x. This process, however, may not lead to an equivalent equation (see Sec. 3.2). A simple illustration will show why. Let $f(x) = 0$ be a given equation and let $g(x)$ be a polynomial. If we multiply each side of $f(x) = 0$ by $g(x)$, we obtain the transformed equation

$$f(x) \cdot g(x) = 0$$

This equation is satisfied by every value of x for which $f(x) = 0$. The equation $f(x) \cdot g(x) = 0$ is also satisfied by every value of x for which $g(x) = 0$, provided $f(x)$ is defined. Hence, the solution set of $f(x) \cdot g(x) = 0$ includes all the elements in the solution set of the equation $g(x) = 0$ in addition to all the elements in the solution set of the equation $f(x) = 0$, provided $f(x)$ is defined. Hence, the solution set of $f(x) \cdot g(x) = 0$ may contain elements that are not in the solution set of the given equation $f(x) = 0$. Such solutions of the transformed equation that are not solutions of the given equation are called *extraneous solutions*, or *extraneous roots*.

Extraneous roots

EXAMPLE 1. The equation $x - 4 = 0$ has only one solution, $x = 4$. If, however, we multiply each side by $x + 7$, the transformed equation

$$(x + 7)(x - 4) = 0$$

has two solutions, $x = 4$ and $x = -7$. Hence, $x = -7$ is an extraneous solution.

The fact that extraneous solutions sometimes appear when we solve a fractional equation requires us to check every solution of the transformed equation by substituting it into the original equation.

Another process that may introduce extraneous roots is that of squaring each side of an equation. We will discuss this case when considering the solution of quadratic equations.

If we divide each side of an equation by a polynomial in the variable, the transformed equation may not have as many solutions as the given equation.

EXAMPLE 2. The equation $(x + 3)(x - 8) = 0$ has two solutions, $x = -3$ and $x = 8$. If we divide each side by $x + 3$, the transformed equation $x - 8 = 0$ has only one solution, $x = 8$.

In the preceding example, we "lost" the solution $x = -3$ when we divided by the polynomial $x + 3$. Solutions lost in this manner cannot be recovered from the transformed equation. Therefore, instead of dividing by a polynomial that is a common factor of each term of an equation, we factor out the polynomial as in the following example.

EXAMPLE 3. Solve for x: $(x + 3)(x - 4) = 5(x + 3)$.

Solution. Write the equation as

$$(x + 3)(x - 4) - 5(x + 3) = 0$$

Then $(x + 3)[(x - 4) - 5] = 0$

Hence, $(x + 3)(x - 9) = 0$

and the solution set is $\{-3, 9\}$. The check is left as an exercise.

EXERCISE SET 6.8

FIND the solution set for each of the following equations. For some of them, the solution set will be the empty set.

1. $\dfrac{y-9}{y-2} + \dfrac{y}{y-2} = \dfrac{y-8}{y-6} + \dfrac{y+1}{y-6}$

2. $\dfrac{1}{x^2+x-12} + \dfrac{1}{x^2-x-6} = \dfrac{2}{x^2-2x-3}$

3. $\dfrac{4}{x^2-1} - \dfrac{5}{x+1} = \dfrac{2}{x-1}$

4. $\dfrac{2y}{y-3} - \dfrac{6}{y-3} = 0$

5. $\dfrac{9-5y}{2y-3} + 3 = \dfrac{3y-3}{2y-3}$

6. $\dfrac{3x-2}{5x+2} = \dfrac{3x}{5x+2}$

7. $\dfrac{10}{x^2-25} + \dfrac{4}{x+5} = \dfrac{1}{x-5}$

8. $\dfrac{4x+2}{x^2-2x-3} + \dfrac{3x-1}{3(x+1)} = \dfrac{2x+1}{2x-6}$

9. $\dfrac{y+10}{2y^2-3y-35} - \dfrac{3y}{(y-5)(2y+7)} = 0$

10. $\dfrac{2x}{(2x+3)(x-2)} = \dfrac{2+x}{2x^2-x-6}$

11. Solve for x in terms of y: $8x + 2y - 7 = 3y - 2x + 5$.
12. Solve for x in terms of y: $12x - 2y - 24 = 0$.
13. Solve for y in terms of x: $10 - 6y + x = 0$.
14. Solve for y in terms of x: $3x + 5(y + 2) - 7 = 0$.
15. Find all values of t between 0 and 2π which are solutions of

$$\cos t = \dfrac{1}{2\tan t}$$

6.10 COMPLEX FRACTIONS

If a fraction has other fractions in its numerator or denominator (or both), it is called a *complex fraction*. To simplify a complex fraction, we may perform the indicated operations in the numerator and in the denominator, and then divide the numerator by the denominator.

EXAMPLE 1.

$$\dfrac{\dfrac{x}{4} + \dfrac{2x}{7}}{\dfrac{x^2}{14}} = \dfrac{\dfrac{7x+8x}{28}}{\dfrac{x^2}{14}} = \dfrac{15x}{28} \div \dfrac{x^2}{14}$$

$$= \dfrac{15x}{28} \cdot \dfrac{14}{x^2} = \dfrac{15}{2x}$$

An alternative method, sometimes more convenient, is to multiply the numerator and the denominator of the complex fraction by a common denominator of all the fractions involved in both numerator and denominator of the complex fraction. The LCD is preferred.

EXAMPLE 2.

$$\frac{x}{1 - \dfrac{1 - x}{1 + x}} = \frac{x(1 + x)}{\left(1 - \dfrac{1 - x}{1 + x}\right)(1 + x)} = \frac{x(1 + x)}{1 + x - (1 - x)}$$

$$= \frac{x(1 + x)}{2x} = \frac{1 + x}{2}$$

EXERCISE SET 6.9

SIMPLIFY the complex fractions:

1. $\dfrac{x - \dfrac{1}{x}}{1 + \dfrac{1}{x}}$

2. $\dfrac{\dfrac{x^2 - 25y^2}{y^3 + 1}}{\dfrac{x - 5y}{y^2 - y + 1}}$

3. $\dfrac{1 - \dfrac{a^2}{b^2}}{1 - \dfrac{a}{b}}$

4. $\dfrac{\dfrac{3}{x^2} - \dfrac{4}{x} + 1}{x - \dfrac{9}{x}}$

5. $\dfrac{\dfrac{2}{x - y} - \dfrac{3}{x + y}}{\dfrac{2}{x^2 - y^2}}$

6. $\dfrac{\dfrac{x - 2}{x - 3} - \dfrac{x - 3}{x - 2}}{\dfrac{1}{x - 2} - \dfrac{1}{x - 3}}$

7. $\dfrac{a + \dfrac{b}{a + b}}{a - \dfrac{a}{a + b} + b}$

8. $\dfrac{x + 1}{\dfrac{1}{x} + 1}$

9. $\dfrac{\dfrac{a + b}{b}}{\dfrac{1}{a} + \dfrac{1}{b}} \div \dfrac{a^2 - b^2}{a + b}$

10. $\left(1 - \dfrac{b^2}{a^2}\right) \div \left(1 - \dfrac{2b}{a} + \dfrac{b^2}{a^2}\right)$

11. $\dfrac{\dfrac{3}{x^2 - y^2}}{\dfrac{1}{x + y} - \dfrac{1}{x - y}}$

12. $\dfrac{\dfrac{x - 3}{x - 4} - \dfrac{x - 4}{x - 3}}{\dfrac{1}{x - 3} - \dfrac{1}{x - 4}}$

13. $2 - \dfrac{1}{2 - \dfrac{1}{2 - \frac{1}{2}}}$

14. $\dfrac{x}{1 - \dfrac{1}{1 + \dfrac{1}{x - 1}}}$

6.11 PROBLEMS LEADING TO FRACTIONAL EQUATIONS

Many types of "story" problems require the solution of fractional equations. We recall that the check is an essential part of the solution of such problems. Each purported solution must satisfy the conditions set forth in the problem.

The exercises of this section are not practical problems in the usual sense of the word, because they do not apply to everyday living. However, for many centuries people have been interested in the "puzzle"-type problem, and for that reason a few of the classical ones are included.

No set rules can be given for finding the solution sets of verbal, or "story," problems. You must "live" with the problem. Read it over and over if necessary and take time to organize the data. You may even calculate seemingly unnecessary quantities, tabulate unnecessary data, draw unnecessary figures, and make some false starts. But you may actually save time by doing all these things. Furthermore, you will better understand the problem and will be more confident that you are correct in your reasoning.

EXAMPLE 1. The numerator of a fraction is 4 less than the denominator. If the numerator is decreased by 2 and the denominator is increased by 15, the resulting fraction is ½. Find the original fraction.

Solution. Let $x =$ the denominator of the original fraction, then $x - 4 =$ the numerator. Hence,

$$\frac{x - 4}{x}$$

is the original fraction. If we decrease the numerator by 2 and increase the denominator by 15, then

$$\frac{(x - 4) - 2}{x + 15}$$

is the resulting fraction. Hence,

$$\frac{x - 6}{x + 15} = \frac{1}{2}$$

giving $x = 27$, the denominator of the original fraction, and $x - 4 = 23$, the numerator. Therefore $^{23}\!/_{27}$ is the original fraction.

Check. $\dfrac{23 - 2}{27 + 15} = \dfrac{21}{42} = \dfrac{1}{2}$

EXERCISE SET 6.10

1. Find two consecutive positive integers such that ¼ the smaller is 3 more than ⅕ the larger.
2. The smaller of two numbers is ¾ the larger. The sum of the multiplicative inverses of the two numbers is ⁷⁄₂₄. Find the numbers.
3. The denominator of a certain fraction is 2 more than the numerator. If the denominator is increased by 3 and the numerator increased by 9, the resulting fraction equals the multiplicative inverse of the original fraction. Find the fraction.
4. The denominator of a fraction is 2 more than the numerator. If the numerator is increased by 30 and the denominator is increased by 40, the value of the fraction is unaltered. Find the fraction.
5. The denominator of a fraction exceeds twice the numerator by 4. If the numerator is increased by 14 and the denominator decreased by 9, the value of the resulting fraction is the multiplicative inverse of ¾. Find the original fraction.
6. The second digit of a two-digit number exceeds the first by 5. If the number is increased by 1 and the result divided by 2 more than the sum of the digits, the quotient is 3. Find the number.

7. A motor boat travels at 35 mi/hr in still water. If it takes ⅔ as many hours to go 35 mi upstream as it does to return, find the rate of the current.

8. With a tail wind of 30 mi/hr an aircraft requires 9/10 as much time to travel 570 mi as it requires to travel the same distance against the wind. Find the rate of the aircraft in still air.

9. The time required for a large intake pipe to fill a water tank is ¾ of the time required by a smaller one. Together the two pipes can fill the tank in 36 min. How many minutes are required for the smaller pipe alone to fill the tank?

10. If Mark can do his chores in 45 min and if Mark and Matt together can do them in 30 min, how long will it take Matt working alone to do them?

11. The larger of two intake pipes can fill a vat in 40 min and the smaller can fill it in 50 min. The two pipes together can fill the vat in 36 min when the vat is being emptied by an outlet pipe. Find the time necessary for the outlet pipe to empty the vat if it is only ¾ full when the intakes are cut off.

12. The tens digit of a two-digit number is 3 less than the units digit. If the number is divided by the sum of its digits, the quotient is 4 and the remainder is 6. Find the number.

13. The width of a rectangle is ⅓ of the length. If each dimension is increased by 3 ft, the area is increased by 117 sq ft. Find the dimensions of the rectangle.

14. The perimeter of a rectangle is 268 in. If each dimension is decreased by 12 in., one of the dimensions becomes ⅗ of the other. Find the dimensions of the rectangle.

15. The digits of a three-digit number are three consecutive integers. The middle digit is the greatest and the first digit is the least. If the number is divided by the sum of its digits, the quotient is 22$\frac{9}{7}$. Find the number.

16. The rate of a small motor boat in still water is 11⅔ mi/hr. It takes just as long to travel 23 mi upstream as it does to travel 47 mi downstream. Find the rate of the current.

7

COMPLEX NUMBERS

There are many equations having real coefficients which do not have real numbers in their solution sets. For example, the equation $x^2 + 1 = 0$ has no real solution. This is due to the fact that if $x \in R$, then x^2 is always positive, and consequently $x^2 + 1 \neq 0$. If we are restricted to real numbers, the solution set of this equation is the empty set $\varnothing$. Hence, if we are to have nonempty solution sets for such equations, it will be necessary to extend the concept of number beyond that of real number. In order to make this extension of the real number system, we construct a set of new numbers, called *complex numbers*. We shall construct these complex numbers in such manner that the set of all real numbers is a proper subset of the set of all complex numbers.

7.1 COMPLEX NUMBERS

A complex number is an ordered pair of real numbers (a,b). We denote the set of all such ordered pairs of real numbers by C. Then

$$C = \{(a,b) \mid a, b \in R\}$$

We define equality, addition, and multiplication of these complex numbers as follows:

DEFINITION. For each (a,b), $(c,d) \in C$,

$$(a,b) = (c,d) \quad \text{if and only if } a = c \text{ and } b = d \tag{7.1}$$

$$(a,b) + (c,d) = (a + c, b + d) \tag{7.2}$$

$$(a,b) \cdot (c,d) = (ac - bd, ad + bc) \tag{7.3}$$

EXAMPLE 1. If $(a,b) = (3, -5)$, then

$$a = 3 \quad \text{and} \quad b = -5$$

EXAMPLE 2. Find the sum of $(5,7)$ and $(-2,3)$.

Solution. $(5,7) + (-2,3) = (5 - 2, 7 + 3) = (3,10)$

EXAMPLE 3. Find the product of $(2,5)$ and $(3,6)$.

Solution. $(2,5) \cdot (3,6) = (6 - 30, 12 + 15) = (-24,27)$

By Definitions (7.2) and (7.3),

$$(a,b) + (0,0) = (a + 0, b + 0) = (a,b)$$

and $\quad (a,b) \cdot (1,0) = (a \cdot 1 - b \cdot 0, a \cdot 0 + b \cdot 1) = (a,b)$

Identity elements

Thus, the complex number $(0,0)$ is the additive identity, and the complex number $(1,0)$ is the multiplicative identity for the set C of complex numbers. It is not difficult to show that

$$(0,0) \cdot (a,b) = (a,b) \cdot (0,0) = (0,0)$$

Additive inverse

We define the additive inverse of the complex number (a,b) to be a number (x,y) such that the sum of (a,b) and (x,y) equals the additive identity. Thus,

$$(a,b) + (x,y) = (0,0)$$

Hence, $\qquad (a + x, b + y) = (0,0) \qquad$ by Def. (7.2)

and by Definition (7.1),

$$a + x = 0$$
$$b + y = 0$$

Therefore, $\qquad\qquad x = -a$
$$y = -b$$

and $\qquad\qquad (x,y) = (-a, -b)$

The difference $(a,b) - (c,d)$ is defined as the sum of (a,b) and the additive inverse of (c,d). Hence,

$$(a,b) - (c,d) = (a,b) + (-c, -d)$$
$$= (a - c, b - d) \qquad (7.4)$$

Multiplicative inverse

The multiplicative inverse of the complex number $(a,b) \neq (0,0)$ is defined to be a number (x,y) such that the product of (a,b) and (x,y) equals the multiplicative identity, that is,

$$(a,b) \cdot (x,y) = (1,0)$$

Then $$(ax - by, ay + bx) = (1,0) \qquad \text{by Def. (7.3)}$$

and by Definition (7.1)

$$ax - by = 1$$
$$bx + ay = 0$$

Since $a^2 + b^2 \neq 0$, these equations can be solved to yield

$$x = \frac{a}{a^2 + b^2} \qquad y = -\frac{b}{a^2 + b^2}$$

Hence, the multiplicative inverse of the complex number (a,b) is the complex number

$$\left(\frac{a}{a^2 + b^2}, -\frac{b}{a^2 + b^2} \right)$$

The quotient of $(a,b) \div (c,d)$ is defined to be the product of (a,b) and the multiplicative inverse of (c,d). Thus,

$$(a,b) \div (c,d) = (a,b) \cdot \left(\frac{c}{c^2 + d^2}, -\frac{d}{c^2 + d^2} \right)$$
$$= \left(\frac{ac + bd}{c^2 + d^2}, \frac{bc - ad}{c^2 + d^2} \right) \qquad (7.5)$$

We omit proof that the complex numbers are associative and commutative under both addition and multiplication and that the distributive law holds. However, we assume these properties and make frequent use of them.

7.2 COMPLEX NUMBERS $(k,0)$ AND $(0,1)$

For complex numbers of the form $(k,0)$, we have

$$(a,0) + (b,0) = (a + b,0)$$
$$(a,0) \cdot (b,0) = (ab,0)$$
$$(a,0) - (b,0) = (a - b,0)$$

and $$\frac{(a,0)}{(b,0)} = \left(\frac{a}{b},0 \right)$$

Thus, the sum, product, difference, or quotient of two complex numbers of the form $(k,0)$ is a complex number that has the same form as $(k,0)$. Furthermore, the first components behave exactly like the real number k, and the second component is always zero.

The conclusions which follow can best be explained by introducing a very important concept of mathematics called an isomorphism between mathematical systems.[1] However, we will not consider isomorphisms here. Instead, we simply make the following statements which our intuition makes reasonable.

Complex number
(k, 0) and real
number k

No contradictions will result if we identify the complex number $(k,0)$ with the real number k. Accordingly, although there is a logical distinction, we shall write

$$(k,0) = k$$

(7.6)

R a subfield of C

By virtue of this identification, we say that the real number field R is a subfield of the complex field (and the complex field is an extension of the real field). When we consider the complex plane in a later section, we shall see that both the complex number $(k,0)$ and the real number k can be made to correspond to the same point in a rectangular coordinate system.

As a result of Eq. (7.6),

$$k \cdot (a,b) = (k,0) \cdot (a,b) = (ka,kb) \qquad (7.7)$$

We define $-(a,b) = -1 \cdot (a,b) = (-a,-b)$.

Imaginary unit

The complex number $(0,1)$ is quite important. It is called the *imaginary unit* and is denoted by i. Thus,

$$i = (0,1) \qquad (7.8)$$

We have, from Definition (7.3),

$$(0,1) \cdot (0,1) = (0 \cdot 0 - 1 \cdot 1, 0 \cdot 1 + 1 \cdot 0)$$

Hence
$$i \cdot i = (-1,0)$$

and by Eq. (7.6)

$$i^2 = -1 \qquad (7.9)$$

Since $i^2 = -1$, we say that $i = \sqrt{-1}$ and then define the *square root* of a negative number as follows:

$$\sqrt{-a^2} = \sqrt{a^2(-1)} = \sqrt{a^2} \cdot \sqrt{-1} = |a|i \qquad (7.10)$$

EXAMPLE 1. $\sqrt{-5^2} = \sqrt{5^2} \cdot \sqrt{-1} = 5i$

$$\sqrt{-(-3)^2} = \sqrt{(-3)^2} \cdot \sqrt{-1} = |-3|i = 3i$$

and $\sqrt{-3} = \sqrt{3} \cdot i = i\sqrt{3}$

---

[1] See, for example, Carl B. Allendoerfer and Cletus O. Oakley, "Principles of Mathematics," pp. 61–64, McGraw-Hill Book Company, New York, 1963.

In writing such imaginary numbers as $\sqrt{5}i$, it is very easy to make the mistake of writing $\sqrt{5i}$. Consequently, it is safer to write $i\sqrt{5}$. It is for this same reason that we have been writing such numbers as $\sqrt{23}$ as $3\sqrt{2}$.

EXAMPLE 2.
$$\sqrt{-3}\cdot\sqrt{-12} = i\sqrt{3}(2i\sqrt{3})$$
$$= 2(i^2)\sqrt{3}\sqrt{3} = 6i^2 = -6$$

It is important to note here that $\sqrt{a}\sqrt{b} \neq \sqrt{ab}$ when a and b are negative.

7.3 RECTANGULAR FORM OF COMPLEX NUMBER (a,b)

The complex number (a,b) can be written in the very useful binomial form $a + bi$. The proof is not difficult: Since

$$(a,b) = (a + 0, b + 0) \qquad \text{why?}$$
$$= (a,0) + (0,b) \qquad \text{by Def. (7.3)}$$
$$= a(1,0) + b(0,1) \qquad \text{by Eq. (7.7)}$$
$$= a\cdot 1 + b\cdot i \qquad \text{why?}$$

Hence, $\qquad\qquad (a,b) = a + bi \qquad\qquad\qquad (7.11)$

Real and imaginary parts We call a the *real part* of the complex number $a + bi$, since if $b = 0$, then $a + bi$ is equivalent to the real number a. We call b the *imaginary part* of the number.

If $b = 0$, then $a + bi$ reduces to the real number a. Thus, the set of real numbers R is a proper subset of the set of complex numbers C; that is,

$$R \subset C$$

Pure imaginary number On the other hand, if $a = 0$ and $b \neq 0$, then $a + bi$ reduces to bi. We call bi a *pure imaginary number*. The set of pure imaginary numbers is also a proper subset of C.

Any real number can be written in the rectangular form $a + bi$. For example, $-3 = -3 + 0\cdot i$. Any pure imaginary number can be written in the rectangular form. For example, $i\sqrt{2} = 0 + i\sqrt{2}$, and $-7i = 0 - 7i$.

The advantage of the rectangular form is that if we write complex numbers (a,b) in the form $a + bi$, we can carry out addition and multiplication as with real binomials, *provided we replace i^2, wherever it occurs, with -1.*

7.4 ALGEBRAIC OPERATIONS ON COMPLEX NUMBERS

Since $i^2 = -1$, $i^3 = i^2\cdot i = (-1)i = -i$.

Also, $\qquad\qquad i^4 = i^2\cdot i^2 = (-1)(-1) = 1$
$$i^5 = i^4\cdot i = (1)i = i$$

and so on. We can derive a simple procedure for writing powers of i as follows. If $n \in P$,

$$i^{4n} = (i^4)^n = (1)^n = 1$$

Hence, $$i^{4n+k} = (i^{4n})(i)^k = (1)(i^k) = i^k$$

Thus, $$i^{17} = i^{4 \cdot 4 + 1} = i$$

$$i^{98} = i^{4 \cdot 24 + 2} = i^2 = -1$$

$$i^{84} = i^{4 \cdot 21} = 1$$

Since the algebraic laws (commutative, associative, and distributive properties) hold for complex numbers, we can use these laws to carry out calculations with numbers of the form $a + bi$ in exactly the same manner as we did with real numbers of the form $a + bx$, except that wherever possible we replace i^2 with -1.

The following statements about complex numbers are in agreement with the definitions of preceding sections:

$$a + bi = c + di$$

if and only if $$a = c \quad \text{and} \quad b = d$$ (7.12)

EXAMPLE 1. If $a + 2i = 3 + bi$, find the values of a and b.

Solution. By (7.12), $a = 3$ and $b = 2$.

$$(a + bi) + (c + di) = (a + c) + (b + d)i$$ (7.13)

EXAMPLE 2. Find the sum of $3 + 8i$ and $2 + 5i$.

Solution. $(3 + 8i) + (2 + 5i) = (3 + 2) + (8 + 5)i$
$$= 5 + 13i$$

$$(a + bi) - (c + di) = (a - c) + (b - d)i$$ (7.14)

EXAMPLE 3. From $\sqrt{2} - 5i$ subtract $3 + 3i$.

Solution. $(\sqrt{2} - 5i) - (3 + 3i) = (\sqrt{2} - 3) + (-5 - 3)i$
$$= (\sqrt{2} - 3) - 8i$$

Complex conjugates   If two complex numbers differ only in the sign of their imaginary parts, each is called the *conjugate* of the other. For example, $a + bi$ is the conjugate of $a - bi$, and $a - bi$ is the conjugate of $a + bi$. The conjugate of the real number a is a itself, since $a + 0 \cdot i = a - 0 \cdot i = a$. Examples of pairs of conjugate complex numbers are $3 + 5i$ and $3 - 5i$; $-7 - \sqrt{5}i$ and $-7 + \sqrt{5}i$; $-6i$ and $6i$.

Since $(a + bi) + (a - bi) = 2a$, the sum of two conjugate complex numbers is a real number. The difference of two conjugate nonreal numbers is a pure imaginary number, since

$$(a + bi) - (a - bi) = (a + bi) + (-a + bi) = 2bi$$

EXAMPLE 4. Find the sum and difference of $6 + 14i$ and $6 - 14i$.

Solution. $(6 + 14i) + (6 - 14i) = 12$
$(6 + 14i) - (6 - 14i) = 28i$

Product

The product $(a + bi)(c + di)$ is equal to $ac + adi + bci + bdi^2$, and since $i^2 = -1$,

$$(a + bi)(c + di) = (ac - bd) + (ad + bc)i \tag{7.15}$$

This is in agreement with (7.3) and (7.11).

EXAMPLE 5. Find the product $(-5 - 8i)(3 - 7i)$.

Solution. $(-5 - 8i)(3 - 7i) = (-15 - 56) + (35 - 24)i$
$= -71 + 11i$

Since $(a + bi)(a - bi) = a^2 + b^2$, we see that the product of two conjugate complex numbers is a nonnegative real number.

EXAMPLE 6. $(3 + 5i)(3 - 5i) = 9 + 25 = 34$
and $(0 + 0 \cdot i)(0 - 0 \cdot i) = 0^2 + 0^2 = 0$

Quotient

Consider the quotient of $a + bi$ divided by $c + di$. If $c + di \neq 0$, we can multiply the numerator and denominator of the fraction

$$\frac{a + bi}{c + di}$$

by the conjugate of the denominator. Thus,

$$\frac{a + bi}{c + di} = \frac{a + bi}{c + di} \cdot \frac{c - di}{c - di} = \frac{(ac + bd) + (bc - ad)i}{c^2 + d^2}$$

$$= \frac{ac + bd}{c^2 + d^2} + \frac{bc - ad}{c^2 + d^2}i \tag{7.16}$$

EXAMPLE 7. Divide i by $1 + i$.

Solution. $\dfrac{i}{1+i} = \dfrac{i}{1+i} \cdot \dfrac{1-i}{1-i} = \dfrac{i-i^2}{1-i^2} = \dfrac{i+1}{2}$

$$= \dfrac{1}{2} + \dfrac{1}{2}i$$

EXAMPLE 8. $\dfrac{3+5i}{2-7i} = \dfrac{3+5i}{2-7i} \cdot \dfrac{2+7i}{2+7i} = \dfrac{-29+31i}{4+49}$

$$= -\dfrac{29}{53} + \dfrac{31}{53}i$$

EXERCISE SET 7.1

FIND the real values of x and y in each of the following equations:

1. $(8x,3y) = (16,6)$ 2. $(2x,-5y) = (-8,15)$
3. $2x + 3yi = 4 + 6i$ 4. $3x - 5yi = 12 + 15i$
5. $4x - 8 = (8 - 2y)i$ 6. $7x - 14 + (2y - 3)i = 0$
7. $x^2 + y^2i = 4$ 8. $(x + 2) + (-3y + 12)i = -x + yi$

PERFORM the indicated operations and reduce all answers to the form $a + bi$:

9. $(5,-3) + (-2,5)$ 10. $(-7,-2) + (5,3)$
11. $(1,-3) - (-2,5)$ 12. $(8,0) - (2,3)$
13. $(3 + 5i) + (7 + 12i)$ 14. $(5 - 3i) + (-5 + 7i)$
15. $(3 + 2i) - (-7 - i) + (8 + 2i)$ 16. $(1 + 2i) + 6 + 4i - (3 + 5i)$
17. $(2 + 3i) - 2 - 5i - (6 - 3i)$ 18. $3 + 8i - (2 + 5i) + (3 - 2i) + 4i$
19. $(1 + \sqrt{3}i) + \left(\dfrac{1}{2} - \dfrac{\sqrt{3}}{2}i\right)$ 20. $(1 - \sqrt{2}i) + \left(\dfrac{3}{2} - \dfrac{\sqrt{2}}{2}i\right)$
21. $\sqrt{-25} + \sqrt{-4} - 6i$ 22. $\sqrt{-49} - \sqrt{-25} + 2i - 3$
23. $(7 + \sqrt{-3}) - (3 + \sqrt{-12}) - 3i$ 24. $5 - \sqrt{-2} - (-7 + \sqrt{-8}) + 4\sqrt{2}i$
25. $(3,2) \cdot (5,2)$ 26. $(-7,-1) \cdot (3,-3)$
27. $(-3,2) \cdot (5,6)$ 28. $(8,-3) \cdot (1,-5)$
29. $(3 + i)(5 + 2i)$ 30. $(5 - 4i)(6 + 2i)$
31. $i(7 - 2i)(5 - 3i)$ 32. $(6 + 2i)(8 - 5i)i$
33. $(\sqrt{5} + \sqrt{2}i)(\sqrt{3} - \sqrt{5}i)$ 34. $(4 + \sqrt{-3})(3 - \sqrt{-12})$
35. $(1 + i)^2$ 36. $(1 - i)^3$
37. $(1 + i)^4$ 38. $(2 + 3i)^2i$
39. $(2 - 3i) \div i$ 40. $(1 + i) \div (1 - i)$
41. $\dfrac{1 - i}{1 + i}$ 42. $\dfrac{1}{1 + i}$
43. $1 \div (1 - i)$ 44. $(\sqrt{3} + i) \div (1 + \sqrt{3}i)$
45. $\dfrac{3}{i^2 - i^3}$ 46. $\dfrac{5}{i^3 - i^2}$
47. $\dfrac{3}{(1 - i)^2}$ 48. $\dfrac{(2 + i)^2}{5 - i}$

49. $\dfrac{3 - 3i}{5 - 4i}$

50. $\dfrac{(1 + i)^2}{3 - i}$

51. If (a,b), (c,d), $(e,f) \in C$, prove the associative law for addition: $[(a,b) + (c,d)] + (e,f) = (a,b) + [(c,d) + (e,f)]$.

52. Prove the commutative law for addition: $(a,b) + (c,d) = (c,d) + (a,b)$.

53. Prove the associative law for multiplication: $[(a,b) \cdot (c,d)] \cdot (e,f) = (a,b) \cdot [(c,d) \cdot (e,f)]$.

54. Prove the commutative law for multiplication: $(a,b) \cdot (c,d) = (c,d) \cdot (a,b)$.

55. Prove the distributive law: $(a,b)[(c,d) + (e,f)] = (a,b) \cdot (c,d) + (a,b) \cdot (e,f)$.

7.5 COMPLEX PLANE

A complex number is an ordered pair (a,b) of real numbers. Consequently, if we consider the ordered pair (a,b) to be the cartesian coordinates of a point in a rectangular coordinate system, every complex number (a,b) corresponds to some point in the coordinate plane. The coordinates of the point are $x = a$ and $y = b$. Conversely, every point (a,b) in the coordinate plane corresponds to some complex number $(a,b) = a + bi$. Thus, we establish a one-to-one correspondence between the set of complex numbers and the set of points in the coordinate plane. This plane is called the *complex plane*.

Axes in complex plane

In the complex plane, the real number $a = a + 0 \cdot i = (a,0)$ corresponds to a point on the x axis. The pure imaginary number $0 + bi = (0,b)$ corresponds to a point on the y axis. We call the x axis the *axis of reals* and the y axis the *axis of imaginaries* (see Fig. 7.1).

7.6 TRIGONOMETRIC FORM OF COMPLEX NUMBERS

We can represent the complex number $(a,b) = a + bi$ by the vector drawn from the origin to the point $P(a,b)$ (see Fig. 7.2). The length of the vector OP is denoted by r. Then, by the distance formula,

$$r = \sqrt{a^2 + b^2} \tag{7.17}$$

FIGURE 7.1 The complex plane

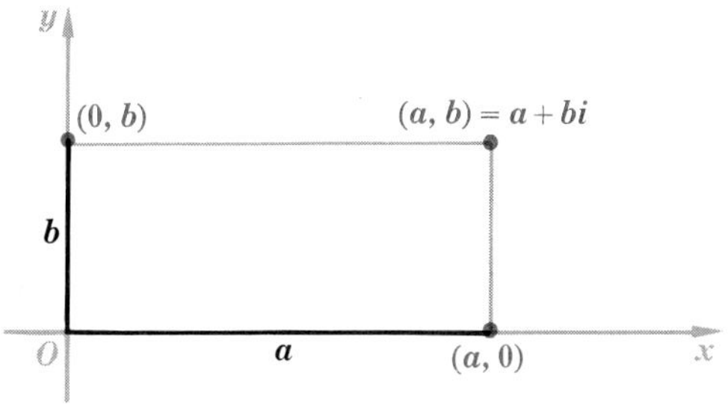

FIGURE 7.2 The complex number (a,b) as a vector in the complex plane

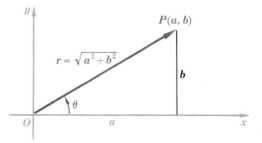

Modulus

We call r the *modulus* or *absolute value* of the complex number $a + bi$. The direction of the vector OP is determined by the angle θ which OP makes with the positive direction of the x axis. Hence, θ is completely defined by the equations

$$a = r \cos \theta \qquad b = r \sin \theta \qquad\qquad (7.18)$$

Amplitude

The angle θ is called an *amplitude* or an *argument* of the complex number $a + bi$.

From Eqs. (7.18) we have

$$\begin{aligned} a + bi &= r \cos \theta + (r \sin \theta) \cdot i \\ &= r(\cos \theta + i \sin \theta) \end{aligned} \qquad (7.19)$$

Polar form

Relation (7.19) is called the *trigonometric form* or the *polar form* of the complex number $a + bi$.

Now, for any integer k, $\cos (\theta + 2k\pi) = \cos \theta$ and $\sin (\theta + 2k\pi) = \sin \theta$. Consequently, we have the very useful relation

$$a + bi = r[\cos (\theta + 2k\pi) + i \sin (\theta + 2k\pi)] \qquad (7.20)$$

EXAMPLE 1. Express $-3 + 3i$ in an equivalent trigonometric form.

Solution. The point $P(-3,3)$ corresponds to the complex number $-3 + 3i$, Fig. 7.3. Hence, $a = -3$, $b = 3$, and the modulus r is found from Eq. (7.17) to be

$$r = \sqrt{(-3)^2 + (3)^2} = 3\sqrt{2}$$

Since $\qquad \cos \theta = \dfrac{-3}{3\sqrt{2}} = \dfrac{-1}{\sqrt{2}} \qquad$ and $\qquad \sin \theta = \dfrac{3}{3\sqrt{2}} = \dfrac{1}{\sqrt{2}}$

θ is in the second quadrant and we have

$$\theta = \frac{3\pi}{4} = 135°$$

Therefore, $\qquad -3 + 3i = 3\sqrt{2}(\cos 135° + i \sin 135°)$

EXAMPLE 2. Express in polar form the complex number $-\frac{1}{2} - (\sqrt{3}/2)i$.

Solution. The point $(-\frac{1}{2}, -\sqrt{3}/2)$ represents the given complex number. Hence,

$$a = -\frac{1}{2} \qquad b = -\frac{\sqrt{3}}{2}$$

(Fig. 7.4). Then

$$r = \sqrt{\left(-\frac{1}{2}\right)^2 + \left(\frac{\sqrt{3}}{2}\right)^2} = 1$$

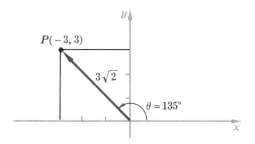

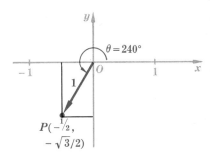

FIGURE 7.3   See Example 1.   FIGURE 7.4   See Example 2.

Since $\cos \theta = -\tfrac{1}{2}$ and $\sin \theta = -(\sqrt{3}/2)$, θ is in the third quadrant and

$$\theta = \frac{4\pi}{3} = 240°$$

Hence, $$-\frac{1}{2} - \frac{\sqrt{3}}{2}i = \cos 240° + i \sin 240°$$

EXAMPLE 3.   Express the complex number $5(\cos 330° + i \sin 330°)$ in the rectangular form.

Solution.   Take angle $\theta = 330°$ in standard position.   Locate point P on the terminal side of θ and at a distance of 5 units from the origin (see Fig. 7.5). Since $P(x,y)$ corresponds to the given complex number, we have

$$x = 5 \cos 330° = 5 \cdot \frac{\sqrt{3}}{2}$$

and $$y = 5 \sin 330° = 5\left(-\frac{1}{2}\right) = -\frac{5}{2}$$

Hence, $$5(\cos 330° + i \sin 330°) = \frac{5\sqrt{3}}{2} - \frac{5}{2}i$$

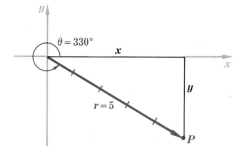

FIGURE 7.5
See Example 3.

EXERCISE SET 7.2

LOCATE the point in the complex plane which corresponds to each of the following numbers. Change each to the trigonometric form.

1. $2 + i$ 2. $-1 + i$ 3. $-4 - 2i$

4. $3 - 2i$ 5. -3 6. $-2i$

7. $-1 + 3i$ 8. $-\dfrac{1}{2} + \dfrac{\sqrt{3}}{2}i$ 9. $\dfrac{1}{2} - \dfrac{\sqrt{3}}{2}i$

10. $1 + i$ 11. $\dfrac{1}{2} + \dfrac{\sqrt{2}}{2}i$ 12. $6 + 8i$

13. $5 + 12i$ 14. $-\dfrac{1}{2} - \dfrac{\sqrt{2}}{2}i$ 15. $\dfrac{3 - 2i}{2 + i}$

16. $\dfrac{1}{3 + 2i}$ 17. $\dfrac{1}{2 - 3i}$ 18. $(2 - 5i)^2$

19. $\dfrac{5 - 2i}{2 - i} + \dfrac{1 - i}{3 + i}$ 20. $\dfrac{1}{i}$ 21. $\dfrac{3}{-i}$

EXPRESS each of the following in the $a + bi$ form:

22. $2(\cos 0° + i \sin 0°)$ 23. $3(\cos 180° + i \sin 180°)$

24. $5(\cos 90° + i \sin 90°)$ 25. $2(\cos 150° + i \sin 150°)$

26. $\cos 225° + i \sin 225°$ 27. $6(\cos 240° + i \sin 240°)$

28. $7(\cos 270° + i \sin 270°)$ 29. $5(\cos 315° + i \sin 315°)$

30. $2[\cos(-120°) + i \sin(-120°)]$ 31. $6[\cos(-240°) + i \sin(-240°)]$

32. $8[\cos(-315°) + i \sin(-315°)]$

7.7 TRIGONOMETRIC FORM IN MULTIPLICATION AND DIVISION

The trigonometric form of complex numbers can be used to great advantage in finding products and quotients. We will find this form especially useful in finding the roots of both real and complex numbers.

Let $r_1(\cos \theta_1 + i \sin \theta_1)$ and $r_2(\cos \theta_2 + i \sin \theta_2)$ be any two complex numbers. Then

$$r_1(\cos \theta_1 + i \sin \theta_1) \cdot r_2(\cos \theta_2 + i \sin \theta_2)$$
$$= r_1 r_2[(\cos \theta_1 \cos \theta_2 - \sin \theta_1 \sin \theta_2) + i(\sin \theta_1 \cos \theta_2 + \cos \theta_1 \sin \theta_2)]$$
$$= r_1 r_2[\cos(\theta_1 + \theta_2) + i \sin(\theta_1 + \theta_2)]$$

$$(7.21)$$

Product

Hence, the modulus of the product of two complex numbers is the product of their moduli, and an amplitude of the product is the sum of the amplitudes.

EXAMPLE 1. Find the product of $3(\cos 60° + i \sin 60°)$ and $2(\cos 120° + i \sin 120°)$.

Solution. By Eq. (7.21), the product

$3(\cos 60° + i \sin 60°) \cdot 2(\cos 120° + i \sin 120°)$
$$= 3 \cdot 2[\cos (60° + 120°) + i \sin (60° + 120°)]$$
$$= 6(\cos 180° + i \sin 180°)$$
$$= 6(-1 + i \cdot 0) = -6$$

To find the quotient

$$\frac{r_1(\cos \theta_1 + i \sin \theta_1)}{r_2(\cos \theta_2 + i \sin \theta_2)}$$

we observe that Eq. (7.21) can be used as follows:

$$r_2(\cos \theta_2 + i \sin \theta_2) \cdot \frac{r_1}{r_2} [\cos (\theta_1 - \theta_2) + i \sin (\theta_1 - \theta_2)] = r_1(\cos \theta_1 + i \sin \theta_1)$$

Therefore,

$$\frac{r_1(\cos \theta_1 + i \sin \theta_1)}{r_2(\cos \theta_2 + i \sin \theta_2)} = \frac{r_1}{r_2} [\cos (\theta_1 - \theta_2) + i \sin (\theta_1 - \theta_2)] \qquad (7.22)$$

Quotient   Hence, the modulus of the quotient of two complex numbers is the quotient of their respective moduli, and an amplitude of the quotient is the amplitude of the numerator minus the amplitude of the denominator.

EXAMPLE 2.   Find the quotient of $8(\cos 75° + i \sin 75°)$ divided by $2(\cos 15° + i \sin 15°)$.

Solution. $8(\cos 75° + i \sin 75°) \div 2(\cos 15° + i \sin 15°)$

$$= \frac{8}{2} [\cos (75° - 15°) + i \sin (75° - 15°)]$$

$$= 4(\cos 60° + i \sin 60°)$$

$$= 4\left(\frac{1}{2} + \frac{\sqrt{3}}{2}i\right) = 2 + 2\sqrt{3}i$$

EXAMPLE 3.   Find the quotient of $1 - i$ divided by $[-\frac{1}{2} - (\sqrt{3}/2)i]$.

Solution.   Express each number in polar form:

$$1 - i = \sqrt{2}(\cos 315° + i \sin 315°)$$

$$-\frac{1}{2} - \frac{\sqrt{3}}{2}i = 1(\cos 240° + i \sin 240°)$$

Hence, $$\frac{1 - i}{-\frac{1}{2} - (\sqrt{3}/2)i} = \frac{\sqrt{2}(\cos 315° + i \sin 315°)}{\cos 240° + i \sin 240°}$$

$$= \sqrt{2}(\cos 75° + i \sin 75°)$$

We can now find the approximate values of $\cos 75°$ and $\sin 75°$ from the table of natural functions. Hence, the required quotient can be written as $\sqrt{2}(0.2588 + 0.9659i)$ or as $0.3659 + 1.366i$. The decimal approximations of complex numbers are not generally useful.

EXERCISE SET 7.3

PERFORM the indicated multiplication or division and express the final result in the $a + bi$ form.

1. $5(\cos 30° + i \sin 30°) \cdot 3(\cos 60° + i \sin 60°)$
2. $3(\cos 90° + i \sin 90°) \cdot 8(\cos 120° + i \sin 120°)$
3. $6(\cos 135° + i \sin 135°) \cdot (\cos 215° + i \sin 215°)$
4. $7[\cos(-120°) + i \sin(-120°)] \cdot 2(\cos 240° + i \sin 240°)$
5. $[3(\cos 30° + i \sin 30°)]^3$
6. $[2(\cos 240° + i \sin 240°)]^2$
7. $8(\cos 80° + i \sin 80°) \div 2(\cos 35° + i \sin 35°)$
8. $3\left(\cos \dfrac{\pi}{3} + i \sin \dfrac{\pi}{3}\right) \div (-12)\left(\cos \dfrac{\pi}{6} + i \sin \dfrac{\pi}{6}\right)$
9. $25\left(\cos \dfrac{2\pi}{3} + i \sin \dfrac{2\pi}{3}\right) \div 5(\cos 30° + i \sin 30°)$
10. $(-21)\left(\cos \dfrac{7\pi}{4} + i \sin \dfrac{7\pi}{4}\right) \div (-7)\left(\cos \dfrac{\pi}{2} + i \sin \dfrac{\pi}{2}\right)$
11. $(\cos 0° + i \sin 0°) \div 6(\cos 240° + i \sin 240°)$

CHANGE each of the following complex numbers to polar form and then perform the indicated operation:

12. $(\sqrt{2} + \sqrt{2}i)(3i)$
13. $(1 - i)^2$
14. $(1 + i)^2$
15. $(1 - i) \div (1 + i)$
16. $(-\sqrt{3} - i) \div (1 - \sqrt{3}i)$
17. $(-1 + i) \div (\sqrt{3} + i)$
18. $(1 + i)(1 + \sqrt{3}i)(-½ + ½ i)$
19. $\dfrac{(\sqrt{3} + i)(-1 + i)}{(1 + \sqrt{3}i)(\sqrt{3} - i)}$

20. Prove that $[2(\cos 30° + i \sin 30°)]^2 = 4(\cos 60° + i \sin 60°)$.
21. Prove that $[r(\cos \theta + i \sin \theta)]^3 = r^3 (\cos 3\theta + i \sin 3\theta)$.
22. Prove that $|a + bi| = |a - bi|$.
23. Show that two complex numbers are equal if and only if (a) they have equal absolute values, and (b) they have amplitudes that differ by $n \cdot 360°$, where n is an integer.

7.8   POWERS AND ROOTS OF NUMBERS

Let $z = a + bi = r(\cos \theta + i \sin \theta)$. Then

$$z^2 = [r(\cos \theta + i \sin \theta)] \cdot [r(\cos \theta + i \sin \theta)]$$

Hence, by Eq. (7.21),

$$z^2 = r^2(\cos 2\theta + i \sin 2\theta)$$

Then, since $z^3 = z^2 \cdot z$, we have

$$z^3 = [r^2(\cos 2\theta + i \sin 2\theta)] \cdot [r(\cos \theta + i \sin \theta)]$$
$$= r^3(\cos 3\theta + i \sin 3\theta)$$

By successive applications of Eq. (7.21), we obtain

$$z^4 = r^4(\cos 4\theta + i \sin 4\theta)$$
$$z^5 = r^5(\cos 5\theta + i \sin 5\theta)$$

and so on. These results are summarized in the following theorem:

THEOREM. If n is any positive integer, then

$$[r(\cos \theta + i \sin \theta)]^n = r^n(\cos n\theta + i \sin n\theta)$$

(7.23)

De Moivre's theorem This theorem is called De Moivre's theorem. It can be proved by the method of mathematical induction, a method of proof that we shall consider in a later chapter. De Moivre's theorem also holds for $n = 0$ and for n a negative integer.

EXAMPLE 1. Find the value of z^6 if $z = 1 - i$.

Solution. Write $z = 1 - i$ in polar form; then

$$z = \sqrt{2}(\cos 315° + i \sin 315°)$$

Hence,
$$z^6 = [\sqrt{2}(\cos 315° + i \sin 315°)]^6$$
$$= (\sqrt{2})^6(\cos 1{,}890° + i \sin 1{,}890°)$$
$$= 8(\cos 90° + i \sin 90°)$$
$$= 8i$$

We define an nth root of a complex number as follows:

DEFINITION. If n is a positive integer, then $p(\cos \theta + i \sin \theta)$ is an nth root of $q(\cos \phi + i \sin \phi)$ if and only if

$$[p(\cos \theta + i \sin \theta)]^n = q(\cos \phi + i \sin \phi)$$

(7.24)

Thus, by De Moivre's theorem, $p(\cos \theta + i \sin \theta)$ is an nth root of $q(\cos \phi + i \sin \phi)$ if and only if

$$q(\cos \phi + i \sin \phi) = [p(\cos \theta + i \sin \theta)]^n$$
$$= p^n(\cos n\theta + i \sin n\theta)$$

Now, two complex numbers are equal if and only if (1) their real parts are equal, and (2) their imaginary parts are equal, by Definition (7.1).

Therefore

$$q \cos \phi = p^n \cos n\theta$$
$$q \sin \phi = p^n \sin n\theta \tag{7.25}$$

If we square each member of the two equations and then add, we have

$$q^2(\cos^2 \phi + \sin^2 \phi) = p^{2n}(\cos^2 n\theta + \sin^2 n\theta)$$

and $$q^2 = p^{2n} \qquad \text{why?}$$

Hence, $$q = p^n$$

and $$p = \sqrt[n]{q} \tag{7.26}$$

Since $p > 0$, p is the principal nth root of q. Hence, Eqs. (7.25) now become

$$\cos \phi = \cos n\theta$$
$$\sin \phi = \sin n\theta \tag{7.27}$$

Equations (7.27) tell us that the angles ϕ and $n\theta$ are coterminal, and therefore either they are equal or they differ by an integral multiple of 2π. Hence,

$$\phi + 2k\pi = n\theta$$

and $$\theta = \frac{\phi + 2k\pi}{n} \qquad k = 0, 1, 2, \ldots \tag{7.28}$$

We have now shown that an nth root of $q(\cos \phi + i \sin \phi)$ has an absolute value equal to the principal nth root of q and an amplitude equal to

$$\frac{\phi + 2k\pi}{n} \qquad k \text{ an integer.}$$

For $k = 0, 1, 2, 3, \ldots, (n - 1)$, the angles are distinct, nonnegative, and less than 2π. Hence, corresponding to each of these n distinct angles, there is one distinct nth root. For $k = n$, the angle is

$$\frac{\phi + 2n\pi}{n} = \frac{\phi}{n} + 2\pi$$

and the corresponding nth root is the same as the root for $k = 0$. In a similar manner, $k = n + 1$ yields the same root as $k = 1$; $k = n + 2$ yields the same root as $k = 2$, and so on.

We conclude, therefore, that the n nth roots of the complex number $q(\cos \phi + i \sin \phi)$ have the modulus $\sqrt[n]{q}$ and amplitudes given by

$$\frac{\phi + 2k\pi}{n} = \frac{\phi + k \cdot 360°}{n} \qquad k = 0, 1, 2, \ldots, (n - 1)$$

EXAMPLE 2.   Find $\sqrt[3]{27(\cos 60° + i \sin 60°)}$.

Solution. $\sqrt[3]{27(\cos 60° + i \sin 60°)}$

$$= \sqrt[3]{27} \left(\cos \frac{60° + k \cdot 360°}{3} + i \sin \frac{60° + k \cdot 360°}{3} \right)$$

$$= 3 \left[\cos (20° + k \cdot 120°) + i \sin (20° + k \cdot 120°) \right]$$

For $k = 0$ $z_0 = 3(\cos 20° + i \sin 20°)$

For $k = 1$ $z_1 = 3(\cos 140° + i \sin 140°)$

For $k = 2$ $z_2 = 3(\cos 260° + i \sin 260°)$

It should be noted that for $k = 3, 4, \ldots ,$

$$z_3 = 3(\cos 380° + i \sin 380°) = 3(\cos 20° + i \sin 20°)$$
$$z_4 = 3(\cos 500° + i \sin 500°) = 3(\cos 140° + i \sin 140°)$$

and so on.

The points in the complex plane which correspond to these three distinct cube roots are equally spaced on a circle whose center is the origin and whose radius is 3 units (see Fig. 7.6).

EXAMPLE 3.   Find $\sqrt{-1 + \sqrt{3}i}$ and express the result in rectangular form.

Solution.   Since $-1 + \sqrt{3}i = 2(\cos 120° + i \sin 120°)$,

$$\sqrt{-1 + \sqrt{3}i} = \sqrt{2(\cos 120° + i \sin 120°)}$$

$$= \sqrt{2} \left(\cos \frac{120° + k \cdot 360°}{2} + i \sin \frac{120° + k \cdot 360°}{2} \right)$$

$$= \sqrt{2}[\cos (60° + k \cdot 180°) + i \sin (60° + k \cdot 180°)]$$

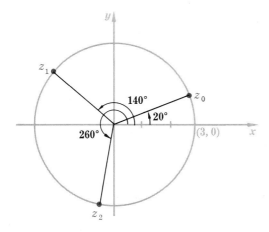

FIGURE 7.6

The three distinct cube roots of $27(\cos 60° + i \sin 60°)$

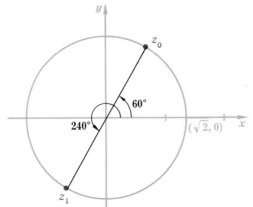

FIGURE 7.7
The two square roots of $-1 + i\sqrt{3}$

For $k = 0$ $\quad$ $z_0 = \sqrt{2}(\cos 60° + i \sin 60°)$

$$= \sqrt{2}\left(\frac{1}{2} + \frac{\sqrt{3}}{2}i\right) = \frac{\sqrt{2}}{2}(1 + \sqrt{3}i)$$

For $k = 1$ $\quad$ $z_1 = \sqrt{2}(\cos 240° + i \sin 240°)$

$$= \sqrt{2}\left(-\frac{1}{2} - \frac{\sqrt{3}}{2}i\right) = -\frac{\sqrt{2}}{2}(1 + \sqrt{3}i)$$

The points in the complex plane corresponding to z_0 and z_1 are shown in Fig. 7.7.

EXERCISE SET 7.4

EVALUATE by using De Moivre's theorem and express the result in rectangular form:

1. $(\cos 15° + i \sin 15°)^3$ $\qquad\qquad$ **2.** $[2(\cos 15° + i \sin 15°)]^6$
3. $(\cos 105° + i \sin 105°)^3$ $\qquad\qquad$ **4.** $(\cos 20° + i \sin 20°)^6$
5. $[2(\cos 315° + i \sin 315°)]^3$ $\qquad$ **6.** $(\cos 50° + i \sin 50°)^9$
7. $(\cos 7.5° + i \sin 7.5°)^8$ $\qquad\qquad$ **8.** $(\cos 5° + i \sin 5°)^{12}$

EXPRESS each of the following numbers in polar form and perform the indicated operations:

9. $(1 + i)^4$ $\qquad\qquad\qquad$ **10.** $(1 - i)^2$
11. $(-\tfrac{1}{2}\sqrt{3} + \tfrac{1}{2} i)^5$ $\qquad\quad$ **12.** $(\tfrac{1}{2}\sqrt{2} + \tfrac{1}{2}\sqrt{2}i)^{200}$
13. $(1 + \sqrt{3}i)^3$ $\qquad\qquad$ **14.** $(\sqrt{3} + \sqrt{3}i)^{10}$
15. $(-\sqrt{2} + \sqrt{2}i)^{10}$ $\qquad$ **16.** $(\tfrac{1}{2}\sqrt{3} - \tfrac{1}{2} i)^{150}$
17. Use De Moivre's theorem to obtain the identities for $\sin 2u$ and $\cos 2u$. HINT:

$$\cos 2u + i \sin 2u = (\cos u + i \sin u)^2 = \cos^2 u + 2(\cos u \sin u)i - \sin^2 u$$

$$= \cos^2 u - \sin^2 u + (2 \sin u \cos u)i$$

Now, use Definition (7.1).

18. Obtain relations for $\sin 3u$ and $\cos 3u$. $\qquad$ **19.** Obtain relations for $\sin 4u$ and $\cos 4u$.
20. Obtain a relation for $\tan 3u$.

21. Find four distinct fourth roots of $25(\cos 48° + i \sin 48°)$.

22. Find three distinct cube roots of $64(\cos 165° + i \sin 165°)$.

23. Find five distinct fifth roots of -32.   **24.** Find four distinct fourth roots of 1.

25. Find the two square roots of $-16i$.   **26.** Find the three distinct cube roots of i.

27. Find the three distinct cube roots of $0.9511 + 0.3090i$.   HINT: Use the table of natural functions and change to trigonometric form.

28. Find the distinct square roots of $0.5000 + 0.8660i$.

FIND the solution set for each of the following equations. Express the results in the $a + bi$ form.

29. $2x^4 = 1 - \sqrt{3}i$ **30.** $x^2 = 4 + 3i$

31. $x^3 - i = 2$ **32.** $x^2 - (2 - i)x - 2i = 0$

33. $x^2 + (3 - i)x - 3i = 0$. HINT: Factor.   **34.** $x^2 - 7ix - 12 = 0$

8

POLYNOMIAL FUNCTIONS

In previous chapters, we have considered trigonometric, exponential, and logarithmic functions and their graphs. We turn our attention now to a study of polynomial functions and their graphs. In particular, we shall be interested in finding the zeros of such functions. The zeros of a polynomial function $f(x)$ are the solutions (roots) of the equation $f(x) = 0$.

8.1 LINEAR FUNCTIONS

We recall from Sec. 3.10 that the function defined by the equation

$$f = \{(x,y) \mid y = mx + b\}$$
or
$$f = \{(x,f(x)) \mid f(x) = mx + b\} \tag{8.1}$$

where m and b are constants, is called a linear function in x. The graph of this function is a straight line.

We recall also that x_0 is a zero of f if and only if $f(x_0) = mx_0 + b = 0$. If we solve the equation $mx + b = 0$ for $x = -b/m$, we see that $f(-b/m) = m(-b/m) + b = 0$. Hence, $-b/m$ is a zero. Thus, to find the zero of f, we solve the equation $f(x) = 0$. The following example shows how this procedure can be used to find the zeros of certain functions that are not linear.

EXAMPLE 1. Find the zeros of the function defined by $f(\theta) = 2 \cos \theta - 1$, where $0 \leq \theta \leq 2\pi$.

Solution. This function is not linear in the variable θ but it is linear in the variable $\cos \theta$.

Set $2 \cos \theta - 1 = 0$.   Then

$$\cos \theta = \frac{1}{2}$$

Hence,
$$\theta = \frac{\pi}{3}, \frac{5\pi}{3}$$

We conclude that the zeros of $f(\theta)$ are $\pi/3$ and $5\pi/3$.

8.2   QUADRATIC FUNCTIONS

The function f defined by the equation

$$f(x) = ax^2 + bx + c \qquad\qquad (8.2)$$

where a, b, c are constants and $a \neq 0$, is called a *quadratic function in x*. The function f is called a real function if a, b, c are real numbers, and a rational function if a, b, c are rational.

The number x_0 is called a zero of f if and only if $f(x_0) = a(x_0)^2 + b(x_0) + c = 0$.   Hence, the zeros of f are the elements in the solution set of the equation

$$ax^2 + bx + c = 0 \qquad a \neq 0 \qquad\qquad (8.3)$$

Quadratic equation

which is called a *quadratic equation*.   If the left member of this equation can be factored by inspection, the solution set (and the zeros of the function) can be determined easily.

EXAMPLE 1.   Find the zeros of $f(x) = 2x^2 - x - 3$.

Solution.   Set $2x^2 - x - 3 = 0$.   Then

$$(2x - 3)(x + 1) = 0$$

We see that if $x = -1$ or if $x = \tfrac{3}{2}$, then the equation is a true statement, but otherwise it is not (why?).   Hence, the zeros of f are -1 and $\tfrac{3}{2}$.

Completing the square

If Eq. (8.3) cannot be solved by factoring, we may employ a device known as *completing the square*.   To illustrate this method, we write the equation in the equivalent form

$$x^2 + \frac{b}{a} x = - \frac{c}{a}$$

and then add the quantity $(\tfrac{1}{2} b/a)^2$ to each side.   Then

$$x^2 + \frac{b}{a} x + \frac{b^2}{4a^2} = - \frac{c}{a} + \frac{b^2}{4a^2} = \frac{b^2 - 4ac}{4a^2}$$

The left-hand side of this equation has now been "completed" so that it is a perfect-square trinomial (an expression of three terms that is the square of a binomial). Hence,

$$\left(x + \frac{b}{2a}\right)^2 = \frac{b^2 - 4ac}{4a^2}$$

and

$$\left(x + \frac{b}{2a}\right) = \frac{+\sqrt{b^2 - 4ac}}{2a}$$

(8.4)

or

$$\left(x + \frac{b}{2a}\right) = \frac{-\sqrt{b^2 - 4ac}}{2a}$$

Each of Eqs. (8.4) yields a value of x that satisfies the equation $ax^2 + bx + c = 0$. Furthermore, these two values of x are the only ones that satisfy the equation. Consequently, the quadratic function has two zeros. These two zeros are equal if $b^2 - 4ac = 0$.

EXAMPLE 2. Solve the equation $3x^2 - 2x + 1 = 0$.

Solution. The factoring method is not useful here, so we complete the square as follows: We write the equation in the equivalent form

$$x^2 - \frac{2}{3}x = -\frac{1}{3}$$

and then complete the square on the left by adding $[\frac{1}{2}(\frac{2}{3})]^2$ to each side. We thus have

$$x^2 - \frac{2}{3}x + \frac{1}{9} = -\frac{1}{3} + \frac{1}{9} = \frac{-2}{9}$$

and

$$\left(x - \frac{1}{3}\right)^2 = \frac{-2}{9}$$

Hence, either

$$x - \frac{1}{3} = \sqrt{\frac{-2}{9}} = \frac{1}{3}i\sqrt{2}$$

or

$$x - \frac{1}{3} = -\sqrt{\frac{-2}{9}} = -\frac{1}{3}i\sqrt{2}$$

The solution set is $\{\frac{1}{3} + \frac{1}{3}i\sqrt{2}, \frac{1}{3} - \frac{1}{3}i\sqrt{2}\}$.

The two linear equations of (8.4) are important in that they lead to the following theorem:

THEOREM. The quadratic polynomial $f(x) = ax^2 + bx + c$ has two zeros (which may be equal):

(8.5)

$$x_1 = \frac{-b + \sqrt{b^2 - 4ac}}{2a} \qquad x_2 = \frac{-b - \sqrt{b^2 - 4ac}}{2a}$$

Quadratic formula   Equations (8.5), called the *quadratic formulas,* give the zeros of

$$ax^2 + bx + c$$

in terms of the coefficients a, b, c.   The equations are frequently combined and written

$$x = \frac{-b \pm \sqrt{b^2 - 4ac}}{2a}$$

EXAMPLE 3.   Solve $3t^2 - 7t - 5 = 0$ by use of the quadratic formulas.

Solution.   Since $a = 3$, $b = -7$, $c = -5$, we have

$$t_1 = \frac{-(-7) + \sqrt{(-7)^2 - 4(3)(-5)}}{2(3)} = \frac{7 + \sqrt{109}}{6}$$

$$t_2 = \frac{-(-7) - \sqrt{(-7)^2 \quad 4(3)(-5)}}{2(3)} = \frac{7 - \sqrt{109}}{6}$$

Zeros as complex numbers   The examples of this section show that even though the coefficients of a quadratic function are rational numbers, the zeros of the function may be nonreal complex numbers, as in Example 2.   Equations (8.5) indicate that when one zero of a rational quadratic function is a nonreal complex number, the other zero is also a nonreal complex number.   Furthermore, the two complex nonreal numbers are conjugates.

If the coefficients a, b, c are nonreal complex numbers, the zeros of the function are found by the same methods as before.

EXAMPLE 4.   Find the zeros of

$$f(x) = x^2 + 2ix - 5$$

Solution.   Set $x^2 + 2ix - 5 = 0$.   Then

$$x_1 = \frac{-2i + \sqrt{-4 + 20}}{2} = \frac{-2i + 4}{2} = 2 - i$$

$$x_2 = \frac{-2i - \sqrt{-4 + 20}}{2} = \frac{-2i - 4}{2} = -2 - i$$

Discriminant   If the coefficients of the quadratic function are real numbers, then the value of $b^2 - 4ac$ in the quadratic formulas gives us the following information about the zeros of the function.   If $b^2 - 4ac < 0$, then x_1 and x_2 are distinct, complex conjugates.   If $b^2 - 4ac = 0$, then x_1 and x_2 are equal real numbers, since each is equal to $-b/2a$.   If $b^2 - 4ac > 0$, x_1 and x_2 are real, and $x_1 \neq x_2$.   The number $b^2 - 4ac$ is called the *discriminant* of the quadratic function $ax^2 + bx + c$ and will be denoted by Δ.   The relation between the discriminant and the zeros of f are exhibited in the following table:

Δ	Zeros of $ax^2 + bx + c$, where $a, b, c \in R$
negative	complex conjugates
zero	real, equal
positive	real, unequal

It is to be noted that this table gives valid results only if a, b, and c are real coefficients. If any of the coefficients are complex, the zeros may be complex even though Δ may be positive. For example, the value of $b^2 - 4ac$ of the function $f(x) = x^2 + 2ix - 5$ is the real number 4, yet the zeros of f are complex numbers.

If the coefficients of a quadratic function are not all real numbers, it may be necessary to find the square roots of a complex number. The following example illustrates a method that is often effective.

EXAMPLE 5. Find $\sqrt{5 - 12i}$.

Solution. Let $\sqrt{5 - 12i} = a + bi$, where a and b are real. Then

$$5 - 12i = (a + bi)^2 = a^2 - b^2 + 2abi$$

Hence, $5 = a^2 - b^2$

and $-12 = 2ab$

To solve this pair of equations, solve the second for b,

$$b = -\frac{6}{a}$$

and substitute into the first equation to obtain

$$a^2 - \left(-\frac{6}{a}\right)^2 = 5$$

Then $a^4 - 5a^2 - 36 = 0$

or $(a^2 - 9)(a^2 + 4) = 0$

Since a is real, the factor $a^2 + 4$ cannot equal zero. Therefore,

$$a^2 - 9 = 0$$

and $a = -3, 3$

If $a = -3$, $b = 2$ and $a + bi = -3 + 2i$

If $a = 3$, $b = -2$ and $a + bi = 3 - 2i$

Hence, the two square roots of $5 - 12i$ are $-3 + 2i$ and $3 - 2i$.

The method of finding roots of complex numbers described in Sec. 7.8 can also be used.

EXERCISE SET 8.1

SOLVE the following equations by factoring if possible:

1. $x^2 - 2x - 3 = x - 3$ **2.** $2x^2 + 11x - 6 = 0$

3. $x^2 + 6x - 27 = 0$ **4.** $2x(4x + 5) = -3$

5. $\tan^2 t - 2 \tan t + 1 = 0$ **6.** $4(\log x)^2 + 4(\log x) + 1 = 0$

7. $2 \sin^2 \theta + 5 \sin \theta - 3 = 0$ **8.** $12 \cos^2 \theta - 25 \cos \theta - 7 = 0$

9. $3x^2 - x = 10$ **10.** $2x^2 - 2 = -4x$

11. $2 \tan^2 \theta + \tan \theta = 0$ **12.** $2 \cos^2 t + 3 \cos t = -1$

13. $2(\log x)^2 - 5(\log x) - 12 = 0$ **14.** $\tan t (2 \sin t - \sqrt{3}) = 0$

SOLVE the following equations by completing the square:

15. $x^2 - 8x = 20$ **16.** $x^2 - 7x - 30 = 0$

17. $2x^2 - 3x - 9 = 0$ **18.** $\tan^2 t + 1 = \sec t + 3$

19. $\tan^2 t - 2 \tan t - 1 = 0$ **20.** $2(\log x)^2 = 3(\log x) + 9$

SOLVE each of the following, using the quadratic formulas. Check all solutions.

21. $2x^2 - 7x + 3 = 0$ **22.** $x^2 - 2x + 5 = 0$

23. $3x^2 - 6x + 2 = 0$ **24.** $\sqrt{3}x^2 = 4x - \sqrt{3}$

25. $x^2 - 5ix = 6$ **26.** $2ix^2 - 3x = -2i$

27. $x^2 + x = 1 - 3i$ **28.** $x^2 + 7(i - 1)x = 25i$

In Probs. 29–34 extraneous solutions may be introduced. Review Sec. 4.4. Check all solutions.

29. $\sqrt{2x + 5} - 3 = -\sqrt{x + 6}$ **30.** $\sqrt{2x - 5} - \sqrt{x - 3} = 1$

31. $\sqrt{5x - 9} = 1 + \sqrt{x + 4}$ **32.** $\dfrac{1}{x + 2} + \dfrac{8}{10} = \dfrac{2x - 1}{4x + 3}$

33. $\dfrac{6}{x + 1} + \dfrac{4}{x - 1} = 2$ **34.** $\dfrac{8}{x - 1} + \dfrac{3}{x + 1} - \dfrac{5}{x - 3} = 0$

35. The product of two consecutive positive integers is 600. Find the integers.

36. The product of two consecutive even integers is 960. Find them.

37. One leg of a right triangle is 12 in. long and the hypotenuse is 3 in. longer than twice the length of the other leg. Find the length of the hypotenuse.

38. The sum of the digits of a two-digit number is 9. The number itself is equal to twice the product of its digits. Find the number.

39. A motorist drove 156 mi at a constant rate. If he had driven 9 mi/hr faster, he would have reduced the driving time by 45 min. Find the rate at which he actually drove.

40. An empty tank can be filled by two pipes in ⅓ hr. The larger pipe alone could fill the tank in 9 min less time than the smaller one alone. Find the time necessary for the larger pipe alone to fill the empty tank.

41. A number is $2\frac{1}{10}$ less than its multiplicative inverse. Find the number.

42. If an object is projected vertically upward, then its height y (measured in feet) above the starting point at the end of t sec is given by

$$y = v_0 t - \frac{1}{2} g t^2$$

where v_0 is the initial speed in feet per second and air resistance is neglected.

If a baseball is thrown vertically upward at 96 ft/sec, how high will it be at the end of 2 sec? Consider $g = 32$.

43. How high will the ball in Prob. 42 be at the end of 4 sec?

44. At the end of 6 sec how high will the ball in Prob. 42 be?

8.3   GRAPH OF THE QUADRATIC FUNCTION

The real zeros of a function are the values of x at points where the graph of the function meets the x axis. Consequently, the graph of a function offers a geometric interpretation of the character of the zeros. For example, the graphs of the quadratic functions

(a) $f(x) = x^2 - 4x + 3$

(b) $f(x) = x^2 - 4x + 4$

(c) $f(x) = x^2 - 4x + 5$

are shown in Fig. 8.1. Note that function (a) has two distinct real zeros, (b) has two equal zeros, and (c) has no real zeros.

The graphs of the three functions shown in Fig. 8.1 have the same general shape. The graph of every quadratic function is of this type and is called a *parabola*. It can be shown that the curve opens upward if $a > 0$ and downward if $a < 0$. To sketch the graph of the function

Parabola

$$f(x) = ax^2 + bx + c$$

we first factor as follows

$$f(x) = a\left[\left(x^2 + \frac{b}{a}x\right) + \frac{c}{a}\right]$$

FIGURE 8.1   Graphs of (a) $f(x) = x^2 - 4x + 3$; (b) $f(x) = x^2 - 4x + 4$; (c) $f(x) = x^2 - 4x + 5$.   What are the zeros of these functions?

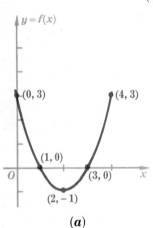

(a)

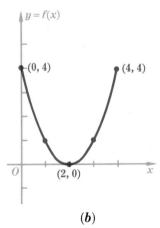

(b)

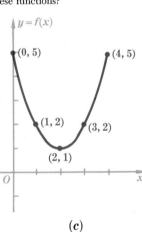

(c)

Then, by adding and subtracting the number $b^2/4a^2$ within the brackets, we have

$$f(x) = a\left[\left(x^2 + \frac{b}{a}x + \frac{b^2}{4a^2}\right) + \frac{c}{a} - \frac{b^2}{4a^2}\right]$$

$$= a\left[\left(x + \frac{b}{2a}\right)^2 + \frac{4ac - b^2}{4a^2}\right]$$

Since $[x + (b/2a)]^2 \geq 0$, the expression within the brackets has its least value when $x + (b/2a) = 0$, that is, when $x = -(b/2a)$. If $a > 0$, then $f(x)$ has its least value at $x = -(b/2a)$. If $a < 0$, then $f(x)$ has its greatest value at $x = -(b/2a)$. Thus, the point whose coordinates are

$$\left(-\frac{b}{2a}, \frac{4ac - b^2}{4a}\right)$$

Minimum and maximum points

is a *minimum point* on the graph of $f(x) = ax^2 + bx + c$ if a is positive and a *maximum point* if a is negative. We call this point the *vertex* of the parabola. The vertex and a few additional points are all that are needed to sketch the graph of the function.

EXAMPLE 1. Find the minimum value of $f(x) = x^2 - 4x + 3$ and sketch the graph of the function.

Solution. $f(x) = x^2 - 4x + 3 = (x^2 - 4x) + 3$
$= (x^2 - 4x + 4) + 3 - 4$
$= (x - 2)^2 - 1$

Hence, the vertex is at the point $(2, -1)$. Since the coefficient of x^2 is positive, the curve opens upward. The minimum value of $f(x)$ is -1. The graph is shown in Fig. 8.1a.

EXERCISE SET 8.2

FIND the values of x for which each of the following functions has a minimum or maximum value. Sketch the graph.

1. $f(x) = x^2 - 4x + 2$ 2. $f(x) = -x^2 + 3x + 2$
3. $f(x) = -3x^2 - 6x + 2$ 4. $f(x) = 2x^2 - 12x + 1$

FIND the minimum or maximum values of the following functions. Sketch the graph.

5. $f(x) = x^2 - 8x + 3$ 6. $f(x) = 2x^2 + 5x - 12$
7. $f(x) = -x^2 + 4x + 1$ 8. $f(x) = 7 - 10x + 2x^2$
9. Find two numbers, whose sum is 40, such that their product is a maximum.
10. Find two numbers, whose sum is 32, such that the sum of their squares is a minimum.
11. A rectangular field next to the straight bank of a river is to be fenced with 160 rd of fence. No fence is required along the riverbank. Find the dimensions of the rectangle which will enclose the maximum area. Find the maximum area.

12. A long sheet of tin 12 in. wide is to be made into a rain gutter by turning up strips vertically along the two sides. How many inches should be turned up at each side to obtain the greatest carrying capacity? HINT: The carrying capacity will be greatest when the area of a cross-section is a maximum.

13. An object is thrown vertically upward from the ground with a velocity of 96 ft/sec. Its height y, measured in feet, after t sec, is given by

$$y = 96t - 16t^2$$

After how many seconds does the body reach its maximum height? What is the maximum height?

14. Suppose a corporation is formed to bring cable television to a fringe area and that there is an annual profit of $20 per installation if there are 1,000 subscribers or less. If there are more than 1,000 subscribers, the annual profit per installation decreases 1 cent for each subscriber over that number. Find the number of subscribers that will yield the maximum annual profit.

8.4   RELATIONS BETWEEN ZEROS AND COEFFICIENTS

As we have seen, the zeros of the quadratic function $f(x) = ax^2 + bx + c$ are x_1 and x_2, where

$$x_1 = \frac{-b + \sqrt{b^2 - 4ac}}{2a}$$

and

$$x_2 = \frac{-b - \sqrt{b^2 - 4ac}}{2a}$$

Hence, the sum of the two zeros is given by

$$x_1 + x_2 = -\frac{b}{a} \tag{8.6}$$

and the product by

$$x_1 \cdot x_2 = \frac{c}{a} \tag{8.7}$$

The proofs of relations (8.6) and (8.7) are left as exercises.

EXAMPLE 1. Find the sum and product of the zeros of $f(x) = 2x^2 - 3x + 5$.

Solution. Here $a = 2$, $b = -3$, $c = 5$. Hence,

$$x_1 + x_2 = -\frac{b}{a} = -\frac{-3}{2} = \frac{3}{2}$$

and

$$x_1 \cdot x_2 = \frac{c}{a} = \frac{5}{2}$$

Since the set of zeros of $f(x) = ax^2 + bx + c$ is precisely the same set as the solution set of the equation $ax^2 + bx + c = 0$, we may use relations (8.6) and (8.7) to obtain a quadratic equation when the solution set $\{x_1, x_2\}$

is given. For from $x_1 + x_2 = -(b/a)$ we have

$$b = -a(x_1 + x_2)$$

and from $x_1 \cdot x_2 = c/a$ we have

$$c = ax_1x_2$$

Hence, $\qquad ax^2 + bx + c = ax^2 - a(x_1 + x_2)x + ax_1x_2$
$$= a[x^2 - (x_1 + x_2)x + x_1x_2]$$
$$= a(x - x_1)(x - x_2)$$

Therefore, the quadratic equation

$$ax^2 + bx + c = 0 \qquad a \neq 0$$

can be written in factored form

$$a(x - x_1)(x - x_2) = 0$$

which is equivalent to the equation

$$(x - x_1)(x - x_2) = 0$$

Thus, if $\{x_1, x_2\}$ is the solution set of a quadratic equation, we can obtain the equation (or its equivalent) by setting the product of the factors $x - x_1$ and $x - x_2$ equal to zero.

EXAMPLE 2. Find a quadratic equation whose solution set is $\{\frac{2}{3}, \frac{3}{4}\}$.

Solution. Set $(x - \frac{2}{3})(x - \frac{3}{4}) = 0$. Then

$$\left(\frac{3x - 2}{3}\right)\left(\frac{4x - 3}{4}\right) = 0$$

and $\qquad (3x - 2)(4x - 3) = 0$

Hence, $12x^2 - 17x + 6 = 0$ satisfies the required conditions.

EXAMPLE 3. The solution set of a quadratic equation is $\{2 + i, 2 - i\}$ and the coefficient of x^2 is 1. Find the equation.

Solution. Set $[x - (2 + i)][x - (2 - i)] = 0$. Then

$$[(x - 2) - i][(x - 2) + i] = 0$$

and $\qquad (x - 2)^2 - (i)^2 = 0$

Therefore, $\qquad x^2 - 4x + 4 + 1 = 0$

Thus, $x^2 - 4x + 5 = 0$ is the required equation.

EXERCISE SET 8.3

1. Find the sum and the product of the solutions of the following quadratic equations:

(a) $x^2 + x - 6 = 0$ (b) $8 + 5x - 2x^2 = 0$

(c) $3x^2 = 35x + 22$ (d) $2x^2 + ix - 1 = 0$

(e) $x^2 - 2ix - (1 + 2i) = 0$ (f) $x^2 + (1 - 8i) = 2x$

2. Find a quadratic equation having the given solution set:

(a) $\{-1,3\}$ (b) $\{\frac{1}{5}, \frac{2}{3}\}$

(c) $\{-7,-5\}$ (d) $\{2 + \sqrt{3}, 2 - \sqrt{3}\}$

(e) $\{2 + 3i, 2 - 3i\}$ (f) $\{-4 + i, -4 - i\}$

(g) $\left\{\dfrac{-1 + \sqrt{5}}{2}, \dfrac{-1 - \sqrt{5}}{2}\right\}$ (h) $\left\{\dfrac{-3 + \sqrt{7}}{4}, \dfrac{-3 - \sqrt{7}}{4}\right\}$

(i) $\{a + b, a - b\}$ (j) $\{a + bi, a - bi\}$

3. Show that if $ax^2 + bx + c = 0$, $a \neq 0$, is written in the form $x^2 + (b/a)x + (c/a) = 0$, then the sum of the roots (solutions) is the negative of the coefficient of x and the product of the roots is the constant term. HINT: Compare the coefficient of x with relation (8.6) and the constant term with relation (8.7).

4. Use the result of Prob. 3 to find an equation whose solution set is

(a) $\{3,5\}$ (b) $\{1,6\}$

(c) $\{1 + \sqrt{3}, 1 - \sqrt{3}\}$ (d) $\{2 + 3i, 2 - 3i\}$

5. Use the result of Prob. 3 to find an equation whose solution set is $\{3 + \sqrt{2}i, 3 - \sqrt{2}i\}$.

6. The sum of the roots of $kx^2 + (5k - 4)x + 2 = 0$ is given by $x_1 + x_2 = -1$. Find the value of k.

7. One of the solutions of $2x^2 + 4x + k^2 + 2k - 8 = 0$ is zero. Find the value of k.

8. Find the values of k if the product of the roots of $kx^2 - 3k + 4x + 2 = 0$ is -5.

9. One solution of $3x^2 - kx + k^2 - k - 18 = 0$ is $x = 1$. Find the value of k.

10. One solution of $4x^2 + kx + 6 = 0$ is $x = -2$. Find the value of k.

11. The sum of the zeros of $f(x) = 3x^2 + kx - 2$ is 6. Find the value of k.

12. The function $f(x) = 4x^2 + 12x + k$ has equal zeros. Find k.

13. The zeros of $f(x) = 9x^2 + (6 - k)x - 81$ are numerically equal but opposite in sign. Find k.

14. One solution of $3x^2 + 6 = 7x + k$ is zero. Find the value of k.

15. Find a quadratic equation whose roots are half the roots of $x^2 + 4x - 5 = 0$.

16. Find a quadratic equation whose solutions are one-third the solutions of $x^2 - 5x - 6 = 0$.

17. Determine the real values of k for which the solutions of $kx^2 + 3x + k = 0$ are complex conjugate numbers.

18. Factor the quadratic function $x^2 + 1$ into a product of linear functions of x.

19. Factor the function $kx^2 + (k^2 - k - 1)x + 1 - k$ into a product of linear functions.

20. One solution of $9x^2 + k = 15x$ is 3 more than the other solution. Find the value of k.

8.5 EQUATIONS IN QUADRATIC FORM

Any equation which can be written in the form

$$au^2 + bu + c = 0 \qquad a \neq 0 \tag{8.8}$$

is a quadratic equation in the variable u and can be solved for u by the methods developed in this chapter. If it happens that u is some function of another variable, then Eq. (8.8) is said to be in *quadratic form*. For example, the equation

$$x^4 - 5x^2 + 4 = 0$$

is in quadratic form. To show that such is the case, let $u = x^2$. The equation then becomes

$$u^2 - 5u + 4 = 0$$

with solutions $\qquad u = 1 \qquad$ and $\qquad u = 4$

Since $u = x^2$, we have

$$x^2 = 1 \qquad \text{or} \qquad x^2 = 4$$

Hence, $\qquad\qquad x = \pm 1 \qquad$ or $\qquad x = \pm 2$

Many times in solving equations that are in quadratic form we do not actually write the substitution, as in the previous example. Instead, we write the equation in a form which exhibits the characteristics of a quadratic equation. The procedure is illustrated in the following example.

EXAMPLE 1. Solve for the real values of x: $x^6 + 7x^3 - 8 = 0$.

Solution. Write the equation in the equivalent form

$$(x^3)^2 + 7(x^3) - 8 = 0$$

and consider the variable to be x^3. Then

$$x^3 = \frac{-7 \pm \sqrt{49 + 32}}{2}$$

and $\qquad\qquad x^3 = -8 \qquad$ or $\qquad x^3 = 1$

Since we are interested in the real values of x only,

$$x = -2 \qquad \text{or} \qquad x = 1$$

The other four roots can be determined by the method for finding the cube roots of a complex number, as in Sec. 7.8. The check is left to the student.

EXAMPLE 2. Find the solution set for the equation $t^{2/3} + 2t^{1/3} - 8 = 0$.

Solution. If we write the equation in the form

$$(t^{1/3})^2 + 2(t^{1/3}) - 8 = 0$$

then, from the quadratic formulas,

$$t^{1/3} = 2 \qquad \text{or} \qquad t^{1/3} = -4$$

Hence, $\qquad\qquad t = 8 \qquad$ or $\qquad t = -64$

The solution set is $\{8, -64\}$. The check is left to the student.

An equation involving radicals is usually best solved by squaring out the radicals as in Sec. 4.4. The following example, however, illustrates another method.

EXAMPLE 3. Solve the equation $x^2 - 7\sqrt{x^2 - 3} + 7 = 0$.

Solution. Add $-3 + 3$ to the left member as follows:

$$x^2 - 3 - 7\sqrt{x^2 - 3} + 7 + 3 = 0$$

Let $u = \sqrt{x^2 - 3}$, then $u^2 = x^2 - 3$ and we have

$$u^2 - 7u + 10 = 0$$

Hence, $u = 2$ or $u = 5$

and $u^2 = 4$ or $u^2 = 25$

Therefore, $x^2 - 3 = 4$ or $x^2 - 3 = 25$

The last two equations yield four values of x, which we write as $\pm\sqrt{7}$, or $\pm\sqrt{28}$. The solution set is the set $\{-\sqrt{28}, -\sqrt{7}, \sqrt{7}, \sqrt{28}\}$.

EXAMPLE 4. Solve for values of t such that $0 \le t \le 2\pi$, given that $4\sin^2 t - 11\sin t + 6 = 0$.

Solution. Let the variable be $(\sin t)$ so that the equation is quadratic in $\sin t$. Then

$$\sin t = \frac{11 \pm \sqrt{121 - 96}}{8}$$

and $\sin t = 2$ or $\sin t = \dfrac{3}{4}$

Now, $\sin t$ must be a real number such that $|\sin t| \le 1$. Hence, we conclude that $\sin t = 2$ is impossible and must be rejected. From $\sin t = \frac{3}{4} = 0.7500$, we conclude that $t = 48°40'$ (approx.) and $131°20'$ (approx.).

EXERCISE SET 8.4

SOLVE and check:

1. $x^4 - 13x^2 + 36 = 0$
2. $x^4 + x^2 - 12 = 0$
3. $x^{-4} - 13x^{-2} + 36 = 0$
4. $8x^{-6} + 7x^{-3} - 1 = 0$
5. $x + x^{1/2} - 20 = 0$
6. $(x^2 + 2x)^2 + (x^2 + 2x) - 12 = 0$. HINT: Let $u = x^2 + 2x$, so that $u^2 + u - 12 = 0$.
7. $(3x - 4)^2 + 6(3x - 4) = -13$
8. $(x^2 + 2)^2 + 3(x^2 + 2) = 4$
9. $3x^{-1/2} + 2x^{1/2} - 2x^{-3/2} = 0$
10. $x^3 - 9x^{3/2} + 8 = 0$
11. $x^2 - 5\sqrt{x^2 - 5} + 1 = 0$
12. $2x - 9\sqrt{x + 2} = -14$
13. $24\sqrt{x} = x^{5/2} + 2x^{3/2}$
14. $z^3 - 9z^{3/2} + 8 = 0$

FIND the smallest nonnegative value of x which satisfies the equation. Express results in degree measure. Check each possible answer.

15. $\sin^2 x + 2 \sin x + 3 = 0$ 16. $\cos^2 x - 3 \sin x + 9 = 0$

17. $2 \sin^2 x - 3\sqrt{3} \cos x - 5 = 0$ 18. $3 \sin^2 x - 2 \cos x - 2 = 0$

19. $\csc^2 x + 3 \cot x + 1 = 0$ 20. $\tan^2 x - \sec x - 2 = 0$

21. $3 \cos x - 2 \cos (x/2) - 5 = 0$ 22. $2 \cos^4 x + 9 \sin^2 x = 5$

23. $2 \tan^4 x + 3 \sec^2 x = 5$ 24. $\sin 2x + \sin x - \cos x - 1 = 0$

8.6 LINEAR AND QUADRATIC INEQUALITIES

If $f(x) = mx + b$, $m, b \in R$, $m \neq 0$, then $f(x) \geq 0$, or $f(x) \leq 0$ is a *linear inequality*. We discussed linear inequalities in Sec. 2.11.

If $f(x) = ax^2 + bx + c$, where a, b, $c \in R$ and $a \neq 0$, then a statement of the form

$$f(x) \geq 0 \qquad \text{or} \qquad f(x) \leq 0$$

is called a *quadratic inequality*. An inequality, like an equation, may or may not be a true statement. For example, the inequality $x^2 + 1 < 0$ is not true for any $x \in R$. The inequality $x^2 - 1 < 0$ is true only for values of x between -1 and $+1$. On the other hand, the inequality $x^2 + 1 > 0$ is true for every real value of x. We define a *conditional inequality* to be an inequality which is not satisfied by every value of x in the domain of f. An *unconditional inequality* is one that is satisfied by every value of x in the domain of f.

The order axioms for the real numbers and the theorems of Sec. 2.9 can be used to establish the following theorems:

THEOREM. If each member of an inequality is increased or decreased by the same number, the sense of the inequality is   **(8.9)** unchanged.

Proof. If $a > b$, then

$$a - b > 0 \qquad \text{why?}$$

Therefore $a - b = x$ where $x > 0$

It follows that $(a + c) - (b + c) = x > 0$

The student can verify the last statement by removing the parentheses and combining like terms. Hence,

$$a + c > b + c \qquad \text{why?}$$

and the sense of the inequality is unchanged. It is left to the student to show that if $a < b$, then $a + c < b + c$.

Thus, $x^2 - x > 1$ is equivalent to $x^2 - x - 1 > 0$.

THEOREM. If each member of an inequality is multiplied or divided by the same positive number, the sense of the inequality   **(8.10)** is unchanged.

The proof is left to the student.

As an example, $(x/2) + 3 > 2$ is equivalent to $x + 6 > 4$.   Also,

$$3x - 9 < 15$$

is equivalent to $x - 3 < 5$

> **THEOREM.**   If each member of an inequality is multiplied or
> divided by the same negative number, the sense of the inequality   (8.11)
> is reversed.

Proof.   If $a < b$, then $b - a = x$, where $x > 0$.   Then, if $c < 0$, $c(b - a)$
is negative.   Hence, $ca - cb$ is positive and

$$ca > cb$$

Therefore, the sense of the inequality is reversed.

It is left to the student to show that if $a > b$ and $c < 0$, then $ca < cb$.

For example, if $3 - x^2 > 4$, then $-(3 - x^2) < -4$ and conversely.
The last inequality can be written $x^2 - 3 < -4$.

To find the solution set of a conditional inequality, we follow a procedure
similar to that used in solving equations.

EXAMPLE 1.   Solve the inequality $x^2 + 3x > 10$.

Solution.   Since $x^2 + 3x > 10$,

$$x^2 + 3x - 10 > 0$$

Hence, $(x - 2)(x + 5) > 0$

The product of two factors is greater than zero if and only if both factors are
positive or both negative.   Hence the inequality is true if and only if

(1)   both $x - 2 > 0$ and $x + 5 > 0$
or (2)   both $x - 2 < 0$ and $x + 5 < 0$

From (1), $x > 2$ and $x > -5$ are both satisfied if $x > 2$.   From (2), $x < 2$
and $x < -5$ are both satisfied if $x < -5$.   Hence, the solution set is the
union

$$\{x \mid x < -5\} \cup \{x \mid x > 2\}$$

In solving the inequality $ax^2 + bx + c < 0$ or $ax^2 + bx + c > 0$, we
find the factors of the function by solving the quadratic equation

$$ax^2 + bx + c = 0$$

If x_1 and x_2 are the solutions of the equation, then $x - x_1$ and $x - x_2$ are
factors of the quadratic function.

EXAMPLE 2.   Find the solution set of the inequality $x^2 - 2x - 1 < 0$.

Solution.   We find the factors of the left member by solving the quadratic equation

$$x^2 - 2x - 1 = 0$$

Thus, $x_1 = 1 + \sqrt{2}$ and $x_2 = 1 - \sqrt{2}$

Hence, $[x - (1 + \sqrt{2})]$ and $[x - (1 - \sqrt{2})]$ are factors of $x^2 - 2x - 1$. The inequality can now be written

$$[x - (1 + \sqrt{2})][x - (1 - \sqrt{2})] < 0$$

We note that this inequality holds if and only if one of the factors is positive and the other negative, that is,

(1) $x - (1 + \sqrt{2}) < 0$ and $x - (1 - \sqrt{2}) > 0$

or (2) $x - (1 + \sqrt{2}) > 0$ and $x - (1 - \sqrt{2}) < 0$

From (1), $x < 1 + \sqrt{2}$ and $x > 1 - \sqrt{2}$. Thus, the solution set contains $\{x \mid x < 1 + \sqrt{2}\}$ and $\{x \mid x > 1 - \sqrt{2}\}$. From (2), $x > 1 + \sqrt{2}$ and $x < 1 - \sqrt{2}$ is impossible, since x cannot be both positive and negative at the same time. Hence, the entire solution set is the intersection

$$\{x \mid x > 1 - \sqrt{2}\} \cap \{x \mid x < 1 + \sqrt{2}\}$$

If the solutions of the real quadratic equation $ax^2 + bx + c = 0$ are nonreal numbers, then the inequality $ax^2 + bx + c > 0$ is satisfied either by all real values of x or by no real value of x. The same statement applies to the inequality $ax^2 + bx + c < 0$. To see why this is so, recall that the real solutions of the equation $f(x) = 0$ correspond to the values of x where the graph of the function crosses (or touches) the x axis. Since the graph has no breaks, this means that if the solutions are not real numbers, the graph lies entirely above or entirely below the x axis. Hence, $f(x) = ax^2 + bx + c$ is always greater than zero or less than zero.

EXAMPLE 3.   Find the solution set of the inequality $x^2 - 2x + 5 > 0$.

Solution.   Since the solution set of the equation $x^2 - 2x + 5 = 0$ is the set $\{1 - 2i, 1 + 2i\}$, we know that the graph of $f(x) = x^2 - 2x + 5$ must lie entirely above or entirely below the x axis. When $x = 0$, $f(x) = 5$, so the point $(0,5)$ is on the graph, and the graph must be above the x axis. Hence the solution set is

$$\{x \mid x \in R\}$$

The properties of the inequality relations $<$ and $>$ together with the properties of the equality relation $=$ makes it possible to apply Theorems (8.9) to (8.11) to the relations $\leq$ and $\geq$.

EXAMPLE 4.   Solve the inequality $|(x/3) + 5| \geq \frac{1}{3}$.

Solution. The given inequality is equivalent to

$$\frac{x}{3} + 5 \geq \frac{1}{3} \qquad \text{or} \qquad -\left(\frac{x}{3} + 5\right) \geq \frac{1}{3}$$

Hence, $x + 15 \geq 1$ or $-x - 15 \geq 1$

and $x \geq -14$ or $-x \geq 16$

Hence, $x \geq -14$ or $x \leq -16$

Note that there is no number that satisfies both inequalities. The solution set is the union

$$\{x \mid x \geq -14\} \cup \{x \mid x \leq -16\}$$

EXAMPLE 5. Solve the inequality $|2x + 3| \leq \frac{1}{2}$.

Solution. The inequality is equivalent to the inequalities

$$2x + 3 \leq \frac{1}{2} \qquad \text{and} \qquad -(2x + 3) \leq \frac{1}{2}$$

Hence, $x \leq -\frac{5}{4}$ and $x \geq -\frac{7}{4}$

The solution set is the intersection

$$\left\{x \mid x \leq -\frac{5}{4}\right\} \cap \left\{x \mid x \geq -\frac{7}{4}\right\}$$

This solution can be written

$$-\frac{7}{4} \leq x \leq -\frac{5}{4}$$

The following example illustrates the solution of inequalities involving trigonometric functions.

EXAMPLE 6. Find the positive values of t less than 2π for which

$$2 \sin^2 t < \sin t$$

Solution. Since $2 \sin^2 t < \sin t$,

$$2 \sin^2 t - \sin t < 0$$

and $(\sin t)(2 \sin t - 1) < 0$

The product of two factors is negative if and only if the two factors have unlike signs. For $\sin t$ positive, t is in the first or second quadrant, and $2 \sin t - 1$ must be negative. Hence, $2 \sin t - 1 < 0$ so that $\sin t < \frac{1}{2}$. This means that $0 < t < \pi/6$, or $5\pi/6 < t < \pi$.

If $\sin t < 0$, t is in the third or fourth quadrant, and $2 \sin t - 1 > 0$ so that $\sin t > \frac{1}{2}$. But this is impossible, for $\sin t$ cannot be both less than zero and greater than $\frac{1}{2}$. Hence, the entire solution set is the union.

$$\left\{t \mid 0 < t < \frac{\pi}{6}\right\} \cup \left\{t \mid \frac{5\pi}{6} < t < \pi\right\}$$

EXERCISE SET 8.5

FIND the real values of x which make each of the following inequalities a true statement:

1. $x + 6 > 0$ 2. $x - 5 < 0$
3. $5x - 25 < 0$ 4. $6x + 3 < 0$
5. $5x < -3x - 16$ 6. $3 - 5x < 2x - 11$
7. $-3 < 6x < 3$ 8. $0 < (x + 1)/2 < \frac{2}{3}$
9. $\frac{3}{8} \leq \frac{1}{8} - x$ 10. $\frac{4}{3} < 1/x$
11. $|x| > 4$ 12. $|x - 3| \leq 5$
13. $|x - m| \leq n$ 14. $x^2 + 2x - 3 > 0$
15. $2x^2 + 3x - 4 > 0$ 16. $2x^2 + 4x + 3 > 0$
17. $x^2 - x + 1 < 0$ 18. $x^2 + 2x + 4 > 0$
19. $1/(x - 2) < \frac{1}{3}$ 20. $|x - 3| > 2$
21. $|(x/5) - 7| \leq 5$ 22. $3x^2 - 3x < -4$
23. $(x + 3)(x + 2)(x + 1) < 0$ 24. $|(x + 1)/x| < 1$
25. $x - 1 < 2/x$

SOLVE the following inequalities for nonnegative values of t less than 2π:

26. $2 \sin^2 t < 1$ 27. $\cos^2 t < \cos t$
28. $\sin^2 t > \sin t$ 29. $2 \sin^2 t + \cos t < 2$
30. $\tan t + \sec^2 t > 1$ 31. $\tan^2 t - 2 \sec t + 1 < 0$
32. $4 \cot t > \sqrt{3} \csc^2 t$

8.7   UNCONDITIONAL INEQUALITIES

Solving unconditional inequalities

Many unconditional inequalities can be proved directly by using Theorems (8.9) to (8.11) to transform the inequality to the required form. However, the method presented here is a "work backward" procedure consisting of three essential steps:

1. Assume that the proposed inequality is true.
2. Reduce the inequality to another inequality which is known to be true.
3. Reverse the steps in 2 above.

Steps 1 and 2 are regarded as an analysis. Step 3 provides the actual proof. Therefore, we must be certain that we can justify each statement in step 3 as we retrace the steps of the analysis.

EXAMPLE 1. Prove that if $a, b \in R$, $a \neq b$, then $a^2 + b^2 > 2ab$.

Solution. (1) Assume that $a^2 + b^2 > 2ab$ is true. Then (2)

$$a^2 + b^2 - 2ab > 0 \qquad \text{why?}$$

Hence, $$(a - b)^2 > 0$$

The last statement is valid since the square of the real number $a - b$ is positive. (3) Now, to reverse the steps in 2, we begin with the inequality

$$(a - b)^2 > 0$$

which we know is true.   Then

$$a^2 - 2ab + b^2 > 0$$

and $$a^2 + b^2 > 2ab \qquad \text{why?}$$

EXAMPLE 2.   If $a \in R$, $a > 0$, then the sum of a and its multiplicative inverse $1/a$ is not less than 2.

Solution.   (1) We want to prove that $a + (1/a) \geq 2$ for every positive real number.   Thus, we assume that

$$a + \frac{1}{a} \geq 2$$

(2) Since $a > 0$, we may multiply each member by a and the sense of the inequality will remain unchanged.   Hence

$$a^2 + 1 \geq 2a$$

and $$a^2 - 2a + 1 \geq 0$$

Therefore $$(a - 1)^2 \geq 0$$

The last statement is true.
(3) Begin with the known inequality

$$(a - 1)^2 \geq 0$$

Then $$a^2 - 2a + 1 \geq 0$$

and $$a^2 + 1 \geq 2a$$

Since $a > 0$, we may divide by a without changing the sense of the inequality.   Hence,

$$a + \frac{1}{a} \geq 2$$

EXERCISE SET 8.6

PROVE each of the following unconditional inequalities.   Each letter represents a real number.

1.   If $a > b > 0$, then $a^2 > b^2$.
2.   If $a > b > 0$, then $\sqrt{a} > \sqrt{b}$.
3.   If $a > b > 0$, then $1/a < 1/b$.
4.   If $a > b$, then $-a < -b$.
5. $a^2 + 2ab \leq 2a^2 + b^2$
6. $|a| \geq a$
7. $a^2 + 1 \geq 2a$
8.   If $a > 0$, $b > 0$, then $(a + b)/2 \geq 2ab/(a + b)$.
9.   If $a > 0$, $b > 0$, $a \neq b$, then $(a/b) + (b/a) > 2$.
10.   If $a > 1$, then $a^2 > a$.
11.   If $0 < a < 1$, then $a^2 < 1$.
12.   If $a \neq b$, then $2ab/(a + b) < \sqrt{ab}$, provided $a > 0$, $b > 0$.
13.   If $a \neq b$, then $a^3b + ab^3 < a^4 + b^4$.
14. $a^2 + b^2 + c^2 < (a + b + c)^2$
15.   If $a \neq 0$, $b \neq 0$, then $a^2/2b^2 \geq 1 - b^2/2a^2$.
16.   If $a \neq 0$, then $(16/a^2) + a^2 \geq 8$.

8.8   GRAPHICAL SOLUTION OF INEQUALITIES

Conditional inequalities may be solved graphically. For example, to find the real solutions of the inequality $f(x) < 0$, we sketch the graph of $y = f(x)$ and note the values of x for which points on the graph are below the x axis. At each of these points y is negative, and therefore $f(x) < 0$. To solve $f(x) > 0$, we observe the values of x for which points on the graph of $y = f(x)$ are above the x axis. Since y is positive at each of these points, $f(x) > 0$.

EXAMPLE 1.   Solve the inequality $3 + 2x - x^2 > 0$.

Solution.   Sketch the graph of $f(x) = 3 + 2x - x^2$, as shown in Fig. 8.2. The real solutions of the inequality are those values of x for which points on the graph are above the x axis.

Since the graph lies above the x axis between $(-1,0)$ and $(3,0)$, we conclude that $y = f(x)$ is positive for all values of x between -1 and 3. The solution set is

$$\{x \mid -1 < x < 3\}$$

The graphical method is frequently used to solve inequalities that involve two variables x and y.

EXAMPLE 2.   Solve the inequality $3x - 2y - 6 < 0$.

Solution.   The inequality can be written in the equivalent form

$$y > \frac{3x - 6}{2}$$

and the graph of $y = f(x) = (3x - 6)/2$ can be constructed, as in Fig. 8.3.

FIGURE 8.3 $y > (3x - 6)/2$ for all points lying above the graph of $y = (3x - 6)/2$

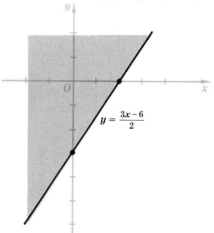

FIGURE 8.2 $f(x) > 0$ for $-1 < x < 3$

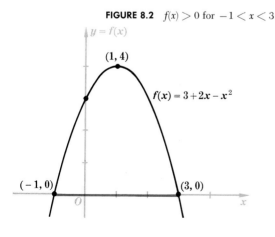

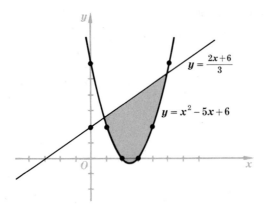

FIGURE 8.4
$y < (2x + 6)/3$ and
$y > x^2 - 5x + 6$ in shaded area

All points lying above the graph have coordinates that satisfy the inequality. Note that the coordinates of points on the graph do not satisfy the inequality. The shaded area of the graph indicates the solution set.

EXAMPLE 3.   Solve the system of inequalities

$$2x - 3y + 6 > 0$$
$$x^2 - 5x + 6 - y < 0$$

Solution.   Here we wish to find all ordered pairs (x,y) which satisfy *both* inequalities.   Write the two inequalities in the forms

$$y < \frac{2x + 6}{3}$$

and $$y > x^2 - 5x + 6$$

and sketch the graphs of the functions as shown in Fig. 8.4.

The coordinates of all points *below* the graph of $y = (2x + 6)/3$ satisfy the linear inequality.   All points *above* the graph of $y = x^2 - 5x + 6$ have coordinates that satisfy the quadratic inequality.   The shaded area indicates the solution set of this system.

$$\{(x,y) \mid 2x - 3y + 6 > 0\} \cap \{(x,y) \mid x^2 - 5x + 6 - y < 0\}$$

EXERCISE SET 8.7

SOLVE each of the following conditional inequalities graphically:

1. $3x + 3 > 6$ 2. $3x - 21 > 0$ 3. $2x + 7 > 3x + 5$
4. $5x - 3 < 7x - 13$ 5. $3x^2 - x < 2$ 6. $2x^2 > 15 - 6x$
7. $6x^2 - 27x > 15$ 8. $x^2 - 4 < 0$ 9. $|x| \leq 1$
10. $|x| \geq 4$ 11. $|2x - 3| > 5$ 12. $5 - 3x \geq x^2$

SOLVE the following systems of inequalities graphically:

13. $3x + 6y > 12$
$\quad\;\; 2x < 3 + y$

14. $x - 2y > 4$
$\quad\;\; x + y < 1$

15. $2x - y > 2$
$\quad\;\; 2x > 2 - y$

16. $y + 1 < x$
$\quad\;\; y + x < 4$

17. $x + 2y \geq 6$
$\quad\;\; 2x - 3y \leq 1$

18. $2x + 3y \leq 12$
$\quad\;\; 2x + 3y \geq 12$

19. $x - y < -2$
$\quad\;\; y - x < -2$

20. $y \leq x$
$\quad\;\; x + y \leq 0$

21. $x^2 - 2x - 1 - y \leq 0$
$\quad\;\; x - y + 2 > 0$

22. $x - 3y \geq 9$
$\quad\;\; 2x + 3y \geq 0$

8.9 POLYNOMIAL FUNCTIONS OF DEGREE n

The function defined by the equation

$$f(x) = a_n x^n + a_{n-1} x^{n-1} + \cdots + a_1 x + a_0 \qquad (8.12)$$

where n is a nonnegative integer and $a_n, a_{n-1}, \ldots, a_0$ are constants, is called a *polynomial function in x*. If $a_n \neq 0$, the polynomial function is said to be of degree n. Since $x^0 = 1$, provided $x \neq 0$, the constant $a_0 \neq 0$ may be written $a_0 x^0$. Thus, we define a nonzero constant to be a polynomial function of degree zero. We also regard the number 0 as a polynomial, but we do not assign a degree to it. Throughout this chapter, the only functions considered will be polynomial functions. The principal problem in this study is that of finding the zeros of $f(x)$. The zeros of $f(x)$ are the numbers in the solution set of the polynomial equation $f(x) = 0$.

Degree

Equal polynomials
Two polynomial functions in x are defined to be equal if and only if the coefficients of the corresponding powers of x are equal. As an example of the definition, we have

$$ax^3 + bx^2 + cx + d = -2x^3 + 5x - 3$$

if and only if $a = -2$, $b = 0$, $c = 5$, and $d = -3$.

In finding the zeros of the polynomial function $f(x)$, we find it convenient to factor $f(x)$ into a product of linear or quadratic factors. Then any value of x that makes at least one of the factors of $f(x)$ equal to zero will be a zero of the function. Linear factors of the form $x - r$ (if there are any) can be found by using *synthetic division*. Synthetic division is an abbreviated form of division that shortens the long-division process.

8.10 SYNTHETIC DIVISION

Long division
In order to understand the useful process called synthetic division, let us first use long division to divide $3x^3 - 11x^2 - 1$ by $x - 3$. Thus,

$$
\begin{array}{r}
3x^2 \quad - \ 2x \quad -6 \\
x - 3 \ \overline{\smash{)}3x^3 \quad -11x^2 \quad +0\cdot x \quad -1} \\
3x^3 \quad - \ 9x^2 \\
\hline
- \ 2x^2 \quad +0\cdot x \quad -1 \\
- \ 2x^2 \quad + \ 6x \\
\hline
-6x \quad - \ 1 \\
-6x \quad +18 \\
\hline
-19
\end{array}
$$

first remainder

second remainder

third remainder

Hence,

$$(3x^3 - 11x^2 - 1) \div (x - 3) = 3x^2 - 2x - 6 + \frac{-19}{x - 3}$$

If we omit writing the variables and write only the coefficients of the terms (writing zero for the coefficient of any missing power), our work appears as follows:

$$
\begin{array}{r}
3 \quad - \ 2 \quad -6 \\
1 - 3 \ \overline{\smash{)}3 \quad -11 \quad +0 \quad -1} \\
3 \quad - \ 9 \\
\hline
-2 \quad +0 \quad -1 \\
-2 \quad +6 \\
\hline
-6 \quad - \ 1 \\
-6 \quad +18 \\
\hline
-19
\end{array}
$$

first remainder

second remainder

third remainder

The numbers in boldface type are repetitions of the numbers immediately above. The numbers in boldface are also repetitions of the coefficients of the variables in the quotient. Omitting these repetitions and also omitting 1, the coefficient of x in the divisor, we can exhibit the process in the form

$$
\begin{array}{r}
\underline{-3)} \ 3 \quad -11 \quad +0 \quad -1 \\
- \ 9 \\
\hline
- \ 2 \quad +0 \quad -1 \\
+6 \\
\hline
-6 \quad - \ 1 \\
+18 \\
\hline
-19
\end{array}
$$

Synthetic division

Moving the scattered terms up near the first line, we can write the entire process in the very compact form

$$
\begin{array}{r}
\underline{-3)} \ 3 \quad -11 \quad +0 \quad - \ 1 \qquad \text{line 1}\\
- \ 9 \quad +6 \quad +18 \qquad \text{line 2}\\
\hline
3 \quad - \ 2 \quad -6 \quad -19 \qquad \text{line 3}
\end{array}
$$

The last number in line 3 is the remainder. The other numbers in line 3, from left to right, are the coefficients of the quotient $3x^2 - 2x - 6$. The numbers in line 3 have been obtained by subtracting the detached coefficients in line 2 from the detached coefficients of the terms having the same degree in line 1.

The same result is obtained if we replace -3 with 3 in the divisor and then *add* at each step instead of subtracting. Thus,

$$\begin{array}{rrrrl}
3 & -11 & +0 & -1 & (3 \quad\ \text{line 1} \\
  & 9 & -6 & -18 & \quad\ \text{line 2} \\
\hline
3 & -\ 2 & -6 & -19 & \quad\ \text{line 3}
\end{array}$$

Note that we placed the divisor 3 on the right. This is not necessary, although many prefer it. The first number 3 on line 3 is brought down from line 1. The product of the 3 in the divisor and the first number in line 3 is 9. This product is written in the next position on line 2. The sum of -11 and 9 is -2, and this sum is written on line 3. The product of 3 and -2 is -6 and is written in the next position on line 2. The sum of 0 and -6 is written on line 3. The product of 3 and -6 is -18. We write this product in the next position on line 2. The sum of -1 and -18 is -19, written on line 3.

Study the following examples carefully.

EXAMPLE 1. Divide $(2x^4 - 3x^2 - 7x + 3)$ by $(x - 2)$.

Solution.
$$\begin{array}{rrrrrl}
2 & 0 & -3 & -7 & 3 & (2 \quad\ \text{line 1} \\
  & 4 & 8 & 10 & 6 & \quad\ \text{line 2} \\
\hline
2 & 4 & 5 & 3 & 9 & \quad\ \text{line 3}
\end{array}$$

The first four numbers on line 3 can be used as coefficients to write a polynomial of degree one less than the degree of the dividend. This polynomial, $2x^3 + 4x^2 + 5x + 3$, is the quotient polynomial. The last number on line 3 is the remainder. Hence, if $x - 2 \neq 0$,

$$(2x^4 - 3x^2 - 7x + 3) \div (x - 2) = 2x^3 + 4x^2 + 5x + 3 + \frac{9}{x - 2}$$

EXAMPLE 2. Divide $(x^3 - 2x^2 + 4x + 1)$ by $(x + 2)$.

Solution. Since $x + 2 = x - (-2)$, we have

$$\begin{array}{rrrrl}
1 & -2 & 4 & 1 & (-2 \\
  & -2 & 8 & -24 & \\
\hline
1 & -4 & 12 & -23 &
\end{array}$$

Hence,

$$(x^3 - 2x^2 + 4x + 1) \div (x + 2) = x^2 - 4x + 12 + \frac{-23}{x + 2}$$

EXAMPLE 3.   Divide $(8x^4 - 4x^3 + 2x - 1)$ by $(x - \frac{1}{2})$.

Solution. $\begin{array}{rrrrr} 8 & -4 & 0 & 2 & -1 \\ & 4 & 0 & 0 & 1 \\ \hline 8 & 0 & 0 & 2 & 0 \end{array}$ $(\frac{1}{2}$

Hence,

$$(8x^4 - 4x^3 + 2x - 1) \div \left(x - \frac{1}{2}\right) = 8x^3 + 0 \cdot x^2 + 0 \cdot x + 2 + \frac{0}{x - \frac{1}{2}}$$

$$= 8x^3 + 2$$

Since the remainder is zero, we know that $x - \frac{1}{2}$ is a factor of

$$8x^4 - 4x^3 + 2x - 1$$

EXERCISE SET 8.8

DIVIDE the first function by the second, using synthetic division.   Write answers in the form

$$f(x) \div (x - r) = q(x) + \frac{R}{x - r}$$

where $q(x)$ is the quotient and R is the constant remainder.

1. $x^2 - 5x + 2, x - 4$
2. $x^3 - 2x^2 + 4x + 1, x - 2$
3. $x^3 + 4x - 7, x - 3$
4. $x^4 - 2x^3 - 3x^2 - 4x - 8, x - 2$
5. $x^3 - 2x + 3x^2 - 5, x + 3$
6. $x^4 + 8x^2 - 5x^3 - 2 + 15x, x - 3$
7. $x^3 + 2x^2 - 25x - 25, x - 5$
8. $9x^3 + 6x^2 + 3x + 9, 3x - 3$.   HINT:

$$\frac{9x^3 + 6x^2 + 3x + 9}{3x - 3} = \frac{3x^3 + 2x^2 + x + 3}{x - 1}$$

9. $4x^3 - 6x^2 + 2x + 1, 2x - 1$
10. $3x^4 + x^3 + 2x^2 + x - 1, x - i$
11. $2x^4 + 6x^3 + 6x^2 - 12x - 20, x - \sqrt{2}$
12. $3x^4 + x^3 - 17x^2 - 5x + 10, x + \sqrt{5}$

8.11   REMAINDER AND FACTOR THEOREMS

If we divide $f(x) = x^3 - 7x^2 + 11x + 2$ by $x - 5$, where $x \neq 5$, we obtain a quotient of $x^2 - 2x + 1$ and a remainder of 7.   Thus

$$\frac{f(x)}{x - 5} = \frac{x^3 - 7x^2 + 11x + 2}{x - 5} = x^2 - 2x + 1 + \frac{7}{x - 5}$$

for every $x \neq 5$.   Now if we multiply each member of the equation by $x - 5$, we have

$$f(x) = x^3 - 7x^2 + 11x + 2 = (x - 5)(x^2 - 2x + 1) + 7$$

Note that the degree of the quotient is 1 less than the degree of the dividend, and the remainder is a constant. This example illustrates the following important theorem which we state without proof.

THEOREM. If $f(x)$ is any real polynomial and r is any real number, then there exists a unique real polynomial $q(x)$ and a real number R such that

$$f(x) = (x - r)q(x) + R$$

and the degree of $q(x)$ is 1 less than the degree of $f(x)$.

(8.13)

The polynomial $q(x)$ is called the quotient, and R is called the remainder. It is understood that R may be zero. The equation in the theorem is an identity, and therefore it is true for every value of x, including $x = r$.

The two theorems which follow are quite useful in finding the zeros of real polynomial functions.

THEOREM (THE REMAINDER THEOREM). If the nonconstant polynomial $f(x)$ is divided by $x - r$, r any constant, until a remainder is obtained which does not involve x, this remainder is equal to $f(r)$.

(8.14)

Proof. By the preceding theorem, there exists a real polynomial $q(x)$ and a real number R such that

$$f(x) = (x - r)q(x) + R \qquad \text{for every value of } x$$

Hence, $\qquad f(r) = (r - r)q(r) + R = 0 \cdot q(r) + R$
$$= R$$

The remainder theorem tells us that the value of the polynomial $f(x)$ at $x = r$ is exactly the same number as the remainder found by dividing $f(x)$ by $x - r$.

EXAMPLE 1. If $f(x) = x^3 + 3x^2 - 2x - 3$, find $f(2)$.

Solution. If we divide $(x^3 + 3x^2 - 2x - 3)$ by $(x - 2)$, the remainder will equal $f(2)$, by the remainder theorem.

$$
\begin{array}{rrrr|}
1 & 3 & -2 & -3 \quad \underline{(2} \\
  & 2 & 10 & 16 \\
\hline
1 & 5 & 8 & 13 \\
\end{array}
$$

Hence, $f(2) = 13$.

The check is easy in this example.   Since

$$f(x) = x^3 + 3x^2 - 2x - 3$$
$$f(2) = (2)^3 + 3(2)^2 - 2(2) - 3 = 13$$

The preceding example illustrates the use of synthetic division in finding the value of $f(x)$ for different values of x.   The advantages of the method will become more apparent as we progress to later sections.   At the moment, we shall consider a second theorem of importance in finding the zeros of a function.

THEOREM (THE FACTOR THEOREM).   The polynomial $x - r$ is a factor of $f(x)$ if and only if   (8.15)

$$f(r) = R = 0$$

Proof.   If $f(r) = R = 0$, then, by the remainder theorem,

$$f(x) = q(x)(x - r) + R = q(x)(x - r)$$

Hence, $x - r$ is a factor of $f(x)$.
   If $x - r$ is a factor of $f(x)$ so that

$$f(x) = q(x)(x - r)$$

then $$f(r) = q(r)(r - r) = q(r) \cdot 0 = 0$$

EXAMPLE 2.   Show that $x - 5$ is a factor of $f(x) = x^3 - x^2 - 25x + 25$.

Solution.   By the factor theorem, $x - 5$ is a factor of

$$f(x) = x^3 - x^2 - 25x + 25$$

if and only if $f(5) = 0$.   To find $f(5)$, we use synthetic division and apply the remainder theorem.

$$
\begin{array}{rrrr|}
1 & -1 & -25 & 25 \quad \underline{(5} \\
  & 5 & 20 & -25 \\
\hline
1 & 4 & -5 & 0
\end{array}
$$

Since the remainder is 0, $f(5) = 0$.   Hence, $x - 5$ is a factor of $f(x)$.

EXAMPLE 3.   Show that $x + 2$ is not a factor of $x^4 - 2x + 1$.

Solution.   Use synthetic division to find $f(-2)$.   If $f(-2) \neq 0$, then

$$x - (-2) = x + 2$$

is not a factor of $f(x)$.

$$
\begin{array}{rrrrr|}
1 & 0 \quad 0 & -2 & 1 & (-2 \\
  & -2 \quad 4 & -8 & 20 & \\
\hline
1 & -2 \quad 4 & -10 & 21 &
\end{array}
$$

Since $f(-2) = 21 \neq 0$, $x + 2$ is not a factor of $f(x)$.

EXAMPLE 4. Show that $x + 1$ is a factor of $x^3 - 4x^2 + x + 6$, and find at least one other factor.

Solution. Using synthetic division,

$$
\begin{array}{rrrr|}
1 & -4 & 1 & 6 & (-1 \\
  & -1 & 5 & -6 & \\
\hline
1 & -5 & 6 & 0 &
\end{array}
$$

we see that $x + 1$ is a factor of $f(x) = x^3 - 4x^2 + x + 6$. The other factor is the quotient obtained by the division of $f(x)$ by $x + 1$ and is $x^2 - 5x + 6$. Hence,

$$
\begin{aligned}
x^3 - 4x^2 + x + 6 &= (x + 1)(x^2 - 5x + 6) \\
&= (x + 1)(x - 2)(x - 3)
\end{aligned}
$$

EXERCISE SET 8.9

FIND $f(r)$, using synthetic division and the remainder theorem:

1. $f(x) = x^3 - 5x + 2$; $r = 1$
2. $f(x) = x^3 - 2x^2 + 4x + 2$; $r = 2$
3. $f(x) = x^3 + 3x^2 + 3x + 1$; $r = -1$
4. $f(x) = 8x^4 + 4x^3 + 2x + 1$; $r = -\frac{1}{2}$
5. $f(x) = x^5 + 32$; $r = 2$
6. $f(x) = 2x^4 - 4x^3 + 9x^2 + 2x - 5$; $r = 1 + 2i$

DETERMINE in each of the following whether or not the second polynomial is a factor of the first. If it is a factor, write the first polynomial in factored form.

7. $x^3 - x^2 - 11x + 15$, $x - 3$
8. $x^4 + 3x^3 - 5x^2 + 2x - 24$, $x - 2$
9. $x^4 - 5x^3 + 8x^2 + 15x - 2$, $x - 3$
10. $x^3 + 2x^2 - 3x - 1$, $x - 1$
11. $2x^4 + 5x^3 + 3x^2 + 8x + 12$, $x + 3$
12. $x^3 - 4x^2 - 18x + 9$, $x + 3$
13. $x^4 - 4x^3 - x^2 + 16x - 12$, $x + 1$
14. $x^4 - 16y^8$, $x - 2y$
15. $x^3 + 2x^2 - 25x - 50$, $x - 5$
16. $x^5 - 10x^4 - 24x$, $x + 2$
17. $2x^4 - 31x^3 + 21x^2 - 17x + 10$, $x + 1$
18. $12x^4 - 40x^3 - x^2 + 111x - 90$, $2x - 3$
19. $2x^4 + 5x^3 + 3x^2 + 8x + 12$, $2x + 3$
20. $9x^3 + 6x^2 + 4x + 2$, $3x + 1$

21. Show that $x - y$ is a factor of $x^5 - y^5$.
22. Show that $x - y$ is a factor of

    (a) $x^6 - y^6$ (b) $x^8 - y^8$ (c) $x^7 - y^7$

23. Show that $x + y$ is a factor of

(a) $x^5 + y^5$ (b) $x^7 + y^7$

24. Show that $x + y$ is not a factor of $x^6 + y^6$.
25. Show that the zeros of the polynomial $f(x) = x^3 - 19x + 30$ are $-5, 2, 3$.
26. Show that $x - b$ is a factor of $x^n - b^n$ for all $n > 0$, $n \in I$.
27. Show that if n is a positive even integer, then $x + y$ is a factor of $x^n - y^n$.
28. Show that if n is a positive odd integer, then $x + y$ is a divisor of $x^n + y^n$.
29. Find the value of k so that $(x^3 - kx^2 + 5) \div (x - 1)$ has remainder 5.
30. Find k so that $(kx^3 + 5x - k) \div (x + 3)$ has remainder 9.

8.12 UPPER AND LOWER BOUNDS FOR ZEROS OF POLYNOMIALS

We can determine an upper bound and a lower bound for the real zeros of a polynomial function as follows:

Let $f(x) = a_n x^n + a_{n-1} x^{n-1} + \cdots + a_1 x + a_0$ be divided by $x - r$, using synthetic division.

Finding upper and lower bounds

1. If $r > 0$ and all the numbers in the third row are positive, then r is an upper bound for the positive zeros of $f(x)$. Thus, no zero is greater than r.

2. If $r < 0$ and all numbers in the third row alternate in sign, then r is a lower bound for the negative zeros of $f(x)$. This means that no zero is less than r.

To prove these statements we reason thus: In case 1, if r is increased, then each number in the third row (except the first) will be increased. Hence the remainder will be increased. This means that the remainder cannot equal zero for any increase in r. In case 2, if r is decreased, each number in the third row is numerically increased. Hence, the remainder is numerically increased and cannot equal zero for any decrease in r.

EXAMPLE 1. Show that $f(x) = x^3 + 13x - 30$ does not have a real zero greater than 2 or less than -1.

Solution. By synthetic division,

$$
\begin{array}{rrrr|l}
1 & 0 & 13 & -30 & \underline{(2} \\
  & 2 & 4 & 34 & \\
\hline
1 & 2 & 17 & 4 &
\end{array}
$$

Since each number in the third row is positive, there can be no real zero greater than 2. Again

$$
\begin{array}{rrrr|l}
1 & 0 & 13 & -30 & \underline{(-1} \\
  & -1 & 1 & -14 & \\
\hline
1 & -1 & 14 & -44 &
\end{array}
$$

Since the successive numbers in the third row alternate in sign, we conclude that -1 is a lower bound of the real zeros of $f(x)$.

We are now certain that if $f(x) = x^3 + 13x - 30$ has any real zeros, they all lie in the interval $-1 < x < 2$.

Bounds of zeros of a polynomial not unique

It is important to note that there is not a unique upper bound or a unique lower bound for the zeros of a polynomial. There are infinitely many upper bounds and infinitely many lower bounds. In the preceding example, -2, -2.5, -8 are lower bounds, and 5, 9.1, 17 are upper bounds for the zeros of the polynomial.

8.13 FUNDAMENTAL THEOREM OF ALGEBRA

Most proofs of the following fundamental theorem require a knowledge of functions of a complex variable, or other advanced topics. We shall assume the theorem here.

THEOREM (THE FUNDAMENTAL THEOREM OF ALGEBRA). If $f(x)$ is a polynomial of degree $n \geq 1$, with real or complex coefficients, then there exists at least one number b (real or complex) such that (8.16)

$$f(b) = 0$$

With this theorem, we can construct an intuitive proof of the following important theorem:

THEOREM. If $f(x)$ is a polynomial of degree $n > 0$, with real or complex coefficients, then $f(x)$ has exactly n zeros (not necessarily distinct). (8.17)

Proof. By Theorem (8.16), $f(x)$ has at least one zero, say r_1. Then by Theorem (8.15) $x - r_1$ is a factor of $f(x)$. Hence,

$$f(x) = (x - r_1) \cdot q_1(x)$$

where $q_1(x)$ is a polynomial of degree $n - 1$ whose leading coefficient is a_n.

Now $q_1(x)$ has at least one zero, say r_2. Hence,

$$q_1(x) = (x - r_2) \cdot q_2(x)$$

where $q_2(x)$ is of degree $n - 2$ with leading coefficient a_n. Therefore, we can write $f(x)$ as

$$f(x) = (x - r_1) \cdot q_1(x)$$
$$= (x - r_1)(x - r_2) \cdot q_2(x)$$

Since $q_2(x)$ has at least one zero, say r_3, we can write

$$q_2(x) = (x - r_3) \cdot q_3(x)$$

where $q_3(x)$ is of degree $n - 3$ with leading coefficient a_n.

We can continue this process until we have

$$f(x) = (x - r_1)(x - r_2)(x - r_3) \cdots (x - r_n) \cdot q_n(x)$$

where $q_n(x)$ is of degree $n - n = 0$ with leading coefficient a_n. This means that $q_n(x)$ must be a_n. Thus, we can write $f(x)$ in the factored form

$$f(x) = a_n(x - r_1)(x - r_2) \cdots (x - r_n)$$

Since $f(x) = 0$ at $x = r_1, r_2, \ldots, r_n$, it follows that $f(x)$ has at least n zeros (not necessarily distinct).

It remains to be proved that $f(x)$ has no other zeros. We do this by assuming that there is another zero, say r, which is not equal to any of the $r_1, r_2, \ldots, r_n$. Then, since

$$f(x) = a_n(x - r_1)(x - r_2) \cdots (x - r_n)$$

it follows that $f(r) = a_n(r - r_1)(r - r_2) \cdots (r - r_n)$

None of the factors of $f(r)$ is equal to zero, because r is not equal to any of the $r_1, r_2, \ldots, r_n$. Hence, $f(r) \neq 0$. This means that r is not a zero of $f(x)$. We conclude that $f(x)$ has no other zeros than $r_1, r_2, \ldots, r_n$.

Multiplicity of roots   If any factor $x - r_i$ of $f(x)$ occurs m times in the factored form of $f(x)$, then r_i is called a *zero of multiplicity m*. For example,

$$f(x) = (x + 3)^2(x - 2)^3(x - 5)(x - 6)$$

is a polynomial of degree 7. The set of zeros of $f(x)$ is the set

$$\{-3, -3, 2, 2, 2, 5, 6\}$$

We say that -3 is a zero of multiplicity 2 (or that -3 is a *double zero*). Also, 2 is a *triple zero* (or a zero of multiplicity 3).

Since $f(x)$ has at least n zeros and not more than n zeros, we conclude that $f(x)$ has exactly n zeros provided a zero of multiplicity m is counted m times.

EXAMPLE 1.   Find a polynomial $f(x)$ which has integral coefficients and has 2 as a double zero and 3 as a triple zero.

Solution.   One polynomial which has the required zeros is

$$f(x) = (x - 2)^2(x - 3)^3$$
$$= x^5 - 13x^4 + 67x^3 - 171x^2 + 216x - 108$$

EXAMPLE 2. Find the zeros of $f(x) = (x^2 - 4)(x^2 - 5x + 6)$.

Solution. $f(x) = (x + 2)(x - 2)(x - 2)(x - 3)$

By inspection, $f(x)$ will be zero for any value of x that makes a factor equal to zero. Hence, the zeros are -2, 2, 2, 3. Note that 2 is a double zero.

EXAMPLE 3. Find the real zeros of $f(x) = 2x^3 - 5x^2 - 14x + 8$.

Solution. If we can factor $f(x)$ into linear factors, we can determine the zeros by inspection. By synthetic division,

$$
\begin{array}{rrrr|r}
2 & -5 & -14 & 8 & \underline{(-2} \\
 & -4 & 18 & -8 & \\
\hline
2 & -9 & 4 & 0 &
\end{array}
$$

Hence, $x + 2$ is a factor of $f(x)$. The other factor is the quotient $2x^2 - 9x + 4$. Thus,

$$2x^3 - 5x^2 - 14x + 8 = (x + 2)(2x^2 - 9x + 4)$$
$$= (x + 2)(2x - 1)(x - 4)$$

The zeros are -2, ½, 4.

EXERCISE SET 8.10

FIND a set of upper and lower bounds of the real zeros (show that the real zeros of each function lie in the indicated interval):

1. $f(x) = 15x^4 + 41x^3 - 13x - 3;\ -3 < x < 1$
2. $f(x) = 32x^5 + x^2 - x - 2;\ -1 < x < 1$
3. $f(u) = u^3 - 2u^2 - 3u - 24;\ -1 < u < 5$
4. $f(v) = 2v^4 + 3v^3 + 2v^2 - 1;\ -2 < v < ½$

FIND an interval in which the real zeros of each of the following functions must lie, i.e., find a set of upper and lower bounds:

5. $f(x) = 3x^3 - 20x^2 - 5x - 50$ 6. $f(x) = x^3 + 2x^2 - 7x - 8$
7. $f(x) = x^4 - 5x^2 + 6x - 9$ 8. $f(x) = x^3 + 16x - 29$

FIND a polynomial function $f(x)$ with integral coefficients having each of the following sets as zeros:

9. 4 is a double zero and 2 is a single zero.
10. 1 is a single zero and 2 is a zero of multiplicity 3.
11. 2 is a zero of multiplicity 2 and 1 is a zero of multiplicity 3.
12. ½ is a zero of multiplicity 2 and ⅔ is a zero of multiplicity 1.

FIND the equation of lowest degree with rational coefficients having each of the following sets for its solution set:

13. $\{3,1,2\}$ 14. $\{-2,4,-1\}$

15. $\{i\sqrt{2}, -i\sqrt{2}\}$

16. $\{-4 + i\sqrt{3}, -4 - i\sqrt{3}, 3\}$

17. $\{\sqrt{2}, -\sqrt{2}, 2i, -2i\}$

18. $\{i\sqrt{7}, -i\sqrt{7}, \sqrt{2}, \sqrt{2}, -\sqrt{2}, -\sqrt{2}\}$

FIND the real zeros of each of the following:

19. $f(x) = (x - 1)^2(x + 3)^3$

20. $f(x) = (x + 2)^3(x - 1)$

21. $f(x) = (2x - 6)^2(x + 1)^3$

22. $f(x) = (3x + 3)^3(x + 2)(x - 3)^2$

23. $f(x) = x^3 + 3x^2 - 4x - 12$. HINT: Factor $f(x)$ by grouping terms.

24. $f(x) = x^3 + 4x^2 - 11x - 30$. Factor $f(x)$ by using synthetic division. Try as factors $x - 1$, $x - 2$, $x - 3$, etc.

8.14 RATIONAL ZEROS WHEN COEFFICIENTS ARE INTEGERS

If all the coefficients of a polynomial $f(x)$ are integers, an application of the following theorem will save a great deal of time in determining the rational zeros (if there are any). To prove this theorem, we must make use of a theorem from the theory of numbers: Let b and c be integers having no common factors other than ± 1. Then, if c is a factor of the product of two integers a and b, c must be a factor of a. For example, if 2 is a factor of $4 \cdot 3$, then, since 2 and 3 have no common factors other than 1, 2 must be a factor of 4.

THEOREM.   If the rational number (reduced to lowest terms) b/c, $c \neq 0$, is a zero of

$$f(x) = a_n x^n + a_{n-1} x^{n-1} + \cdots + a_0 \qquad (8.18)$$

where the a_i are integers, then b is a factor of a_0, and c is a factor of a_n.

Proof.   Since b/c is a zero of $f(x)$, it follows that

(1) $\qquad a_n \left(\dfrac{b}{c}\right)^n + a_{n-1} \left(\dfrac{b}{c}\right)^{n-1} + \cdots + a_1 \left(\dfrac{b}{c}\right) + a_0 = 0$

Multiplying each side of this equation by c^n, we obtain

(2) $\qquad a_n b^n + a_{n-1} b^{n-1} c + \cdots + a_1 b c^{n-1} + a_0 c^n = 0$

Hence, $\qquad\qquad a_n b^n = -c(a_{n-1} b^{n-1} + \cdots + a_0 c^{n-1})$

Thus, c is a factor of $a_n b^n$. Since b and c have no common factors except ± 1, it follows that c is a factor of a_n.

From Eq. (2) above,

$$a_0 c^n = -b(a_n b^{n-1} + \cdots + a_1 c^{n-1})$$

Hence, b is a factor of $a_0 c^n$, and since b and c have no factors in common except ± 1, we conclude that b is a factor of a_0.

EXAMPLE 1. Find all rational zeros (if there are any) of the polynomial $f(x) = 2x^3 - 3x^2 - 11x + 6$.

Solution. If b/c is a rational zero of $f(x)$, b must be a factor of 6 and c must be a factor of 2. The possible choices for b and c are

$$b \qquad \pm 1, \ \pm 2, \ \pm 3, \ \pm 6$$
$$c \qquad \pm 1, \ \pm 2$$

(Note that it is not necessary to list both positive and negative values of c.) Then the possible values of b/c are

$$\frac{\pm 1}{1}, \ \frac{\pm 1}{2}, \ \frac{\pm 2}{1}, \ \frac{\pm 2}{2}, \ \frac{\pm 3}{1}, \ \frac{\pm 3}{2}, \ \frac{\pm 6}{1}, \ \frac{\pm 6}{2}$$

When duplications are removed, we have the set of all possible rational zeros of $f(x)$

$$\pm \frac{1}{2}, \ \pm 1, \ \pm \frac{3}{2}, \ \pm 2, \ \pm 3, \ \pm 6$$

listed in order of increasing magnitude. We arrange them in this order because we usually test them in the order of increasing magnitude.

Synthetic division shows that $-\frac{1}{2}$ is not a zero of $f(x)$. For $\frac{1}{2}$,

$$
\begin{array}{rrrr}
2 & -3 & -11 & 6 \quad (\frac{1}{2} \\
 & 1 & -1 & -6 \\
\hline
2 & -2 & -12 & 0
\end{array}
$$

Depressed function

Since the remainder is zero, $\frac{1}{2}$ is a zero of $f(x)$ and $x - \frac{1}{2}$ is a factor of $f(x)$. The other factor is $2x^2 - 2x - 12$, which is called a *depressed function* with respect to the given function. Hence,

$$f(x) = \left(x - \frac{1}{2} \right)(2x^2 - 2x - 12)$$

Two additional zeros can now be found by finding the zeros of the depressed function, which is a quadratic function. The two zeros of the depressed function are $-2, 3$. Hence, the set of zeros of $f(x)$ is the set

$$\left\{ -2, \frac{1}{2}, 3 \right\}$$

EXAMPLE 2. Show that $f(x) = x^4 + x^2 + 2x + 6$ has no rational zeros.

Solution. If b/c is a rational zero of $f(x)$, then b is a factor of 6, and c is a factor of 1. The possible choices for b are $\pm 1, \pm 2, \pm 3, \pm 6$ and the possible choices for c are ± 1. Thus, the possible choices for b/c are the same as the possible choices for b. By synthetic division and the remainder and factor theorems, we find that none of the numbers $\pm 1, \pm 2, \pm 3, \pm 6$ is a zero of $f(x)$. Hence, $f(x)$ has no rational zeros. Since $f(x)$ has four zeros, by Theorem (8.17), we conclude that they are either irrational or complex or a combination of irrational and complex numbers. The possibilities are: two pairs of complex roots; one pair of complex roots and two irrational roots; or four irrational roots.

EXERCISE SET 8.11

FIND all zeros of each of the following functions. Find all the rational zeros first, then use the quadratic formulas to find the zeros of any depressed function which may be quadratic.

1. $f(x) = x^3 + 3x^2 - 5x - 39$
2. $f(x) = x^3 - 4x^2 + x + 6$
3. $f(x) = 4x^3 - 11x^2 + x + 1$
4. $f(x) = 10x^4 - 13x^3 + 17x^2 - 26x - 6$
5. $f(v) = 2v^4 + 5v^3 - 11v^2 - 20v + 12$
6. $f(u) = u^4 - 4u^3 + 4u - 1$
7. $f(w) = w^4 - 6w^3 - w^2 + 34w + 8$
8. $f(x) = 12x^4 + 5x^3 + 10x^2 + 5x - 2$
9. $f(z) = 12z^3 - 52z^2 + 61z - 15$
10. $f(z) = z^4 - 4z^3 + 6z^2 - 4z + 1$
11. $f(x) = 20x^5 - 9x^4 - 74x^3 + 30x^2 + 42x - 9$
12. $f(w) = 6w^4 - 13w^3 + 2w^2 - 4w + 15$

FIND all the positive zeros which are less than 2π for each of the following functions:

13. $f(\theta) = \cos^4 \theta - 4 \cos^3 \theta + 4 \cos \theta - 1$
14. $f(t) = \sin^3 t - 4 \sin^2 t + \sin t + 6$
15. $f(t) = 2 \sin^4 t + 5 \sin^3 t + 11 \cos^2 t - 20 \sin t + 1$
16. $f(t) = 2 \sin^3 t - 8 \cos^2 t - \sin t + 4$
17. Show that $\sqrt{3}$ is irrational. HINT: Try to find the rational zeros of $f(x) = x^2 - 3$.
18. Show that $\sqrt[3]{4}$ is irrational.

8.15   GRAPHS OF POLYNOMIAL FUNCTIONS

To sketch the graph of the polynomial $f(x)$, we use synthetic division and the remainder theorem to determine some of the ordered pairs $[x,f(x)] = (x,y)$ of the function. By taking these ordered pairs as coordinates of points on the graph and by assuming that the graph is a smooth continuous curve, we can obtain a fair approximation to the graph.

Since the real zeros of $f(x)$ are the abscissas of points where the graph of $f(x)$ touches or crosses the x axis, the graph can be of distinct service in determining the number and approximate values of the real zeros of the function.

EXAMPLE 1.   Sketch the graph of $f(x) = 12x^3 - 28x^2 - 9x + 10$, and find approximate values of the zeros of the function.

Solution. By use of synthetic division and the remainder theorem,

$$f(-1) = -21$$
$$f\left(-\frac{1}{2}\right) = 16$$
$$f(0) = 10$$
$$f\left(\frac{3}{2}\right) = -26$$
$$f(1) = -15$$
$$f(2) = -24$$
$$f(3) = 55$$

Hence, a subset of ordered pairs of f is the set

$$\left\{(-1,-21),\left(-\frac{1}{2},16\right),(0,10),(1,-15),\ldots,(3,55)\right\}$$

The points whose coordinates are these ordered pairs have been plotted and the graph sketched in Fig. 8.5, where the scale on the y axis is not the same as the scale on the x axis.

On the assumption that the graph of $y = f(x)$ is a smooth, continuous curve, we conclude that there is at least one real zero between -1 and $-\frac{1}{2}$;

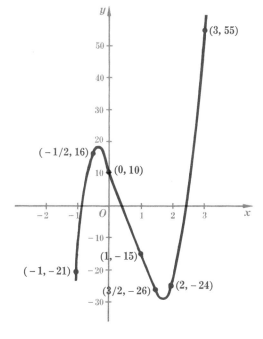

FIGURE 8.5

Graph of
$f(x) = 12x^3 - 28x^2 - 9x + 10$.
What are the approximate values
of the zeros?

at least one zero between 0 and 1; and at least one zero between 2 and 3. Since there can be not more than three zeros in all, we have approximately located all the real zeros of $f(x) = 12x^3 - 28x^2 - 9x + 10$.

In the preceding example, we made use of a theorem which we shall now discuss.

For any real number c, the point $[c, f(c)]$ is on the graph of $y = f(x)$. If $f(c) > 0$, then this point is above the x axis. If $f(c) < 0$, the point is below the x axis. Hence, if a and b are real numbers such that $f(a)$ and $f(b)$ have opposite signs, then one of the points $[a, f(a)]$, $[b, f(b)]$ is above the x axis and the other is below the x axis. Since it can be proved that the graph of $y = f(x)$, where $f(x)$ is a polynomial in x, is a smooth curve (has no gaps or breaks), it follows that the graph must cross the x axis between the point where $x = a$ and the point where $x = b$ (see Fig. 8.6a). We summarize the discussion in the following theorem:

> **THEOREM.** If $f(x)$ is a polynomial function in x with real coefficients, and if a and b are real numbers such that $f(a)$ and $f(b)$ have opposite signs, then there exists a real number c between a and b such that $f(c) = 0$.   (8.19)

Note that Theorem (8.19) does not rule out the possibility of real zeros between $x = a$ and $x = b$ if $f(a)$ and $f(b)$ have the same sign (see Fig. 8.6b). If $f(a)$ and $f(b)$ do have the same sign, however, we conclude from the geometric consideration of the figures that there are an even number of real zeros between $x = a$ and $x = b$, or there are no real zeros between $x = a$ and $x = b$ (see Fig. 8.6c).

The graph of $y = f(x)$ gives us no information about the nonreal zeros of the function.

FIGURE 8.6   If $f(a)$ and $f(b)$ have opposite signs, then the graph of $f(x)$ crosses the x axis between the points $x = a$ and $x = b$.

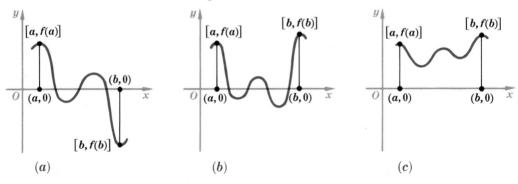

8.16   APPROXIMATING REAL ZEROS OF POLYNOMIALS

There are several procedures that may be used to approximate the real zeros of a polynomial function. The one we shall consider in this book is called the method of successive approximations. It relies heavily upon geometric consideration of the graph of $f(x)$.

If we can find two numbers a and b such that $f(a)$ and $f(b)$ have opposite signs, then the points on the graph of $f(x)$ whose coordinates are $[a,f(a)]$ and $[b,f(b)]$ lie on opposite sides of the x axis. Hence, the graph of $f(x)$ crosses the x axis between $x = a$ and $x = b$. Thus, there is a real zero between a and b.

First approximation   We take as an approximate value of this zero the abscissa of the point where the chord joining the points $[a,f(a)]$ and $[b,f(b)]$ crosses the x axis (see Fig. 8.7).

By using properties of similar triangles, we can deduce from the figure that

$$\frac{h}{b - a} = \frac{|f(a)|}{|f(a)| + |f(b)|}$$

Hence, h can be determined. The value of $a + h$ (assuming that $a < b$) is an approximation of a real zero lying between a and b and may be a better **Second approximation** approximation than either a or b. We can use this new approximation to find two points on the graph of $f(x)$ which are closer together than the original two points and between which a zero must lie. For example, the point $[a + h, f(a + h)]$ and one or the other of the points $[a,f(a)]$, $[b,f(b)]$ lie on opposite sides of the x axis. Hence, there is a zero between $x = a$ and $x = a + h$, or there is a zero between $x = a + h$ and $x = b$.

By continuing the process of finding shorter intervals on the x axis, each

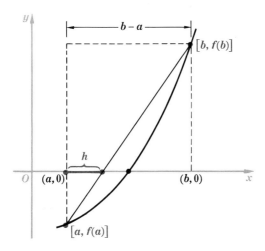

FIGURE 8.7
Finding an approximate value for a real zero of $f(x)$

Further
approximations

of which contains the desired zero, we can approximate this zero to any number of decimal places.

EXAMPLE 1. Find correct to two decimal places the positive zero of

$$f(x) = x^4 + 4x^3 + 2x^2 - 12x - 15$$

Solution. Since $f(0) = -15$, $f(1) = -20$, and $f(2) = 17$, we know that $f(x)$ has a real zero between 1 and 2. By synthetic division, we find that $f(1.5)$ is negative. Thus, the zero we seek is between 1.5 and 2. We find, also, that $f(1.6)$, $f(1.7)$ are negative but that $f(1.8)$ is positive. Specifically,

$$f(1.7) = -1.6159$$

and

$$f(1.8) = 3.7056$$

Hence, the zero lies between 1.7 and 1.8.

By enlarging that portion of the graph of $y = f(x)$ which lies between $x = 1.7$ and $x = 1.8$, as in Fig. 8.8, we see that

$$\frac{h}{0.1} = \frac{|-1.6159|}{|-1.6159| + |3.7056|}$$

Hence, $h = 0.03$, correct to two decimal places. Therefore,

$$1.7 + 0.03 = 1.73$$

is an approximate value of the zero.

We can find a more accurate approximation of the zero of the preceding example as follows. By synthetic division, $f(1.73) = -0.1358$, to four decimal places. This means that the zero is slightly greater than 1.73, as indicated by the fact that the graph is below the x axis at this point. Since

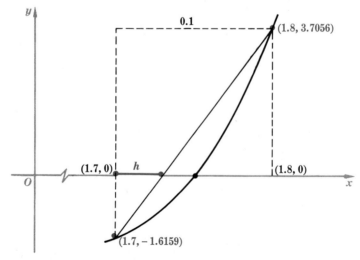

FIGURE 8.8
See Example 1.

$f(1.74) = 0.4136$, to four decimal places, the zero is between 1.73 and 1.74. Again,

$$\frac{h}{0.01} = \frac{|-0.1358|}{|-0.1358| + |0.4136|}$$

Hence, $h = 0.002$, correct to three decimal places, and

$$1.73 + 0.002 = 1.732$$

is an approximate value of a zero of $f(x)$ lying between 1.73 and 1.74.

The method of successive approximations can be applied to functions which are not polynomials.

EXERCISE SET 8.12

SKETCH the graph of each of the following functions showing (between consecutive integer values of x) the location of all the real zeros:

1. $f(x) = x^3 - 5x - 3$
2. $f(x) = x^4 - 9x^2 + 3x + 4$
3. $f(x) = x^3 - 4x^2 - 4x - 7$
4. $f(x) = x^4 + 4x^3 - 37$

SKETCH the graph of each of the following functions and use the method of successive approximations to find the indicated zeros correct to two decimal places:

5. $f(x) = x^3 + 3x - 5$; the positive zero.
6. $f(x) = x^4 - 5x^3 + 2x^2 + x + 7$; the smaller of the positive zeros.
7. $f(x) = x^4 - 2x^3 - 3x^2 - 2x - 4$; the negative zero.
8. $f(x) = x^3 + 6x - 23$; the positive zero.
9. $f(x) = x^3 + 2x + 47$; the negative zero.
10. $f(x) = x^3 - 6x^2 - x + 23$; all real zeros.
11. $f(x) = 2x^4 + 4x^3 + 1$; all irrational zeros.
12. $f(x) = x^5 - 20x + 8$; all irrational zeros.

FIND correct to two decimal places each of the following irrational numbers:

13. $\sqrt[3]{6}$. HINT: $\sqrt[3]{6}$ is the positive zero of $f(x) = x^3 - 6$.
14. $\sqrt[3]{7}$
15. $\sqrt[4]{2}$
16. $\sqrt[5]{-9}$

8.17 COMPLEX ZEROS OF POLYNOMIALS

If the polynomial $f(x)$ has real coefficients, some of the zeros may not be real. For example, the two zeros of $f(x) = x^2 - 5x + 25\frac{1}{2}$ are the complex numbers $\frac{5}{2} + \frac{5}{2} i$ and $\frac{5}{2} - \frac{5}{2} i$. We shall prove the theorem that any complex zeros of a polynomial with real coefficients always occur in conjugate pairs.

THEOREM. If $f(x)$ is a polynomial with real coefficients and if $a + bi$, $b \neq 0$, is a zero of $f(x)$, then $a - bi$ is also a zero of $f(x)$. **(8.20)**

Proof. Since $a + bi$ is a zero of $f(x)$, $f(a + bi) = 0$. Hence, $x - (a + bi)$ is a factor of $f(x)$. We now show that $x - (a - bi)$ is also a factor of $f(x)$.

Divide $f(x)$ by the product

$$[x - (a + bi)][x - (a - bi)] = x^2 - 2ax + (a^2 + b^2)$$

until a remainder of the form $Rx + S$ is obtained. Now R and S are real numbers, since the coefficients of both dividend and divisor are real. Then we can write

$$f(x) = q(x) \cdot [x^2 - 2ax + a^2 + b^2] + Rx + S$$

for all values of x.

Since $f(a + bi) = 0$, we have

$$f(a + bi)$$
$$= q(a + bi) \cdot [(a + bi)^2 - 2a(a + bi) + a^2 + b^2] + R \cdot (a + bi) + S = 0$$

Thus, $\qquad\qquad\qquad q(a + bi) \cdot 0 + R \cdot (a + bi) + S = 0$

and $\qquad\qquad\qquad\qquad (Ra + S) + (Rb)i = 0$

Since two complex numbers are equal if and only if their real parts are equal and their imaginary parts are equal,

$$Ra + S = 0 \qquad \text{and} \qquad Rb = 0$$

Now, we are given that $b \neq 0$. Therefore

$$R = 0 \qquad \text{from} \qquad Rb = 0$$

and $\qquad\qquad\qquad S = 0 \qquad \text{from} \qquad Ra + S = 0$

Thus, we conclude that when $f(x)$ is divided by

$$[x - (a + bi)][x - (a - bi)]$$

the remainder is 0. Hence, $x - (a - bi)$ is a factor of $f(x)$, and $a - bi$ is a zero of $f(x)$.

EXAMPLE 1. Find a polynomial function with real coefficients having 5 and $3 - 2i$ as zeros.

Solution. Since $3 - 2i$ is a zero, it follows from Theorem (8.20) that $3 + 2i$ is also a zero. Hence, the factors of $f(x)$ are

$$x - 5, \ x - (3 - 2i), \ x - (3 + 2i)$$

and $\qquad f(x) = (x - 5)[x - (3 - 2i)][x - (3 + 2i)]$
$$= (x - 5)(x^2 - 6x + 13) = x^3 - 11x^2 + 43x - 65$$

EXAMPLE 2. If $2 + 3i$ is a zero of the function $f(x) = x^3 - 2x^2 + 5x + 26$, find all the zeros.

Solution.   Since $2 + 3i$ is a zero, $2 - 3i$ is also a zero, and

$$[x - (2 + 3i)][x - (2 - 3i)] = x^2 - 4x + 13$$

is a factor of $f(x)$.   By long division, we find that the quotient of $f(x)$ divided by $x^2 - 4x + 13$ is $x + 2$.   Thus, $x + 2$ is also a factor of $f(x)$.   Hence,

$$f(x) = (x + 2)[x - (2 + 3i)][x - (2 - 3i)]$$

The set of zeros is the set

$$\{-2, 2 + 3i, 2 - 3i\}$$

EXERCISE SET 8.13

FIND another zero, if possible, if $f(x)$ is a polynomial with real coefficients and has the given number as a zero:

1. $2i$ 2. $-3i$ 3. $1 + 3i$
4. $(-3 - 2i)/5$ 5. 7 6. $(2 + 7i)/3$

FIND a polynomial with integral coefficients having the given numbers as zeros:

7. $2, 1 - i$ 8. $-3, 2 + i$
9. $1, 2, 1 + i$ 10. $2, 2, -1, -1 - i$

FIND all the other zeros of the functions having the indicated zeros:

11. $f(x) = 2x^3 + 3x^2 + 4x + 6$; $-\frac{3}{2}$ 12. $f(x) = 2x^4 + x^3 - 2x - 1$; $-\frac{1}{2}, 1$
13. $f(x) = x^3 - 4x^2 + 9x - 36$; $3i$ 14. $f(x) = 2x^3 + 9x^2 + 14x + 5$; $-2 - i$
15. $f(x) = x^4 - 3x^3 + 3x^2 - 2$; $1 - i$
16. $f(x) = x^4 + 4x^3 + 14x^2 + 20x + 25$; $-1 - 2i$

FIND all zeros of each of the following:

17. $f(x) = 2x^3 + 3x^2 + 2x + 3$ 18. $f(x) = x^3 - 8x^2 + 23x - 22$
19. Let a and b be rational numbers such that $\sqrt{b}$ is irrational and let $f(x)$ be a polynomial with rational coefficients.   Prove that if $a + \sqrt{b}$ is a zero of $f(x)$, then $a - \sqrt{b}$ is also a zero of $f(x)$.
HINT: Develop a proof similar to the proof of Theorem (8.20).

FIND another zero if possible (each of the following numbers is a zero of a polynomial with rational coefficients):

20. $\sqrt{5}$ 21. $\frac{1}{2}\sqrt{13}$ 22. 3
23. $(2 + \sqrt{3})/2$ 24. $-1 - 2\sqrt{2}$ 25. $-3 + 5\sqrt{2}$

FORM a polynomial with integral coefficients having the following numbers as zeros:

26. $\sqrt{3}, i$ 27. $\frac{1}{2}, 1 - i\sqrt{3}$
28. $-\sqrt{5}, 2 + i\sqrt{3}$ 29. $1 + 2\sqrt{3}, i$

FIND the real zeros of the following functions of x where $0 \le x < 2\pi$:

30. $f(x) = \sin^3 x + 3 \sin^2 x - 4$ 31. $f(x) = 2 \cos^3 x - 3 \cos^2 x - 3 \cos x + 2$
32. $f(x) = \cos^3 x - 3 \sin^2 x - \cos x$ 33. $f(x) = \sec^3 x - 7 \tan^2 x + 15 \sec x - 16$

LOGARITHMIC FUNCTIONS

We have considered in some detail (Sec. 4.5) the exponential function $f(x) = b^x$, $x \in R$. Is the inverse of this relation also a function? If it is a function, what are some of its properties and what are some of its applications? We shall attempt to answer these questions in the following discussions.

9.1   LOGARITHMIC FUNCTION

The exponential function defined by the equation $y = b^x$, $b > 0$, $b \neq 1$, is a subset of $R \times R$, that is, $f = \{(x, b^x) \mid x \in R\}$. Some examples of the ordered pairs of f are

$$(1, b), (2, b^2), \left(\frac{3}{2}, b^{3/2}\right), (\sqrt{10}, b^{\sqrt{10}})$$

Since no two distinct ordered pairs of f have the same second coordinate, by Theorem (4.13), the inverse relation $f^{-1} = \{(b^x, x) \mid x \in R\}$ is a function. Some of the ordered pairs of f^{-1} are

$$(b, 1), (b^2, 2), \left(b^{3/2}, \frac{3}{2}\right), (b^{\sqrt{10}}, \sqrt{10})$$

Logarithmic function to base b   We call f^{-1} the *logarithmic function of the base b*.

Let us now consider the problem of determining the ordered pairs of f^{-1}. We recall (Sec. 3.8) that the equation which defines the inverse function may be found by interchanging the first and second coordinates of each ordered pair of f. Hence, for the exponential function defined by $y = b^x$, we have

$$x = b^y$$

as the equation which defines the inverse function. It is important to note that for the function $x = b^y$ there is exactly one value of y corresponding to any positive value of x. We call this unique value of y the *logarithm of x to the base b*, and write it as $\log_b x$.

DEFINITION. If $x = b^y$, $b > 0$, $b \neq 1$, then

$$y = \log_b x \tag{9.1}$$

We read the symbol $\log_b x$ as "the logarithm of x to the base b."

The two equations $x = b^y$ and $y = \log_b x$ are equivalent, and we may use whichever is most convenient for our purposes. In the following arrangement, the exponential equation on the left is equivalent to the corresponding logarithmic equation on the right.

Exponential Form	Equivalent Logarithmic Form
$2^4 = 16$	$\log_2 16 = 4$
$9^{3/2} = 27$	$\log_9 27 = \frac{3}{2}$
$8^{-2} = \frac{1}{64}$	$\log_8 \frac{1}{64} = -2$
$b^0 = 1$	$\log_b 1 = 0$
$b^1 = b$	$\log_b b = 1$

EXAMPLE 1.   Find the value of b if $\log_b 9 = \frac{2}{3}$.

Solution.   Write $\log_b 9 = \frac{2}{3}$ in the equivalent form

$$b^{2/3} = 9$$

Then

$$(b^{2/3})^{3/2} = 9^{3/2}$$

and

$$b = 27$$

EXAMPLE 2.   Find the value of $\log_{1/4} 32$.

Solution.   Let $x = \log_{1/4} 32$.

Then

$$\left(\frac{1}{4}\right)^x = 32$$

since $\frac{1}{4} = 2^{-2}$, we have

$$(2^{-2})^x = 32 = 2^5$$

Hence

$$-2x = 5 \qquad \text{and} \qquad x = -\frac{5}{2}$$

A useful property of the logarithmic function can be derived from Definition (9.1) as follows: If $y = \log_b x$, then

$$b^y = x$$

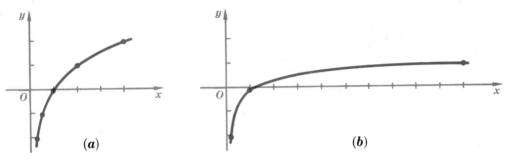

FIGURE 9.1   Graphs of (a) $y = \log_2 x$; (b) $y = \log_{10} x$

Hence, $$b^{\log_b x} = x$$ (9.2)

For example, $$5^{\log_5 N} = N$$
$$13^{\log_{13} p} = p$$
$$8^{\log_8 (x^2 - 2x - 3)} = x^2 - 2x - 3$$

**Graphing
logarithmic functions**

To sketch the graph of a logarithmic function, for example, $y = \log_2 x$, we sketch the graph of the equivalent exponential function, in this case $x = 2^y$. A convenient subset of the ordered pairs $(x, 2^y)$ of the exponential function are found to be

$$\left\{ \left(\tfrac{1}{4}, -2\right), \left(\tfrac{1}{2}, -1\right), (1,0), (2,1), (4,2) \right\}$$

When the points whose coordinates are these ordered pairs are plotted and connected by a smooth curve, the approximate graph is obtained, Fig. 9.1a.

In a similar manner, the graph of $y = \log_{10} x$ has been constructed, Fig. 9.1b.

The graph of the logarithmic function to any base b, $b > 1$, has the same general shape and characteristics as the graphs shown in Fig. 9.1. Two properties of this function illustrated by the graphs are the following:

1.   If $x > 1$, then $\log_b x > 0$.   For example, $\log_2 3 > 0$.
2.   If $x < 1$, then $\log_b x < 0$.   For example, $\log_2 (\tfrac{1}{2}) < 0$.

9.2   PROPERTIES OF LOGARITHMS

We now consider three basic properties of logarithms which are often called the *laws of logarithms:*

1.   **THE LOGARITHM OF A PRODUCT**

$$\log_a (MN) = \log_a M + \log_a N$$

Proof. Let $x = \log_a M$ and let $y = \log_a N$. Then

$$M = a^x \quad \text{and} \quad N = a^y \qquad \text{by Def. (9.1)}$$

and
$$MN = a^x \cdot a^y = a^{x+y} \qquad \text{why?}$$

Hence
$$\log_a (MN) = x + y \qquad \text{Def. (9.1)}$$
$$= \log_a M + \log_a N \qquad \text{by substitution}$$

EXAMPLE 1. Find $\log_2 (8)(32)$.

Solution. $\log_2 (8)(32) = \log_2 8 + \log_2 32 = 3 + 5 = 8$

EXAMPLE 2. If $\log_a 5 = 0.6990$ and $\log_a 6 = 0.7782$, find $\log_a 30$.

Solution. $\log_a 30 = \log_a (5)(6) = \log_a 5 + \log_a 6$
$$= 0.6990 + 0.7782$$
$$= 1.4772$$

2. THE LOGARITHM OF A QUOTIENT

$$\log_a \frac{M}{N} = \log_a M - \log_a N$$

Proof. The proof is left as an exercise.

EXAMPLE 3. Find the value of $\log_4 100 - \log_4 50$.

Solution. $\log_4 100 - \log_4 50 = \log_4 {}^{100}\!/_{50} = \log_4 2 = \frac{1}{2}$

EXAMPLE 4. If $\log_a 24 = 1.3802$ and $\log_a 6 = 0.7782$, find $\log_a 4$.

Solution. $\log_a 4 = \log_a {}^{24}\!/_{6} = \log_a 24 - \log_a 6$
$$= 1.3802 - 0.7782 = 0.6020$$

3. THE LOGARITHM OF A POWER

$$\log_a M^p = p \log_a M \qquad M > 0$$

Proof. Let $x = \log_a M$. Then

$$M = a^x \qquad \text{why?}$$

and
$$M^p = (a^x)^p = a^{px} \qquad \text{why?}$$

Therefore, by Definition (9.1),

$$\log_a M^p = px$$
$$= p \log_a M \qquad \text{why?}$$

EXAMPLE 5. $\log_{10} (15)^{20} = 20 \log_{10} 15$

EXAMPLE 6.   Express $3 - \log_4 9$ as the logarithm of a single number.

Solution.   Since $3 = 3 \log_4 4 = \log_4 4^3$,

$$3 - \log_4 9 = \log_4 4^3 - \log_4 9 = \log_4 \frac{4^3}{9} = \log_4 \frac{64}{9}$$

EXAMPLE 7.   Express $3 \log_{10} 5 - 4 \log_{10} x + \log_{10} y^2$ as the logarithm of a single number.

Solution.

$$3 \log_{10} 5 - 4 \log_{10} x + \log_{10} y^2 = \log_{10} 5^3 + \log_{10} y^2 - \log_{10} x^4$$
$$= \log_{10} (5^3 \cdot y^2) - \log_{10} x^4$$
$$= \log_{10} \left(\frac{5^3 \cdot y^2}{x^4} \right) = \log_{10} \left(\frac{125 y^2}{x^4} \right)$$

We have always assumed that $b > 0$, $b \neq 1$.   Since $b^x > 0$ for each real number x, the equation $b^y = x$ does not have a solution if $x \leq 0$.   Hence, neither the logarithm of zero nor the logarithm of a negative number is defined over the set of real numbers.

Additional properties of the logarithmic function are contained in problems to be found in the next set of exercises.   These important properties include the following:
1.   If $x = 1$, then $\log_b x = 0$.
2.   If $x = b$, then $\log_b x = 1$.

EXERCISE SET 9.1

1.   Write the logarithmic equation which is equivalent to each of the following exponential equations:

(a) $10^2 = 100$ (b) $3^5 = 243$ (c) $7^{-1} = \frac{1}{7}$
(d) $16^{1/2} = 4$ (e) $8^{-2/3} = \frac{1}{4}$ (f) $81^{3/4} = 27$
(g) $x = 4^y$ (h) $10^y = x$ (i) $10^{\log_{10} y} = x$

2.   Write the exponential equation which is equivalent to each of the following logarithmic equations:

(a) $\log_5 125 = 3$ (b) $\log_2 128 = 7$ (c) $\log_8 4 = \frac{2}{3}$
(d) $\log_2 \frac{1}{64} = -6$ (e) $\log_4 8 = \frac{3}{2}$ (f) $\log_7 343 = 3$
(g) $\log_{10} 0.001 = -3$ (h) $\log_9 729 = 3$

3.   Find the rational values of each of the following:

(a) $\log_2 8$ (b) $\log_4 2$ (c) $\log_6 6$
(d) $\log_{10} 0.01$ (e) $\log_3 81$ (f) $\log_4 \frac{1}{16}$

4. In each of the following equations find the value of x:

(a) $\log_3 81 = x$ (b) $\log_2 \frac{1}{8} = x$ (c) $\log_x 64 = 3$

(d) $\log_2 x = 8$ (e) $\log_{1/4} x = 2.5$ (f) $\log_x 0.1 = -1$

(g) $\log_x x = 2$ (h) $\log_{64} x = \frac{7}{6}$

5. If $\log_b 2 = 0.3010$, $\log_b 3 = 0.4771$, $\log_b 5 = 0.6990$, and $\log_b 7 = 0.8451$, find the value of each of the following:

(a) $\log_b 9$ (b) $\log_b 25$ (c) $\log_b 18$

(d) $\log_b 10$ (e) $\log_b 1.5$ (f) $\log_b 7.5$

(g) $\log_b 1.2$ (h) $\log_b 3.6$ (i) $\log_b 10b$

(j) $\log_b \sqrt{6}$ (k) $\log_b \sqrt{\frac{4}{5}}$ (l) $\log_b \sqrt[3]{2.5}$

(m) $\log_b \sqrt{50}$ (n) $\log_b 7^6$ (o) $\log_b 75^{7/4}$

(p) $\log_b \dfrac{\sqrt[3]{63}}{7^{2/5}}$ (q) $\log_b \sqrt[7]{98}$ (r) $\log_b (2^{2/3})(15^{1/2})$

WRITE each of the following as a single logarithm, using the laws of logarithms:

6. $\log_b 3 + \log_b 4 - 2 \log_b 5$ **7.** $\log_b 4 + \log_b \pi + 3 \log_b r - \log_b 3$

8. $\log_b \pi + 2 \log_b r$ **9.** $\frac{1}{3} \log_b 8 + \frac{2}{3} \log_b 5 + \frac{2}{3} \log_b 4$

10. $-2 \log_b 5 + \log_b 13 + \log_b 2^3$ **11.** $-2 \log_b 4 + \frac{1}{3} \log_b 9 + \frac{1}{2} \log_b 7$

12. $\log_b x - \frac{1}{2} \log_b (x + \sqrt{x^2 - a^2}) + \frac{1}{2} \log_b (x - \sqrt{x^2 - a^2})$

PROVE the following, assuming that $b > 0$, $b \neq 1$:

13. That $\log_b (M/N) = \log_b M - \log_b N$, $N \neq 0$.   **14.** That $\log_b \sqrt[q]{x^p} = p/q \log_b x$.

15. That $\log_b 1 = 0$.   HINT: Since $b^0 = 1$, the desired result follows from Definition (9.1).

16. That $\log_b b = 1$.   **17.** That if $x_1 = x_2$, then $\log_b x_1 = \log_b x_2$.

18. That if $\log_b x_1 = \log_b x_2$, then $x_1 = x_2$.   **19.** That if $x_2 > x_1$, then $\log_b x_2 > \log_b x_1$ if $b > 1$.

9.3 COMMON LOGARITHMS

Any positive real number except 1 can be used as the base for a system of logarithms. In general practice, however, only two numbers are in use, the rational number 10 and the irrational number $e = 2.718. \ldots$ The log-

Common logarithm arithm of a number to the base 10 is called the *common logarithm* of the number. In this book, when the base is not indicated in writing the logarithm of a number, we shall understand that the base is 10. That is, we shall consider the symbol log x to mean $\log_{10} x$. Hence, from Definition (9.1),

$$\text{If } 10^y = x \qquad \text{then} \qquad y = \log x \qquad (9.3)$$

We shall also use the word "log" to mean "common logarithm."

From Eq. (9.3) we can deduce that

$$\log 10 = 1$$

and

$$\log 1 = 0$$

Natural logarithm

The logarithm of a number to the base e is called the *natural logarithm* of the number and will be considered in a later section. For the present, let us consider the problem of finding the logarithm of any positive number to base 10.

Every real number r can be written in the form $10^c \cdot M$ where $c \in I$ and $1 \leq M < 10$. For example,

$$0.000484 = 10^{-4}(4.84)$$
$$0.00484 = 10^{-3}(4.84)$$
$$0.0484 = 10^{-2}(4.84)$$
$$0.484 = 10^{-1}(4.84)$$
$$4.84 = 10^0(4.84)$$
$$48.4 = 10^1(4.84)$$
$$484 = 10^2(4.84)$$
$$4,840 = 10^3(4.84)$$

and so on. This notation for writing numbers is called *scientific notation*. Now if $r > 0$ and r is expressed in scientific notation, so that

$$r = 10^c \cdot M \qquad c \in I, 1 \leq M < 10$$

then
$$\log r = \log 10^c + \log M$$
$$= c \log 10 + \log M$$
$$= c + \log M$$

If we let $m = \log M$, we can then write

$$\log r = c + m \tag{9.4}$$

Characteristic and mantissa

We call the integer c the *characteristic* of $\log r$ and the number m the *mantissa* of $\log r$. When $\log r$ is written in the form given in Eq. (9.4), it is said to be in *standard form*.

Since $1 \leq M < 10$, it can be shown that

$$\log 1 \leq \log M < \log 10$$
Hence,
$$0 \leq m < 1$$

The characteristic c of $\log r$ can be found by inspection, as the following examples will show. The mantissa m generally must be obtained from a table of approximations of mantissas. Table III gives the approximate mantissa of the logarithm of each number from 1.00 to 9.99 (at intervals of 0.01), correct to four decimal places.

It is important to note that the mantissa m is irrational in most cases and that *tabulated values of m are only rational approximations*.

The use of Table III is illustrated in the following examples.

EXAMPLE 1. Find log 4,840.

Solution. Since $4,840 = 10^3 (4.84)$,

$$\log 4,840 = \log 10^3 + \log 4.84 = 3 + \log 4.84$$

Hence, $\qquad\qquad c = 3$

To find $m = \log 4.84$ from Table III, locate 4.8 in the column headed n and move to the right to the column headed 4. The entry found there is .6848 and is the required mantissa. Therefore,

$$m = .6848 \text{ (approx.)}$$

and we can write

$$\log 4,840 = 3 + .6848$$
$$= 3.6848 \text{ (approx.)}$$

The following example illustrates a useful method of writing the logarithm of a number between 0 and 1. Logarithms of such numbers have negative characteristics.

EXAMPLE 2. Find $\log 0.00484$.

Solution. Since $0.00484 = 10^{-3} (4.84)$,

$$\log 0.00484 = -3 + \log 4.84$$
$$= -3 + .6848 \qquad \text{from Table III}$$

We could now combine the two terms of the binomial and write $-3 + .6848$ as -2.3152. However, the decimal part of -2.3152 is certainly not the mantissa of $\log 0.00484$. To preserve the identity of the mantissa (and to avoid other difficulties), we resort to the standard practice of simultaneously adding and subtracting some multiple of 10 to the binomial form of the logarithm and using the associative law for real numbers. Thus

$$\log 0.00484 = -3 + .6848 + 10 - 10$$
$$= (10 - 3) + .6848 - 10$$
$$= 7.6848 - 10$$

We could also write

$$\log 0.00484 = 17.6848 - 20$$
$$= 27.6848 - 30$$

and so on.

EXAMPLE 3. Find $\frac{1}{3} \log 0.0768$.

Solution. Since $0.0768 = 10^{-2}(7.68)$,

$$\log 0.0768 = -2 + .8854 \text{ (approx.)} \qquad \text{from Table III}$$
$$= 28.8854 - 30 \text{ (approx.)}$$

Hence, $\dfrac{1}{3}$ log 0.0768 = 9.6285 − 10 (approx.)

We chose the form 28.8854 − 30 so that the −30 is exactly divisible by 3, which is necessary in order to avoid a fractional characteristic.

Note that the mantissa of the preceding example was rounded off to four decimal places when we divided by 3. This was done because the tabulated values are correct to four places only. Any figure in the fifth decimal place resulting from the division is unreliable.

EXAMPLE 4. Find the characteristic and mantissa of log $(0.000935)^3$.

Solution.   log $(0.000935)^3$ = 3 log 0.000935 = 3(−4 + .9708)
$$= 3(6.9708 − 10)$$
$$= 20.9124 − 30 \text{ (approx.)}$$

Hence, the characteristic is −10 and the mantissa is 0.9124.

Antilogarithm

The number corresponding to a given logarithm is called the *antilogarithm* of the given logarithm. Thus, if log $r = c + m$, then r is the antilogarithm of $c + m$. For example, since log 50 = 1.6990, it follows that 50 is the antilogarithm of 1.6990, and we write

$$50 = \text{antilog } 1.6990$$

The use of Table III in finding a number when its logarithm is given is illustrated by the following example.

EXAMPLE 5. Find x if log x = 2.8669, that is, find antilog 2.8669.

Solution. We look for .8669, the given mantissa, in the body of Table III. This entry is located in the column headed 6 and to the right of 7.3 in the column headed n. This tells us that log 7.36 = 0.8669 (approx.), and since the characteristic is 2, we reason that

$$\log x = 2.8669 = 2 + .8669$$
$$= 2 \log 10 + .8669$$
$$= \log 10^2 + \log 7.36$$
$$= \log 10^2(7.36)$$

Hence,   log x = log 736
and x = 736 (approx.)

EXAMPLE 6. Find x if log x = 8.5988 − 10.

Solution. The mantissa of log x is .5988 and corresponds to the number 3.97, from Table III. Since the characteristic is -2, we have

$$\log x = -2 + .5988 = \log 10^{-2} + \log 3.97$$
$$= \log (10^{-2})(3.97) = \log 0.0397$$

Hence, $x = 0.0397$

If the mantissa of a given logarithm is not an entry in Table III, we may obtain a rough estimate of the antilogarithm (to three significant figures) by taking the antilogarithm of the entry closest to the given mantissa. For example, in finding antilog 3.9453, we find the given mantissa is closer to .9455 than it is to .9450 and is between the two. Hence,

$$\text{antilog } 3.9453 = \text{antilog } 3.9455 \text{ (approx.)}$$
$$= 8,820 \text{ (approx.)}$$

The use of logarithms to find approximate solutions to exponential equations is illustrated by the following example.

EXAMPLE 7. Find the approximate value of x if $10^{3x-2} = 40.7$.

Solution. Since $10^{3x-2} = 40.7$,

$$(3x - 2) \log 10 = \log 40.7$$

Hence, $3x - 2 = 1.6096$

$$3x = 3.6096$$

and $x = 1.2032 \text{ (approx.)}$

EXERCISE SET 9.2

1. Find the characteristic of the common logarithm of each of the numbers:

   (a) 27.3   (b) 4.58   (c) 758
   (d) 7240   (e) 97,600   (f) 52,374
   (g) 92,786   (h) 0.0794   (i) 0.305
   (j) 0.0008   (k) 0.842   (l) 0.28569
   (m) sin 30°   (n) cos 45°   (o) tan 60°
   (p) tan 89°50′   (q) cot 3°10′   (r) sec 0°
   (s) log 3,240   (t) log 934,000   (u) log 3,898,000

2. If log 5.76 = 0.7604, find the logarithm of each of the following:

   (a) 57.6   (b) 57,600   (c) 0.576
   (d) 0.000576   (e) 5,760,000   (f) 0.00576

3. Place a decimal point in the sequence of digits 4265 so the logarithm of the resulting number will have a characteristic of

(a) 1

(b) 3

(c) -2

(d) 0

(e) $9 - 10$

(f) $7 - 10$

(g) $5 - 10$

(h) $4 - 10$

(i) -6

4. Use Table III to find the logarithm of each of the following numbers:

(a) 273

(b) 4.53

(c) 0.951

(d) 0.379

(e) 87.5

(f) 7,210

(g) 0.00539

(h) 0.0008

(i) 101

(j) $\cos 60°$

(k) $\tan 45°$

(l) $\sin 30°$

(m) $27(380)$

(n) $425/275$

(o) $(47)^3$

(p) $(24)^{1/3}$

(q) $(16)^{2/5}$

(r) $(48)^{1/4}$

5. Use Table III to find the antilogarithm of each of the following logarithms:

(a) 0.6021

(b) 1.6170

(c) 2.7218

(d) 3.6609

(e) $9.5539 - 10$

(f) $9.4857 - 10$

(g) $8.8876 - 10$

(h) $7.9004 - 10$

(i) $9.7973 - 10$

(j) $9.8756 - 10$

(k) $8.8899 - 10$

(l) 1.0043

(m) 3.0170

(n) 2.0086

(o) 0.0212

6. Find the approximate value of x in each of the following equations. If an antilogarithm is to be found, take the entry in Table III which is closest to the given mantissa.

(a) $10^x = 5$

(b) $10^x = 75$

(c) $10^{3x} = 4.02$

(d) $10^{4x} = 32$

(e) $x = 10^{2/5}$

(f) $x = 10^{1/4}$

(g) $x = 10^{1+\sqrt{3}}$

(h) $x = 10^{\pi}$

(i) $x = 10^{1.4}$

(j) $10^{2x+1} = 8.89$

(k) $10^{3x-2} = 5.12$

(l) $10^{-x} = 0.0074$

(m) $10^{-x} = 0.82$

(n) $10^{-x/2} = 969$

(o) $10^{-x/3} = 74,600$

9.4   INTERPOLATION

Table III is a four-place table of mantissas, and from it we can read directly the approximate value of the mantissa of the logarithm of any positive number of three significant digits or less. We do not define the logarithm of a negative number (Sec. 9.2). When a number contains four significant digits, the mantissa of its logarithm may be approximated by interpolation. The principle is similar to that used in finding values of the trigonometric functions by interpolation.

EXAMPLE 1.   Find log 0.04786.

Solution.   Since $0.04786 = 10^{-2}(4.786)$,

$$\log 0.04786 = -2 + \log 4.786$$

Hence, $c = -2$

To find $m = \log 4.786$ by interpolation, we reason that since 4.786 is $.006/.010 = {}^6\!/_{10}$ of the way from 4.780 to 4.790, then $\log 4.786$ is ${}^6\!/_{10}$ of the way between $\log 4.780$ and $\log 4.790$. Hence,

$$\log 4.786 = \log 4.780 + \frac{6}{10}(\log 4.790 - \log 4.780)$$

$$= 0.6794 + \frac{6}{10}(0.6803 - 0.6794)$$

$$= 0.6799 \text{ (approx.)}$$

Since $c = -2$ and $m = .6799$, we see that

$$\log 0.04786 = -2 + .6799$$
$$= 8.6799 - 10 \text{ (approx.)}$$

The following tabular arrangement may be used if preferred:

$$\begin{array}{ccc} & \text{Number} & \text{Mantissa} \\ & \left.\begin{array}{c} 4.780 \\ 4.786 \\ 4.790 \end{array}\right. & \left.\begin{array}{c} .6794 \\ .6794 + d \\ .6803 \end{array}\right. \end{array}$$

$.010\left\{ .006\left\{ \begin{array}{c} 4.780 \quad .6794 \\ 4.786 \quad .6794 + d \end{array}\right\} d \\ 4.790 \quad .6803 \end{array}\right\} .0009$

Assume
$$\frac{.006}{.010} = \frac{d}{.0009}$$

$$d = .00054 = .0005 \qquad \text{(to four decimal places)}$$

Hence, $\qquad m = .6794 + .0005 = .6799 \text{ (approx.)}$

and $\qquad \log 0.04786 = -2 + .6799 = 8.6799 - 10 \text{ (approx.)}$

In the preceding example, the assumption $.006/.010 = d/.0009$ is close enough to give $\log 0.04786 = 8.6799 - 10$ almost correct to four decimal places *but not more.* The result of computing with approximate numbers cannot be more accurate than the numbers used. That is why we round off $d = .00054$ to $d = .0005$.

Any mantissa found from Table III by interpolation must be rounded off to four decimal places, and any antilogarithm must be rounded off to not more than four significant digits.

To find the logarithm of a number having more than four significant figures, we first round off the number to four significant figures and then interpolate. For example, to find $\log 71{,}492$ we round off the number to 71,490 and then interpolate as in the example above.

EXAMPLE 2. Find $\log 56{,}984$.

Solution. Round off 56,984 to 56,980. Then

$$\log 56{,}980 = \log (10^4)(5.698) = 4 + \log 5.698$$

Hence, $\qquad c = 4$

To find log 5.698, we may arrange our work thus:

$$\begin{array}{cc} & \text{Number} \quad \text{Mantissa} \end{array}$$

$$.010 \left\{ .008 \begin{cases} 5.690 & .7551 \\ 5.698 & .7551 + x \\ 5.700 & .7559 \end{cases} x \right\} .0008$$

$$\frac{.008}{.010} = \frac{x}{.0008}$$

$$x = .0006 \qquad \text{(to four decimal places)}$$

Hence, $m = .7551 + .0006 = .7557$

and $\log 56{,}984 = \log 56{,}980 \text{ (approx.)}$

$$= 4.7557 \text{ (approx.)}$$

Antilogarithms whose mantissas are not entered in Table III are also found by interpolation.

EXAMPLE 3.   Find N if $\log N = 8.9720 - 10$.

Solution.   Here, $c = -2$, $m = .9720$.   The given mantissa, .9720, is be-tween the entries .9717 and .9722 of Table III.   Now, since antilog .9717 = 9.370 and antilog .9722 = 9.380, we reason that $M = $ antilog .9720 is between 9.370 and 9.380.   We arrange our work thus:

$$\begin{array}{cc} & \text{Number} \quad \text{Mantissa} \end{array}$$

$$.010 \left\{ x \begin{cases} 9.370 & .9717 \\ 9.370 + x & .9720 \\ 9.380 & .9722 \end{cases} .0003 \right\} .0005$$

$$\frac{x}{.010} = \frac{.0003}{.0005}$$

$$x = .006$$

Hence, $M = 9.370 + .006 = 9.376$

and $N = 10^{-2}(9.376) = 0.09376 \text{ (approx.)} = 10^c \cdot M$

9.5   COMPUTING WITH LOGARITHMS

The importance of logarithms in calculating products, quotients, powers, and roots of numbers has decreased with the development of high-speed electronic computers.   The logarithmic function, however, has not diminished in importance in mathematics.   We can increase our under-standing of this function by applying its properties to problems that in-volve somewhat complicated computations, as the following examples show.

In computation problems we frequently must decide on how many digits to retain in the answers.   Let us assume here that the accuracy of the data in a problem is such that an answer to four digits is warranted.

EXAMPLE 1.   Find the approximate value of $(^{34}\!/_{27})^{5.8}$.

Solution.   Let $N = (^{34}\!/_{27})^{5.8}$; then

$$\log N = 5.8 \log \frac{34}{27} = 5.8(\log 34 - \log 27)$$

$$\log 34 = 1.5315$$
$$\log 27 = \underline{1.4314}$$
$$0.1001 \qquad \text{difference}$$

Hence,
$$\log N = 5.8(0.1001)$$
$$= 0.5806 \qquad \text{to four decimal places}$$

Therefore,
$$N = 3.807 \text{ (approx.)} \qquad \text{by interpolation}$$

EXAMPLE 2.   Find N if $N = \dfrac{2.78(3.41)^2}{\sqrt[3]{7.84}}$.

Solution.
$$\log N = \log 2.78 + 2 \log 3.41 - \tfrac{1}{3} \log 7.84$$

$$\log 2.78 = 0.4440$$
$$2 \log 3.41 = \underline{1.0656}$$
$$1.5096 \qquad \text{log of numerator}$$

$$\frac{1}{3} \log 7.84 = \underline{0.2981} \qquad \text{log of denominator}$$
$$1.2115 \qquad \text{difference}$$

Hence,
$$\log N = 1.2115$$
and
$$N = 16.27 \text{ (approx.)} \qquad \text{after interpolation}$$

EXERCISE SET 9.3

1.   Use interpolation to find the logarithm of each of the following:

(a) 1,256

(b) 275.4

(c) 57.62

(d) 0.8976

(e) 0.04827

(f) 215,600

2.   Use interpolation to find the antilogarithm of each of the following:

(a) 0.4978

(b) 1.6004

(c) 9.3936 − 10

(d) 8.9312 − 10

(e) 9.9765 − 10

(f) 7.9210 − 10

3. Use logarithms to find the value of each of the following:

(a) $(426)^5$ (b) $(72.8)^3$ (c) $(85.4)^2(48.7)^3$

(d) $(3.673)^4(3.14)^2$ (e) $(21.7)^{1/3}$ (f) $(66.3)^{1/5}$

4. Evaluate each of the following:

(a) $\sqrt{720}$ (b) $\sqrt{2350}$ (c) $\sqrt{0.0947}$

(d) $\sqrt{0.00429}$ (e) $\sqrt[3]{18.7}$ (f) $\sqrt[3]{75.8}$

(g) $\sqrt[4]{4.78}$ (h) $\sqrt[5]{-28.9}$ (i) $(25.7)^{-0.2}$

(j) $(4.23)^{-0.3}$ (k) $(0.196)^{-0.6}$ (l) $(7.36)^{1.4}$

(m) $(56.72)^{2.8}$ (n) $(3.1416)^{3.7}$

5. Use logarithms to perform the indicated operations:

(a) $\sqrt{9183}(\sqrt[3]{5.27})$ (b) $\sqrt[3]{-0.42}(\sqrt[4]{8.1})$

(c) $\dfrac{\sqrt[4]{81}}{\sqrt{35.6}}$ (d) $\dfrac{\sqrt{0.0351}}{\sqrt[5]{-20.3}}$

(e) $\sqrt[3]{\log 2}$ (f) $\sqrt[5]{\log 9.9}$

(g) $\log 7.96 + \sqrt{348}$ (h) $(28.1)^{-1.3} + (0.084)^{2/3}$

6. Find the area of a circle whose radius is 34.6 in. Let $\pi = 3.14$; use logarithms.

7. The volume of a sphere is given by $V = \frac{4}{3}\pi r^3$. Find the volume of a sphere whose radius is 25.8 ft.

8. A spherical shell whose outside diameter is 6.34 in. is made of metal 0.0375 in. thick. If the metal weighs 489 lb/cu ft, find the weight of the shell.

9. If a, b, c represent the lengths of the sides of a triangle, the area of the triangle is given by $T = \sqrt{s(s-a)(s-b)(s-c)}$, where s equals one-half the perimeter. Find the area of a triangle whose sides are 72.4 ft, 85.6 ft, and 34.8 ft respectively.

10. The approximate time for one complete oscillation of a simple pendulum is given by

$$t = 2\pi \sqrt{\frac{l}{32}}$$

where l is the length of the pendulum in feet and t is the time in seconds. Find the period (time for one complete oscillation) of a pendulum 6.25 ft in length.

11. Assuming that the annual rate r of depreciation of a certain machine is constant, the scrap value S after n yr is given by $S = C(1 - r)^n$, where C is the initial cost. If the machine originally cost \$11,250 and its annual depreciation rate is 15 percent, find its scrap value after 8 yr.

12. When an inductance of L henrys and a resistance of R ohms are connected in series with an electromotive force of E volts, the current I (in amperes) flowing after t sec is given by

$$I = \frac{E(1 - e^{-Rt/L})}{R}$$

If $L = 0.1$ henry, $E = 110$ volts, $R = 6$ ohms, find the current flowing when $t = 0.01$ sec. Take $\log e = 0.4343$.

9.6   CHANGE OF BASES

Table III permits us to find the logarithm of any positive number to any base b, $b > 0$, $b \neq 1$.   The following examples illustrate the procedure.

EXAMPLE 1.   Find $\log_6 420$.

Solution.   Let $x = \log_6 420$.   Then

$$6^x = 420 \qquad \text{by Def. (9.1)}$$

and $\qquad\qquad\qquad x \log 6 = \log 420 \qquad\qquad \text{why?}$

Therefore $\qquad\qquad\qquad x = \dfrac{\log 420}{\log 6}$

$$= \dfrac{2.6232}{0.7782}$$

The value of x can be found either by ordinary division or by the use of logarithms.   [Note that $\log 420/\log 6 \neq \log (420 \div 6)$.]   By division, $x = 3.371$.

EXAMPLE 2.   Take $\pi = 3.14$ and find $\log_\pi 8$.

Solution.   Let $x = \log_\pi 8$.   Then

$$\pi^x = 8$$

and $\qquad\qquad\qquad x \log \pi = \log 8$

$$x = \frac{\log 8}{\log \pi} = \frac{0.9031}{0.4969} = 1.813$$

If the logarithm of a number N to any base b is given, then the logarithm of N to any other base a can be found by applying the following theorem:

THEOREM.   If a and b are bases and $N > 0$,

$$\log_a N = \frac{\log_b N}{\log_b a} \qquad\qquad\qquad \textbf{(9.5)}$$

Proof.   Let $x = \log_b N$ and $y = \log_a N$.   Then

$$b^x = N \qquad \text{and} \qquad a^y = N$$

Hence, $\qquad\qquad\qquad\qquad a^y = b^x$

and $\qquad\qquad\qquad y \log_b a = x \log_b b \qquad \text{why?}$

$$= x \qquad \text{why?}$$

By substitution, we have

$$(\log_a N)(\log_b a) = \log_b N$$

so that
$$\log_a N = \frac{\log_b N}{\log_b a}$$

EXAMPLE 3. Find $\log_5 7$.

Solution. $\log_5 7 = \dfrac{\log_{10} 7}{\log_{10} 5} = \dfrac{\log 7}{\log 5} = \dfrac{0.8451}{0.6990} = 1.209$

In the decimal system of notation, multiplying any number by 10^c, $c \in I$, serves to change the location of the decimal point but does not alter the sequence of digits in the number. This property of our number system enables us to find separately the characteristic and mantissa of the common logarithm of the number. When the base of the logarithm is different from 10, this convenient property is not available. We then make no effort to preserve the mantissa as a positive number.

EXAMPLE 4. Find $\log_{0.8} 3$.

Solution. $\log_{0.8} 3 = \dfrac{\log 3}{\log 0.8} = \dfrac{0.4771}{9.9031 - 10}$

We keep in mind that the logarithm of a number is also a number, and we write $9.9031 - 10$ as -0.0969. Hence,

$$\log_{0.8} 3 = \frac{0.4771}{-0.0969} = -4.924$$

9.7 NATURAL LOGARITHMS

The system of logarithms whose base is the irrational number

$$e = 1 + \frac{1}{1} + \frac{1}{1 \cdot 2} + \frac{1}{1 \cdot 2 \cdot 3} + \frac{1}{1 \cdot 2 \cdot 3 \cdot 4} + \cdots$$

which is approximately equal to 2.718, is called the system of *natural logarithms*. These natural logarithms are used extensively in both applied and theoretical problems in advanced mathematics. They are convenient in the study of the calculus.

We shall use the abbreviation $ln\ N$ for the natural logarithm of N. Thus, $\log_e N$ is written $\ln N$. By Definition (9.1),

$$\text{If } e^y = N \qquad \text{then} \qquad y = \ln N \tag{9.6}$$

Although tables of natural logarithms of numbers are available, we shall continue to use Table III. To understand how this can be done, we make use of Theorem (9.5). Since

$$\log_a N = \frac{\log_b N}{\log_b a}$$

we may take $a = e$ and $b = 10$. Then

$$\ln N = \frac{\log N}{\log e} \qquad (9.7)$$

From Table III,

$$\log e = \log 2.718 \text{ (approx.)} = 0.4343$$

Hence, $\qquad \ln N = \dfrac{\log N}{0.4343} = \dfrac{1}{0.4343} \log N$

$$= 2.303 \log N \text{ (approx.)} \qquad (9.8)$$

EXAMPLE 1.   Find $\ln 10$.

Solution. $\ln 10 = 2.303 \log 10 = 2.303(1) = 2.303$

EXAMPLE 2.   Find $\ln 25$.

Solution. $\ln 25 = 2.303 \log 25 = 2.303(1.3979)$
$$= 3.219 \qquad \text{to four significant figures}$$

The reason for choosing the irrational number e as a base for a system of logarithms is too lengthy to present here. In the calculus, the number e is defined as a limit:

$$e = \lim_{x \to \infty} \left(1 + \frac{1}{x}\right)^x = 2.71828 \cdots$$

and the application of certain concepts of differential calculus to the logarithmic function leads naturally to this limit and to its adoption as a convenient base.

EXERCISE SET 9.4

FIND each of the following, using Table III:

1. $\log_4 75$	2. $\log_5 \pi$	3. $\log_\pi 64$
4. $\log_6 e$	5. $\log_\pi e$	6. $\log_e e$
7. $\log_{100} 64$	8. $\log_{0.1} 10$	9. $\log_{0.2} 48$
10. $\log_{\sqrt{2}} 4$	11. $\log_\pi 1$	12. $\log_\pi \pi$
13. $\ln 2.3$	14. $\ln 0.83$	15. $\ln 0.076$

16. $\ln \dfrac{1}{7.2}$ $\qquad\qquad$ 17. $\ln \dfrac{1}{0.45}$ $\qquad\qquad$ 18. $\ln \dfrac{1}{0.023}$

FIND the approximate value of x in each of the following:

19. $x = \log_4 7$	20. $x = \log_{12} 9$	21. $x = \log_5 10$
22. $x = \log_\pi 3$	23. $\log_3 x = 1.73$	24. $\log_5 x = 2.14$
25. $\ln x = 1.1053$	26. $\ln x = 2.0819$	

9.8 EXPONENTIAL AND LOGARITHMIC EQUATIONS

Logarithms are quite useful in determining the solution sets (approximate values) of certain types of equations that are not easily solved by other algebraic methods. The following examples illustrate the method to be used.

EXAMPLE 1. Solve the equation $4^x = 5^{2x+1}$ for x.

Solution. Since $4^x = 5^{2x+1}$,

$$\log (4^x) = \log (5^{2x+1}) \qquad\qquad\qquad \text{why?}$$

and $\qquad\qquad x \log 4 = (2x + 1)\log 5 = 2x \log 5 + \log 5$

Hence, $\qquad\qquad x \log 4 - 2x \log 5 = \log 5$

and $\qquad\qquad x(\log 4 - 2 \log 5) = \log 5$

Therefore, $\qquad x = \dfrac{\log 5}{\log 4 - 2 \log 5} = \dfrac{0.6990}{0.6021 - 1.3980}$

$$= -0.878 \text{ (approx.)}$$

EXAMPLE 2. If $\log_4 (x^2 - 12x) = 3$, find x.

Solution. By Definition (9.1),

$$4^3 = x^2 - 12x \qquad \text{and} \qquad x^2 - 12x - 64 = 0$$

Hence, $\qquad\qquad\qquad (x - 16)(x + 4) = 0$

and $\qquad\qquad\qquad\qquad x = -4 \qquad \text{or} \qquad x = 16$

EXAMPLE 3. Find x if $\log (x - 9) - \log (x + 3) = 1$.

Solution. Use the property $\log M/N = \log M - \log N$ to change the form of the left member of the equation. Then

$$\log \frac{x - 9}{x + 3} = 1$$

Hence, $\qquad\qquad\qquad 10^1 = \dfrac{x - 9}{x + 3} \qquad$ by Def. (9.1)

and $\qquad\qquad\qquad\qquad x = -\dfrac{13}{3}$

This value of x does not satisfy the given equation because we do not define the logarithm of a negative real number.

Hence, $x = -13\frac{3}{}$ is an extraneous root. The equation has no solution.

EXAMPLE 4. Solve the inequality $4^x > \frac{5}{8}$.

Solution.   Since $4^x > \frac{5}{3}$,

$$\log 4^x > \log \frac{5}{3}$$

(See Prob. 19 of Exercise Set 9.1.)   Then

$$x \log 4 > \log 5 - \log 3$$

and

$$x > \frac{\log 5 - \log 3}{\log 4}$$

Therefore,

$$x > \frac{0.6990 - 0.4771}{0.6021}$$

or

$$x > 0.3685$$

EXERCISE SET 9.5

SOLVE for x.   (Some of the equations may not have solutions.)

1. $2^{x+1} = 128$
2. $3^{4x-1} = 729$
3. $30^{x+4} = 810,000$
4. $4^x = \frac{5}{2}$
5. $(0.01)^x = 3$
6. $7^{3x}(5^{x-1}) = 35$
7. $4^{2x+1} = 7^{2x-1}$
8. $4^{-x} = 6^{x+2}$
9. $3^{2x+3} = 8^{3x-4}$
10. $2^{3x-1} = 7.2^{3x+1}$
11. $\log (x + 3)^2 = 2$
12. $\log (2x - 3)^3 = 3$
13. $\log (13 + 2x)^7 = 14$
14. $\log (52 - x)^6 = 6$
15. $\log (3x + 4) - \log (2x - 2) = 2$
16. $\log (x - 1) - \log (4x + 7) = 1$
17. $\log (x^2 - 16) - \log (x - 4) = 2$
18. $\log (x^2 - 4) - \log (x + 2) = 3$
19. $\log_2 (x + 1) + \log_2 (x - 1) = 3$
20. $\log_3 (x + 9) + \log_3 (x + 3) = 3$
21. $\log x - \log (x + 4) = 0.3010$
22. $\log (x - 6) - \log (x + 6) = 1.4771$
23. $(0.5)^x > \frac{3}{2}$
24. $(0.08)^{2x} > (\frac{4}{3})^2$
25. $x^{\log x} = 10$
26. $x^{\log x} = 100x$

CHANGE each of the following equations into an equation free of logarithms:

27. $\log x - 1 = 2 \log y$
28. $3 \log x = 2 \log (y + 2)$
29. $\ln x = 3 \ln y + 2y$
30. $\ln (\ln x) = \ln x + \ln y$

FIND the real zeros of the following functions of x:

31. $f(x) = e^{2x} + 4e^x - 1 - 4e^{-x}$
32. $f(x) = e^{-4x} - 4e^{-3x} - 5e^{-2x} + 36e^{-x} - 36$
33. $f(x) = \log (3x + 5) + \log (2x - 5) + \log (x + 5) - 3$
34. $f(x) = 2 \log (3x - 5) + \log (3x + 4) - 1$
35. $f(x) = \log (10 - x^2) + \log (2x - 5)$

10

SOLUTION
OF TRIANGLES.
VECTORS

We turn our attention to some of the basic relations that exist between the lengths of the sides and the trigonometric functions of the angles of a triangle. These relations enable us to solve certain types of geometrical problems involving triangles.

We shall usually denote the measure of the angles at the vertices of the triangle ABC by α, β, and γ (alpha, beta, and gamma) respectively. We shall also denote the lengths of the sides opposite the vertices A, B, C by a, b, c respectively, as shown in Fig. 10.1.

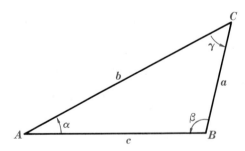

FIGURE 10.1
Lettering a triangle

10.1 LAW OF SINES

Let ABC be a given triangle. Place a rectangular coordinate system in the plane of the triangle so that the vertex A is at the origin, and so that side c lies on the positive x axis. This puts the angle α in standard position as shown in Fig. 10.2a. Hence, the vertex C has the coordinates $(b \cos \alpha, b \sin \alpha)$.

If the coordinate system is now placed so that the vertex B is at the origin, with side c again lying on the positive x axis, the vertex C has the coordinates $(a \cos \beta, a \sin \beta)$ as shown in Fig. 10.2b.

In either case, the ordinate of the vertex C is the altitude of the triangle from the vertex C. Since $a \sin \beta$ and $b \sin \alpha$ are merely different expressions for the ordinate of C, it follows that

$$a \sin \beta = b \sin \alpha$$

and

$$\frac{a}{\sin \alpha} = \frac{b}{\sin \beta}$$

By a similar procedure, with the origin at C and with b lying on the positive x axis, we find that the ordinate of the vertex B is $a \sin \gamma$. When the coordinate system is then taken so that the origin is at A and b again lies on the positive x axis, the ordinate of B is $c \sin \alpha$. Hence, we have

$$a \sin \gamma = c \sin \alpha$$

and

$$\frac{a}{\sin \alpha} = \frac{c}{\sin \gamma}$$

Thus, if α, β, and γ are the measures of the angles of a triangle and if a, b, and c are the lengths of the sides opposite the respective angles, then

$$\frac{a}{\sin \alpha} = \frac{b}{\sin \beta} = \frac{c}{\sin \gamma} \tag{10.1}$$

Law of sines

The relations (10.1) are called the *law of sines* and may be used to solve a triangle when (1) two angles and a side are given or (2) two sides and an angle opposite one of them are given. The following examples illustrate the method.

EXAMPLE 1. Find the side a of the triangle ABC if $b = 30$, $\alpha = 36°$, $\gamma = 48°$.

FIGURE 10.2 (*a*) $h = b \sin \alpha$; (*b*) $h = a \sin \beta$

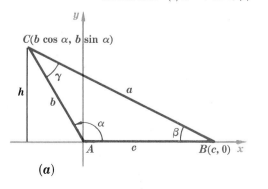

(*a*)

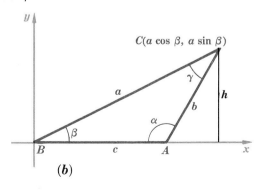

(*b*)

Solution. $\beta = 180° - (36° + 48°) = 96°$

By the law of sines,

$$\frac{a}{\sin \alpha} = \frac{b}{\sin \beta}$$

Therefore,

$$\frac{a}{\sin 36°} = \frac{30}{\sin 96°}$$

and

$$a = \frac{30(\sin 36°)}{\sin 96°}$$

The computation can now be done in either of two ways. We can substitute values from Table II for sin 36° and for sin 96°. Then

$$a = \frac{30(0.5878)}{0.9945} = 17.73$$

Since the result cannot be more accurate than the given measurements, we round off the answer. Hence,

$$a = 18 \text{ (approx.)}$$

We need to know how the accuracy in the measure of angles should compare with the accuracy in the measure of distances. For example, if the sides of a triangle are expressed to two significant figures, then the angles should be expressed to the nearest degree. In solving triangles, we shall use as a working rule the following corresponding accuracies:

Significant figures	2	3	4	5
Accuracy of angle (nearest)	1°	10′	1′	0.1′

Thus, angles should be expressed to the nearest degree, nearest multiple of 10′, nearest minute, or nearest tenth of a minute, accordingly as the sides are expressed to two, three, four, or five significant figures. For angles near quadrantal angles, the accuracy may vary from the above. According to this working rule, the answer in the preceding example must be rounded off to two significant figures. Hence,

$$a = 18 \text{ (approx.)}$$

Another way to compute the value of *a* is by the use of common logarithms. Since

$$a = \frac{30(\sin 36°)}{\sin 96°}$$

$$\log a = \log 30 + \log \sin 36° - \log \sin 96°$$

Use of Table IV   Table IV contains the logarithms (common) of the trigonometric functions correct to four decimal places. To understand the use of this table, we recall that if $0 < \theta < 90°$, then $0 < \sin \theta < 1$, and therefore the logarithm of $\sin \theta$ has a negative characteristic. The same statement can be made about $\cos \theta$. Also, if $0 < \theta < 45°$, then log tan θ has a negative characteristic. To simplify determination of characteristics, Table IV has been constructed so that -10 must be added to each entry in the body of the table. Hence, we have the following rule:

In using Table IV, add -10 to each printed entry. Since

$$
\begin{array}{lll}
\log 30 = & 1.4771 & \text{Table III} \\
\log \sin 36° = & \underline{9.7692 - 10} & \text{Table IV} \\
& 11.2463 - 10 & \text{log numerator} \\
\log \sin 96° = & \underline{9.9976 - 10} & \text{log denominator} \\
& 1.2487 & \text{log fraction}
\end{array}
$$

Hence, $\log a = 1.2487$

and $a = 17.73$

                      $= 18$ to two significant figures

If we know the lengths of two sides of a triangle, say a and b, and the measure of an angle opposite one of them, say α, we may find the remaining unknown parts of the triangle by use of the law of sines. In this case, however, complications frequently arise due to the fact that there may be two, one, or no triangles having the given sides a and b and the angle α. That is why we call this case the *ambiguous case*.

Ambiguous case

EXAMPLE 2. In triangle ABC, $a = 14$, $b = 17$, and angle $\alpha = 32°$. Find the angle γ and the side c.

Solution. Since $a/\sin \alpha = b/\sin \beta$, we have

$$\frac{14}{\sin 32°} = \frac{17}{\sin \beta}$$

and $\sin \beta = \dfrac{17 \sin 32°}{14}$

Hence, $\log \sin \beta = \log 17 + \log \sin 32° - \log 14$

                      $= 9.8085 - 10$

Now, there are two values of β between $0°$ and $180°$ for which

$$\log \sin \beta = 9.8085 - 10$$

namely $40°03'$ and $139°57'$.

In accordance with our agreement as to the accuracy of results, we round off to the nearest degree. Hence,

$$\beta = 40° \quad \text{or} \quad \beta = 140°$$

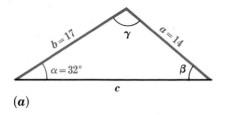

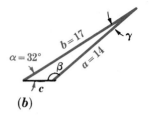

FIGURE 10.3   The ambiguous case (see Example 2)

If $\beta = 40°$, then $\gamma = 180° - (40° + 32°) = 108°$. The law of sines now gives us

$$c = \frac{14 \sin 108°}{\sin 32°}$$

$$= 25 \qquad \text{to two significant figures}$$

This triangle is shown in Fig. 10.3a.

However, if we take $\beta = 140°$, then

$$\gamma = 180° - (140° + 32°) = 8°$$

and
$$c = \frac{14 \sin 8°}{\sin 32°}$$

$$= 3.7 \qquad \text{to two significant figures}$$

This triangle is shown in Fig. 10.3b.

To understand why there may be two, one, or no triangles determined when the angle α and the sides a and b of the triangle ABC are given, let us construct the triangle as follows: Assume that α is acute and draw the angle α. On one of the sides of α take point C so that $AC = b$.   With C as a center and a radius of length a, construct the arc of a circle as in Fig. 10.4.   This arc meets the other side of angle α in two points, one point, or no point.   Thus, there are two, one, or no triangles having the given angle α and the given sides a and b.

The interested student can analyze the possibilities when $\alpha > 90°$.

FIGURE 10.4   The ambiguous case: there may be two, one, or no triangles determined when the angle α and the sides a and b of triangle ABC are given.

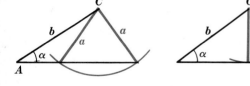

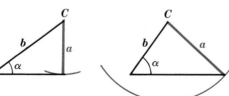

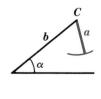

EXERCISE SET 10.1

SOLVE the triangle ABC, given that:

1. $\alpha = 46°$, $\gamma = 59°$, $b = 10$
2. $\beta = 65°$, $\gamma = 35°$, $c = 24$
3. $\alpha = 120°$, $\beta = 18°$, $a = 90$
4. $\beta = 30°$, $\gamma = 48°$, $a = 12$
5. $\alpha = 98°20'$, $\gamma = 45°10'$, $c = 155$
6. $\beta = 115°40'$, $\gamma = 16°50'$, $c = 4.43$
7. $\alpha = 27°10'$, $\beta = 132°20'$, $c = 0.758$
8. $\beta = 30°$, $b = 50$, $c = 30$
9. $\alpha = 60°$, $a = 75$, $b = 80$
10. $\beta = 29°20'$, $a = 200$, $b = 500$
11. $\gamma = 59°50'$, $b = 800$, $c = 500$
12. $\beta = 31°10'$, $a = 225$, $b = 115$

13. Two points A and B on the bank of a stream are 36 ft apart. Point C on the other side of the stream is located so that angle CAB is $72°$ and angle ABC is $80°$. Find the approximate width of the stream.

14. The side of a regular polygon of seven sides is 125.0 in. long. Find the radius of the circumscribed circle.

15. Two points A and B are on opposite banks of a farm pond. A third point C is located so that angle ABC is $42°$. If BC and AC are measured and found to be 260 and 320 ft respectively, what is the distance AB?

16. The longer side of a parallelogram is 172.3 in. and the shorter side is 99.4 in. The longer diagonal makes an angle of $31°40'$ with the longer side. Find the length of the longer diagonal.

SKETCH figures to illustrate the following, given the angle α and the sides a and b of the triangle ABC:

17. If α is acute and $a \geq b$, one triangle is determined.
18. If α is acute and $b \sin \alpha < a < b$, two triangles are determined.
19. If α is obtuse and $a \leq b$, no triangle is determined.
20. If α is obtuse and $a > b$, one triangle is determined.

10.2 SOLUTION OF RIGHT TRIANGLES

Suppose the angle γ of the triangle ABC is a right angle (Fig. 10.5). Then, by the law of sines,

$$\frac{a}{\sin \alpha} = \frac{c}{\sin 90°} = \frac{c}{1} = c$$

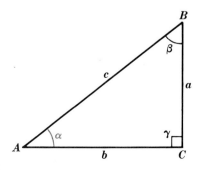

FIGURE 10.5
The right triangle

Hence,
$$\sin \alpha = \frac{a}{c}$$

Since $\sin^2 \alpha + \cos^2 \alpha = 1$, we have

$$\cos \alpha = \sqrt{1 - \sin^2 \alpha} = \sqrt{1 - \frac{a^2}{c^2}} = \frac{b}{c}$$

Also,
$$\tan \alpha = \frac{\sin \alpha}{\cos \alpha} = \frac{a/c}{b/c} = \frac{a}{b}$$

Thus, for any triangle ABC with $\gamma = 90°$,

$$\sin \alpha = \frac{a}{c} = \frac{\text{side opposite angle } \alpha}{\text{hypotenuse}}$$

$$\cos \alpha = \frac{b}{c} = \frac{\text{side adjacent to angle } \alpha}{\text{hypotenuse}} \qquad (10.2)$$

$$\tan \alpha = \frac{a}{b} = \frac{\text{side opposite angle } \alpha}{\text{side adjacent angle } \alpha}$$

These are the only equations needed to solve right triangles. However, since $b/\sin \beta = c/\sin 90° = c$, we can deduce that

$$\sin \beta = \frac{b}{c} = \frac{\text{side opposite } \beta}{\text{hypotenuse}}$$

$$\cos \beta = \frac{a}{c} = \frac{\text{side adjacent } \beta}{\text{hypotenuse}}$$

and
$$\tan \beta = \frac{b}{a} = \frac{\text{side opposite } \beta}{\text{side adjacent } \beta}$$

The fundamental identities of Sec. 5.5 can now be used to show that $\cot \alpha = b/a$, $\sec \alpha = c/b$, $\csc \alpha = c/a$, and so on.

In solving right triangles, it is always helpful to sketch a figure and indicate the two parts which, in addition to the right angle, have been given. The procedure then requires the writing of an equation involving the two given parts and one of the unknown parts. It is usually helpful to select an equation that requires a multiplication rather than a division.

EXAMPLE 1. In the right triangle ABC, $\gamma = 90°$, $\beta = 32°$, and $b = 21$. Find side a.

Solution. Sketch the triangle and indicate the known parts (Fig. 10.6). Since $\beta = 32°$, $\alpha = 58°$. Now,

$$\tan \alpha = \frac{\text{side opposite } \alpha}{\text{side adjacent } \alpha}$$

Hence,
$$\tan 58° = \frac{a}{21}$$

and
$$a = 21 \tan 58°$$

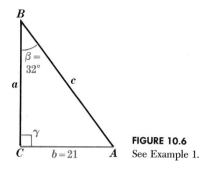

FIGURE 10.6
See Example 1.

The computation can be done by finding the value of tan 58° from Table II or by using logarithms as follows:

$$\log a = \log 21 + \log \tan 58°$$

$$\log 21 = 1.3222$$
$$\log \tan 58° = \frac{10.2042 - 10}{11.5264 - 10}$$

Hence, $\qquad \log a = 1.5264$

and $\qquad\qquad a = 33.61 = 34 \qquad$ to two significant figures

EXERCISE SET 10.2

FIND the lengths of the unknown sides and the measures of the unknown angles in each of the following triangles. In each case $\gamma = 90°$.

1. $\alpha = 36°$, $a = 24$
2. $\alpha = 64°$, $b = 80$
3. $\beta = 52°20'$, $c = 9.36$
4. $\beta = 55°10'$, $a = 825$
5. $a = 32$, $b = 62$
6. $a = 54$, $c = 63$
7. $b = 355$, $c = 484$
8. $a = 327$, $b = 764$
9. $a = 43.2$, $c = 63.7$
10. $a = 36.8$, $\beta = 57°40'$
11. $a = 4.32$, $\alpha = 67°30'$
12. $b = 318$, $\alpha = 25°50'$
13. $\beta = 48°22'$, $b = 7.255$
14. $b = 35.72$, $c = 48.46$
15. A ladder 13 ft long rests against a vertical wall. The foot of the ladder is 5 ft from the base of the wall. Find the angle which the ladder makes with the ground.
16. A wire stretching from a point on level ground to the top of a vertical pole touches the ground at a point 16 ft from the foot of the pole. If the wire makes an angle of 62° with the horizontal, find the approximate height of the pole and the approximate length of the wire.
17. A rectangular city lot is 110 by 140 ft. Find the angle which a diagonal makes with the longest side.
18. A regular hexagon is inscribed in a circle whose radius is 10 cm. Find the radius of a circle inscribed in the hexagon.
19. One of the base angles of an isosceles triangle is 36°10', and one of the equal sides is 5.75 in. Find the length of the base.
20. From a point 150 ft from the foot of a flagpole, the angle of elevation of the top of the pole is 28°10'. Find the height of the flagpole. (If a point Y is above the level of a point X, the

angle which the line XY makes with a horizontal line through X in the same vertical plane as XY is the *angle of elevation* of Y from X. The angle which the line YX makes with a horizontal line through Y in the same vertical plane as YX is the *angle of depression* of X from Y.)

21. From an observation balloon 950 ft above level ground, the angle of depression of an object is $12°40'$. Find the distance of the object from a point directly under the balloon.

22. From a point A the angle of elevation of a mountain top is $28°40'$. From a point B 8,600 ft nearer and in the same horizontal plane, the angle of elevation of the top is $36°50'$. Find the height of the mountain top above the horizontal plane.

23. Consider the Earth a sphere of radius 3,960 mi. Find the radius of the parallel of latitude which passes through a city whose latitude is $37°23'$.

24. Two radar installations A and B are 75 mi apart, and B is due east of A. An aircraft modification center is located at point C due north of A. The bearing of BC is N$44°10'$W. Find the distance from A to C. (The bearing of a line is the acute angle which its direction makes with a north-south line. In describing the bearing of a line, the measure of the acute angle is preceded by the letter N or S to indicate that the angle is measured from the north half-line or the south half-line. The measure of the angle is followed by the letter E or the letter W to indicate that the terminal side of the angle lies to the east or to the west of the north-south line. The bearing N$44°10'$W is shown in Fig. 10.7.)

25. A ship sails due west a distance of 25 naut mi, then turns due north and continues a distance of 30 naut mi. Find its bearing from the starting point.

26. Two ships leave the same port at the same time. If one sails in the direction N$38°$E at 20 knots and the other travels S$42°$E at 30 knots, find the distance between them after 5 hr. Assume the surface of the water is a plane.

27. A lookout tower 90 ft high is on top of a hill. From the tower a forest ranger looking down the hill observes that the angle of depression of the top of a certain tree is $20°30'$, while the angle of depression of its base is $36°40'$. If the base of the tree is 215 ft down the hill from a point on the ground directly beneath the observer, how high is the tree?

28. At a point 225 ft from a building, the angle of elevation of the top of the building is $27°50'$. The angle of elevation of the top of a flagpole on the top of the building is $33°20'$. Find the height of the flagpole.

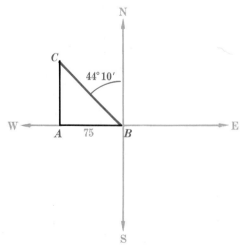

FIGURE 10.7
See Prob. 24.

10.3   VECTORS

A vector quantity is a quantity that requires both magnitude and direction for its determination.   Examples of vector quantities are forces, displacements, velocities, and accelerations.   We represent a vector quantity geometrically by an arrow whose length is proportional to the magnitude of the vector quantity and whose direction coincides with (or is parallel to) the direction of the vector quantity.   For example, the arrow from the initial point O to the terminal point A (shown in Fig. 10.8a) represents a vector quantity and is called a *vector*.   We denote this vector by the symbol $\overrightarrow{OA}$.

Resultant

Experiments show that the combined effect of any two vector quantities can be represented by a single vector called the *vector sum,* or *resultant,* of the two vectors representing the vector quantities.

> **DEFINITION.**   The vector sum (or resultant) of the two vectors $\overrightarrow{OA}$ and $\overrightarrow{OB}$ is the diagonal vector $\overrightarrow{OC}$ of the parallelogram   **(10.3)** $OACB$ constructed with $\overrightarrow{OA}$ and $\overrightarrow{OB}$ as sides.

An illustration of this definition is shown in Fig. 10.8a where the vector $\overrightarrow{OC}$ is the resultant, or sum, of the vectors $\overrightarrow{OA}$ and $\overrightarrow{OB}$.

Two vectors whose sum is a given vector are called *components* of the given vector.   Thus, $\overrightarrow{OA}$ and $\overrightarrow{OB}$ are components of the vector $\overrightarrow{OC}$, Fig. 10.8a. Finding the components of a given vector is called *resolving* the vector.   We usually find it convenient to resolve a vector into components parallel to the

Rectangular components

x and y axes.   Such components are called the *rectangular components* of the vector.   In Fig. 10.8b, **h** and **k** are the rectangular components of the vector **u**.   We call **h** the *horizontal component* and **k** the *vertical component*.

FIGURE 10.8   (*a*) Vector sum: $\overrightarrow{OC} = \overrightarrow{OA} + \overrightarrow{OB}$; (*b*) Vector components: $\mathbf{u} = \mathbf{h} + \mathbf{k}$

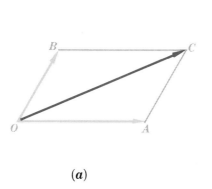

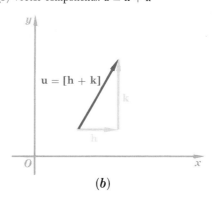

(*a*)   (*b*)

Note that when we use a single letter to represent a vector, we set it in bold-face type.

A vector **u** in the coordinate plane is the sum of its rectangular components **h** and **k**. If **h** and **k** are given, the vector **u** can be determined, and conversely. This means that we can describe the vector **u** by describing its rectangular components.

For the moment, consider that the vector **u** has nonzero length and is not parallel to either axis. Then **h**, the horizontal component (or x component) of **u**, is a vector parallel to the x axis. Let the length of **h** be denoted by h, then we can represent **h** by the single real number h if **h** has the same direction as the positive x axis and by $-h$ if **h** has the opposite direction. Similarly, the vertical component (or y component) **k** of **u** is parallel to the y axis. Let the length of **k** be denoted by k. We can then represent **k** by the real number k if the vector **k** has the same direction as the positive y axis and by $-k$ if **k** has the opposite direction.

We can now describe the vector **u** by the ordered pair of real numbers $[h,k]$ which specify the rectangular components of **u**, and we write

$$\mathbf{u} = [h,k]$$

As examples, the vector $\mathbf{u} = [-3,2]$ has a horizontal component 3 units long in the direction of the negative x axis and a vertical component 2 units long in the direction of the positive y axis. The vector $[2,0]$ has a horizontal component 2 units long in the direction of the positive x axis and, since the vertical component is zero, $[2,0]$ is parallel to the x axis. The vector $[0,-5]$ is in the direction of the negative y axis and has length 5.

We define equality of vectors as follows:

$$[h_1,k_1] = [h_2,k_2]$$

if and only if $\qquad h_1 = h_2 \qquad$ and $\qquad k_1 = k_2$

Sum of two vectors   The sum of two vectors $[h_1,k_1]$ and $[h_2,k_2]$ is defined as follows:

DEFINITION.   If $[h_1,k_1]$ and $[h_2,k_2]$ are vectors, then $[h_1,k_1] +$ $[h_2,k_2] = [h_1 + h_2, k_1 + k_2]$.   **(10.4)**

Thus the sum of any two vectors has horizontal and vertical components that are the sums, respectively, of the horizontal and vertical components of the given vectors. For example, the sum vector of $[3,5]$ and $[6,-1]$ is given by

$$[3,5] + [6,-1] = [9,4]$$

If $\mathbf{u} = [h,k]$ is any vector, then the vector $-\mathbf{u} = [-h,-k]$ is defined to be the negative of **u**. This vector has the same magnitude (length) as the

Difference of two vectors

vector **u** but has the opposite direction. It follows that $\mathbf{u} = -(-\mathbf{u})$. This relation suggests a method for subtracting one vector from another:

$$\mathbf{u} - \mathbf{v} = \mathbf{u} + (-\mathbf{v})$$

Thus, $[h_1,k_1] - [h_2,k_2] = [h_1,k_1] + [-h_2,-k_2] = [h_1 - h_2, k_1 - k_2]$.

The symbol $c[h,k]$ is used to denote the vector $[ch,ck]$. If c is positive, the new vector has the same direction as $[h,k]$; if c is negative, the new vector has the direction opposite to that of $[h,k]$.

Zero vector

The vector $[0,0]$ is defined to be the *zero vector*, even though it has no unique direction. Thus, the vector $[h,k] - [h,k] = [0,0]$ is the zero vector.

Vector magnitude

The magnitude of the vector $\mathbf{u} = [h,k]$ is denoted by $|\mathbf{u}|$ and is equal to $\sqrt{h^2 + k^2}$. For example, the magnitude of the vector $[3,7]$ is

$$\sqrt{3^2 + 7^2} = \sqrt{58}$$

EXAMPLE 1. Two forces are acting on a body, one of 30 lb acting horizontally to the right, the other of 40 lb acting vertically upward. Find the resultant of these two forces.

Solution. Let R be the resultant force, Fig. 10.9. Then

$$|R| = \sqrt{30^2 + 40^2} = 50$$

Let α be the angle between R and the horizontal force. Then

$$\tan \alpha = \frac{40}{30} = 1.333$$

and $\alpha = 53°$ (to agree with the number of significant digits in the measure of the forces).

Hence, the resultant force has magnitude 50 lb and makes an angle of $53°$ with the horizontal force.

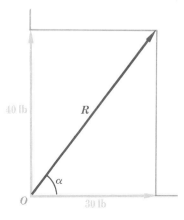

FIGURE 10.9
See Example 1.

EXAMPLE 2.   A force of 225 lb is applied at an angle of $47°50'$ with the horizontal.   Find two forces, one horizontal and the other vertical, which will produce the same result.

Solution.   Let v_x and v_y be the lengths of the horizontal and vertical components respectively of the force **v**, Fig. 10.10.   Then

$$v_x = 225 \cos 47°50' = 151$$

and $$v_y = 225 \sin 47°50' = 167$$

EXAMPLE 3.   An inclined plane makes an angle of $20°10'$ with the horizontal. Assume there is no friction and find the force required to hold a 100-lb weight stationary on the plane.

Solution.   Let the weight be at point R, Fig. 10.11.   Then the force due to the weight may be represented by the vector RS directed downward.   Let RT and RQ be the components of the vector RS parallel to and perpendicular to the inclined plane, respectively.   Then the required force F must be equal to the force represented by the vector RT and must be directed *up* the plane.

Since angle RST is $20°10'$ (angle RST = angle of inclination of the plane),
$$RT = 100 \sin 20°10' = 34.48$$
$$= 34.5 \qquad \text{to three significant figures}$$
Hence, $F = 34.5$ lb   directed up the plane

EXAMPLE 4.   Add the vectors $[2,3]$ and $[6,-8]$, then find the magnitude of the sum vector.

Solution. $[2,3] + [6,-8] = [8,-5]$
$$|[8,-5]| = \sqrt{8^2 + (-5)^2} = \sqrt{89}$$

FIGURE 10.10
See Example 2.

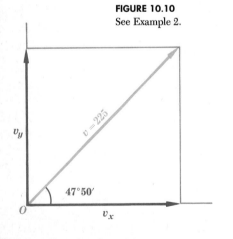

FIGURE 10.11
See Example 3.

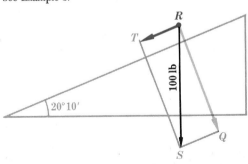

EXERCISE SET 10.3

1. A force of 50 lb acts at an angle of 53° with the horizontal. Find the horizontal and vertical components of the force.
2. A force of 155 lb makes an angle of 38°20′ with the horizontal. Find its rectangular components.
3. A force of 100 lb has a vertical component of 60 lb. Find the horizontal component of the force.
4. The magnitude of a force is 384 lb. Find the rectangular components of the force if the horizontal component is three times the vertical component.
5. A barrel weighing 128 lb is rolled up an incline which makes an angle of 15°30′ with the horizontal. Find the force parallel to the incline necessary to keep the barrel in motion.
6. A boat sails 54.6 mi in the direction S28°E. How far south and how far east has it gone?
7. A river 2.4 mi wide flows due south at 1.1 mi/hr. A swimmer who can swim at the rate of 1.8 mi/hr in still water heads east across the river. Find the direction in which the swimmer is actually moving.
8. The motor of a boat can propel it at the rate of 15 mi/hr in still water. If the boat is moving directly across a river whose current is 4 mi/hr, find the angle in which the boat is heading upstream.
9. A jet transport travels with a speed of 500 mi/hr in calm air. If the wind is blowing with a velocity of 30 mi/hr from N28°E and the aircraft is headed in the direction S20°E, find the magnitude of the speed of the aircraft and its direction with reference to the ground.
10. A ship is sailing due east at 25 mi/hr. A passenger walks directly across the deck from starboard to port at 4 mi/hr. Find the velocity of the passenger relative to the surface of the water.
11. An aircraft has a speed of 600 mi/hr in calm air. The wind is blowing from the west at 20 mi/hr. In what direction must the aircraft head in order to travel directly north?
12. An automobile weighs 3,500 lb and is parked on a street inclined 7°50′ with the horizontal. Find the force which tends to pull the automobile downhill.
13. In preparing for a probe of the planet Mars, scientists observe that the lines from Earth to the sun and from Earth to Mars make an angle of 120°. If the distances from Earth to the sun and from Mars to the sun are, respectively, 9.3×10^7 miles and 1.41×10^8 miles, find the distance between Earth and Mars at the time of the observation.
14. In each of the following add the vectors and find the magnitude of the sum vector:

    (a) [2,7] and [1,3]   (b) [−5,2] and [−3,2]   (c) [−3,−5] and [4,5]
    (d) [3,−6] and [2,−1]   (e) [5,6] and [3,−8]   (f) [−3,6] and [1,−5]

15. A *unit vector* is defined to be a vector whose magnitude is 1 unit. Show that if [h,k] is a nonzero vector, then

$$\left[\frac{h}{\sqrt{h^2 + k^2}}, \frac{k}{\sqrt{h^2 + k^2}} \right]$$

    is a unit vector.
16. Show that any nonzero vector can be expressed as the magnitude of the vector times a unit vector. HINT: Since $c[h,k] = [ch,ck]$, it follows that

$$[h,k] = \sqrt{h^2 + k^2} \left[\frac{h}{\sqrt{h^2 + k^2}}, \frac{k}{\sqrt{h^2 + k^2}} \right]$$

17. Express each of the following as the magnitude of the vector times a unit vector:

    (a) [4,3]   (b) [2,−3]   (c) [−3,4]   (d) [−2,−5]

18. In each of the following let the vector from the origin to the point P meet the unit circle at the point Q. Find the coordinates of Q, given the point P to be

(a) $(1,3)$ (b) $(3,0)$ (c) $(-3,2)$ (d) $(0,-4)$

10.4 LAW OF COSINES

If two sides and the included angle of a triangle are given, or if the three sides are given, the law of sines will not apply directly. In order to solve such triangles, we develop a second basic relation between the sides and angles of any triangle.

Place a rectangular coordinate system in the plane of any triangle ABC so that the vertex A is at the origin, with side c lying on the positive x axis. Then angle α is in standard position, and the coordinates of B are $(c,0)$. The coordinates of the vertex C are $(b \cos \alpha, b \sin \alpha)$, as shown in Fig. 10.12. By the distance formula,

$$a^2 = (b \cos \alpha - c)^2 + (b \sin \alpha - 0)^2$$
$$= b^2 \cos^2 \alpha - 2bc \cos \alpha + c^2 + b^2 \sin^2 \alpha$$
$$= b^2(\cos^2 \alpha + \sin^2 \alpha) + c^2 - 2bc \cos \alpha$$

Hence $a^2 = b^2 + c^2 - 2bc \cos \alpha$ (10.5)

Similarly, by placing the coordinate system so that angle β is in standard position, we can obtain

$$b^2 = a^2 + c^2 - 2ac \cos \beta \qquad (10.6)$$

By placing the coordinate system so that angle γ is in standard position, we can show also that

$$c^2 = a^2 + b^2 - 2ab \cos \gamma \qquad (10.7)$$

Law of cosines   Relations (10.5) to (10.7) are referred to as the *law of cosines*.

EXAMPLE 1. Given triangle ABC in which $a = 30$, $b = 20$, and $\gamma = 108°$, find side c.

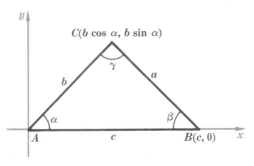

FIGURE 10.12
Law of cosines

Solution. By (10.7), $c^2 = a^2 + b^2 - 2ab \cos \gamma$. Hence,

$$c^2 = 30^2 + 20^2 - 2(30)(20) \cos 108°$$
$$= 900 + 400 - (1{,}200)(-0.3090)$$
$$= 1{,}670.8$$

Then $c = 40.87$

$$= 41 \qquad \text{to two significant figures}$$

EXAMPLE 2. Given triangle ABC in which $a = 15$, $b = 20$, $c = 28$. Find the angle γ.

Solution. Since $c^2 = a^2 + b^2 - 2ab \cos \gamma$,

$$\cos \gamma = \frac{a^2 + b^2 - c^2}{2ab}$$

$$= \frac{15^2 + 20^2 - 28^2}{2(15)(20)} = -0.2650$$

Therefore, $\gamma = 105°22'$

$$= 105° \qquad \text{to agree with measurement of } a,\ b,\ c$$

EXERCISE SET 10.4

SOLVE the following triangles:

1. $a = 110$, $b = 205$, $\gamma = 41°50'$
2. $b = 51.5$, $c = 81.6$, $\alpha = 24°40'$
3. $b = 73.4$, $c = 67.2$, $\alpha = 106°10'$
4. $a = 220$, $c = 180$, $\beta = 54°30'$
5. $a = 350$, $c = 40$, $\beta = 105°20'$
6. $b = 0.125$, $c = 0.350$, $\alpha = 135°40'$
7. $a = 76.1$, $b = 33.4$, $\beta = 82°50'$
8. $a = 43.75$, $c = 35.12$, $\beta = 151°32'$
9. $a = 43$, $b = 37$, $c = 55$
10. $a = 51$, $b = 52$, $c = 40$
11. $a = 522$, $b = 873$, $c = 717$
12. $a = 255$, $b = 290$, $c = 425$
13. The points A and B are on opposite sides of a hangar. A third point C is chosen so that $CA = 212$ ft, $CB = 350$ ft. The angle BCA then measures $66°10'$. Find the distance from A to B.
14. Two forces act on a body, one of 84.5 lb, the other of 32.1 lb. Their directions make an angle of $72°30'$ with each other. Find the magnitude and direction of their resultant.
15. Two forces have directions which make an angle of $118°50'$ with each other. The magnitudes of the forces are 280 and 370 lb respectively. Find the magnitude and direction of their resultant.
16. A pilot takes off from city A to fly to city B 400 mi away. After flying 120 mi he discovers that he has been traveling $8°50'$ off course. At this point, how far from city B is he?
17. Two sides of a parallelogram are 12.5 in. and 16.2 in. If the angle between them is $42°20'$, find the length of the shorter diagonal.
18. The angle of elevation of the top of a building at a point due north of it is $28°30'$. At a point 145 ft due west of the first observation point the angle of elevation is $22°20'$. How high is the building?

10.5   AREA OF A TRIANGLE

If we place a rectangular coordinate system in the plane of a triangle ABC so that the vertex A and the origin coincide, and so that the side c lies on the positive x axis, then the angle α will be in standard position.   The coordinates of B are $(c,0)$, and the coordinates of C are $(b \cos \alpha, b \sin \alpha)$, as shown in Fig. 10.13.   This means that the altitude h of the triangle is the y coordinate of C.   Hence,

$$h = b \sin \alpha$$

Let T be the area of the triangle ABC.   Then

$$T = \frac{1}{2} ch = \frac{1}{2} c(b \sin \alpha)$$

or $$T = \frac{1}{2} bc \sin \alpha \qquad (10.8)$$

Additional relations for finding the area T of the triangle can be developed. For example, from the law of sines,

$$c = \frac{b \sin \gamma}{\sin \beta}$$

and if we substitute this value of c into Eq. (10.8), we have

$$T = \frac{b^2 \sin \alpha \sin \gamma}{2 \sin \beta} \qquad (10.9)$$

Equation (10.8) is used when two sides and the included angle are given, and Eq. (10.9) can be applied when two angles and a side of the triangle are given.

EXAMPLE 1.   Find the area of triangle ABC if $b = 20$, $c = 80$, and $\alpha = 64°$.

Solution. $T = \frac{1}{2}(20)(80) \sin 64° = 800(0.8988) = 719.04$
$= 720$ sq units (approx.)

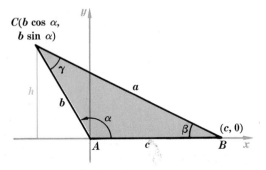

$C(b \cos \alpha,$
$b \sin \alpha)$

FIGURE 10.13
Area of a triangle

EXAMPLE 2. Find the area of the triangle ABC if $b = 60$, $\alpha = 54°$, $\beta = 110°$.

Solution. $\gamma = 180° - (54° + 110°) = 16°$

Then $\qquad T = \dfrac{b^2 \sin \alpha \sin \gamma}{2 \sin \beta} = \dfrac{60^2 \sin 54° \sin 16°}{2 \sin 110°}$

$$= \dfrac{1{,}800 \sin 54° \sin 16°}{\sin 110°}$$

$$\log T = \log 1{,}800 + \log \sin 54° + \log \sin 16° - \log \sin 110°$$

$$\log 1{,}800 = 3.2553$$
$$\log \sin 54° = 9.9080 - 10$$
$$\log \sin 16° = \underline{9.4403 - 10}$$
$$22.6036 - 20$$
$$\log \sin 110° = \underline{9.9730 - 10}$$
$$12.6306 - 10$$

Thus $\quad \log T = 2.6306$

and $\qquad T = 427.2 = 430$ sq units (approx.)

EXERCISE SET 10.5

FIND the area of triangle ABC in each of the following:

1. $b = 12$, $c = 7$, $\alpha = 36°$
2. $a = 15$, $b = 24$, $\gamma = 155°$
3. $a = 50$, $c = 72$, $\beta = 73°$
4. $b = 15$, $c = 40$, $\alpha = \pi/5$ rad
5. $a = 155$, $b = 255$, $\gamma = 2\pi/5$ rad
6. $b = 12$, $\beta = 1.2$ rad, $\alpha = 0.72$ rad
7. $c = 12.5$, $\alpha = 150°30'$, $\beta = 12°20'$
8. $a = 7$, $b = 9$, $c = 13$. HINT: Use the law of cosines to find one angle.
9. $a = 30$, $b = 38$, $c = 49$
10. The sides of a triangular field measure 23, 29, and 46 rd respectively. If 160 sq rd = 1 acre, find the number of acres in the field.
11. Prove that $T = \dfrac{a^2 \sin \beta \sin \gamma}{2 \sin \alpha}$.
12. Prove that $T = \dfrac{c^2 \sin \alpha \sin \beta}{2 \sin \gamma}$.
13. Show that if $\gamma = 90°$, the law of cosines is equivalent to the Pythagorean theorem.
14. Let ABC be an acute triangle (for convenience only). Then show that $a = b \cos \gamma + c \cos \beta$, that $b = c \cos \alpha + a \cos \gamma$, and that $c = a \cos \beta + b \cos \alpha$.
15. Use the three equations of the preceding problem to derive the law of cosines. HINT: Multiply each member of the first equation by a, of the second by b, and of the third by c.
16. A light airplane travels in calm air with a speed of 160 mi/hr. If the wind is blowing at 25 mi/hr from 20° (measured clockwise from due north) and the airplane heads in the direction 150° (measured from due north), find the magnitude and direction of its velocity with reference to the ground.
17. A ship travels with a speed of 30 mi/hr in still water. If an ocean current is flowing at the rate of 8 mi/hr in the direction N10°E and the ship heads in the direction N42°E, find the magnitude and direction of its velocity with reference to the surface of the water.
18. Show that if $a^2 + b^2 = c^2$, the triangle ABC is a right triangle. This is the converse of the Pythagorean theorem.

10.6   DOT PRODUCT

A real number is called a *scalar*. In Sec. 10.3 we defined the product of a scalar times a vector as follows: If $u = [h,k]$ and c is any scalar, then $cu = c[h,k] = [ch,ck]$. The length of the vector cu is $|c|$ times the length of u. If c is positive, then cu has the same direction as u, and if c is negative, then cu has a direction opposite to that of u. For example, $3[-2,1]$ is a vector which is three times as long and has the same direction as $[-2,1]$, whereas $-3[-2,1]$ is three times as long as $[-2,1]$ and has the opposite direction.

If $u = [h_1,k_1]$ and $v = [h_2,k_2]$, the sum $h_1h_2 + k_1k_2$ is a real number that has a special name in mathematics. It is called the *dot* product (or *scalar* product) of u and v and is denoted by $u \cdot v$.

> **DEFINITION.**   If $u = [h_1,k_1]$
>
> and $v = [h_2,k_2]$ **(10.10)**
>
> then $u \cdot v = h_1h_2 + k_1k_2$

We read the symbol $u \cdot v$ as "u dot v."

EXAMPLE 1.   Find the dot product of u and v if $u = [5,-2]$ and $v = [3,6]$.

Solution.   Since $u = [5,-2]$

$$v = [3,6]$$

and $u \cdot v = 15 - 12 = 3$ by Def. (10.10)

From Definition (10.10) the following properties of the dot product can be established.

(1) $u \cdot v = v \cdot u$ the commutative property

(2)   If t is a scalar,

$$(tu) \cdot v = t(u \cdot v)$$ the associative property

(3) $u \cdot (v + w) = u \cdot v + u \cdot w$ the distributive property

The proofs of these properties are left as exercises at the end of this section.

We have defined the length of a vector $u = [h,k]$, denoted by $|u|$ to be $|u| = \sqrt{h^2 + k^2}$, Sec. 10.3. This definition can be stated in terms of the dot product as

$$|u| = \sqrt{u \cdot u} \tag{10.11}$$

This definition is easily justified because $u \cdot u = [h,k] \cdot [h,k] = h^2 + k^2$. An important thing to remember here is that

$$|u|^2 = u \cdot u \tag{10.12}$$

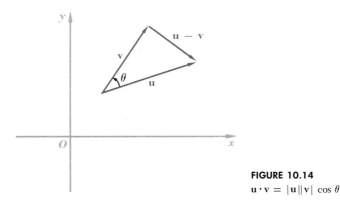

FIGURE 10.14

$\mathbf{u} \cdot \mathbf{v} = |\mathbf{u}||\mathbf{v}| \cos \theta$

Let $\mathbf{u}$ and $\mathbf{v}$ be plane vectors of nonzero length having a common initial point and let θ be the angle between them as in Fig. 10.14. Then the vector from the terminal point of $\mathbf{v}$ to the terminal point of $\mathbf{u}$ is the vector $\mathbf{u} - \mathbf{v}$. From the law of cosines, we have

$$|\mathbf{u} - \mathbf{v}|^2 = |\mathbf{u}|^2 + |\mathbf{v}|^2 - 2|\mathbf{u}||\mathbf{v}| \cos \theta$$

Therefore, by (10.12) we obtain

$$(\mathbf{u} - \mathbf{v}) \cdot (\mathbf{u} - \mathbf{v}) = |\mathbf{u}|^2 + |\mathbf{v}|^2 - 2|\mathbf{u}||\mathbf{v}| \cos \theta$$

and by the distributive property

$$(\mathbf{u} - \mathbf{v}) \cdot \mathbf{u} - (\mathbf{u} - \mathbf{v}) \cdot \mathbf{v} = |\mathbf{u}|^2 + |\mathbf{v}|^2 - 2|\mathbf{u}||\mathbf{v}| \cos \theta$$

From this, after using the commutative and distributive properties, we obtain

$$\mathbf{u} \cdot \mathbf{u} - 2\mathbf{u} \cdot \mathbf{v} + \mathbf{v} \cdot \mathbf{v} = |\mathbf{u}|^2 + |\mathbf{v}|^2 - 2|\mathbf{u}||\mathbf{v}| \cos \theta$$

Then by (10.12)

$$|\mathbf{u}|^2 - 2\mathbf{u} \cdot \mathbf{v} + |\mathbf{v}|^2 = |\mathbf{u}|^2 + |\mathbf{v}|^2 - 2|\mathbf{u}||\mathbf{v}| \cos \theta$$

and

$$\mathbf{u} \cdot \mathbf{v} = |\mathbf{u}||\mathbf{v}| \cos \theta \qquad (10.13)$$

This interpretation of the dot product of two vectors is of particular importance to the engineer.

If Eq. (10.13) is solved for $\cos \theta$, we have

$$\cos \theta = \frac{\mathbf{u} \cdot \mathbf{v}}{|\mathbf{u}||\mathbf{v}|}$$

EXAMPLE 2. Find the cosine of the angle θ between [2,3] and [3,2].

Solution. $\cos \theta = \dfrac{[2,3] \cdot [3,2]}{|[2,3]||[3,2]|}$

$= \dfrac{6 + 6}{\sqrt{4 + 9} \; \sqrt{4 + 9}}$

$= \dfrac{12}{13}$

EXAMPLE 3.   Find the cosine of the angle between $[5, -1]$ and $[2,7]$.

Solution. $\cos \theta = \dfrac{3}{\sqrt{26} \; \sqrt{53}} = \dfrac{3}{\sqrt{1,378}}$

If $\mathbf{u}$ and $\mathbf{v}$ are perpendicular, then $\theta = 90°$ and from (10.13) $\mathbf{u} \cdot \mathbf{v} = 0$. Conversely, if $\mathbf{u} \cdot \mathbf{v} = 0$, then $\cos \theta = 0$ and $\theta = 90°$. We understand that both $\mathbf{u}$ and $\mathbf{v}$ have nonzero length. We conclude that the vectors $\mathbf{u}$ and $\mathbf{v}$ are perpendicular if and only if $\mathbf{u} \cdot \mathbf{v} = 0$.

EXAMPLE 4.   Show that $\mathbf{u} = [2,6]$ and $\mathbf{v} = [6, -2]$ are perpendicular.

Solution.   Here $\mathbf{u} \cdot \mathbf{v} = 12 - 12 = 0$. Hence $\mathbf{u}$ and $\mathbf{v}$ are perpendicular.

The slope of a line in the coordinate plane was defined in Sec. 3.11. Now if a vector $\mathbf{u} = [h_1,k_1]$ lies on a line L_1, then $k_1/h_1 = m_1$ is the slope of the line L_1. If the vector $\mathbf{v} = [h_2,k_2]$ lies on a line L_2, then $m_2 = k_2/h_2$ is the slope of L_2. Let θ be the angle between L_1 and L_2. Now L_1 and L_2 are perpendicular if $\mathbf{u}$ and $\mathbf{v}$ are perpendicular; that is, if $\mathbf{u} \cdot \mathbf{v} = h_1 h_2 + k_1 k_2 = 0$. Hence, L_1 is perpendicular to L_2 if

$$h_1 h_2 = -k_1 k_2$$

so that $$\dfrac{k_1}{h_1} = -\dfrac{1}{k_2/h_2}$$

or $$m_1 = -\dfrac{1}{m_2}$$

This agrees with Theorem (3.14) of Sec. 3.11 which states that two lines are perpendicular if and only if their slopes are negative reciprocals of each other.

EXAMPLE 5.   Show that the line through the points $P_1(3,2)$ and $P_2(5,6)$ is perpendicular to the line through the points $P_3(5,2)$ and $P_4(1,4)$.

Solution.   The vector from $(3,2)$ to $(5,6)$ is the vector $[2,4]$. Hence, the slope of the line through P_1 and P_2 is $\frac{4}{2} = 2$. The vector from P_3 to P_4 is the vector $[-4,2]$. Therefore the slope of the line through these two points is $2/-4 = -\frac{1}{2}$. Since the slope of one line is the negative reciprocal of the slope of the other, the lines are perpendicular.

EXERCISE SET 10.6

1. Find the dot product of $[5,3]$ and $[2,7]$.
2. Find the product $[6,2] \cdot [7,-1]$.
3. Find $\mathbf{u} \cdot \mathbf{v}$ if $\mathbf{u} = [-2,-3]$ and $\mathbf{v} = [-1,-5]$.
4. Find the cosine of the angle between $[5,7]$ and $[7,5]$.
5. Find the cosine of the angle between $[3,-2]$ and $[5,1]$.
6. Show that $[2,3]$ and $[-3,2]$ are perpendicular.
7. Show that $[-5,7]$ and $[7,5]$ are perpendicular.
8. If $\mathbf{u} = [3,5]$ and $\mathbf{v} = [5,k]$ and $\mathbf{u}$ is perpendicular to $\mathbf{v}$, find k.
9. Find the value of h so that $[h,7]$ is perpendicular to $[3,6]$.
10. Find the acute angle between the vectors $[1,\sqrt{3}]$ and $[-1,\sqrt{3}]$.
11. Find the acute angle between the vectors $[\sqrt{3},1]$ and $[3,3\sqrt{3}]$.

In the following assume that all vectors have nonzero length:

12. Prove that $\mathbf{u} \cdot \mathbf{v} = \mathbf{v} \cdot \mathbf{u}$.
13. Prove that $(t\mathbf{u}) \cdot \mathbf{v} = t(\mathbf{u} \cdot \mathbf{v})$, t a scalar.
14. Prove that $\mathbf{u} \cdot (\mathbf{v} + \mathbf{w}) = \mathbf{u} \cdot \mathbf{v} + \mathbf{u} \cdot \mathbf{w}$.

15. The work done by a constant force $\vec{f}$ producing a displacement $\vec{s}$ in the direction of $\vec{f}$ is the product $|\vec{f}\,\|\vec{s}|$. Show that if the force $\vec{f}$ makes an angle θ with $\vec{s}$, the work w done in producing a displacement s is given by

$$w = \vec{f} \cdot \vec{s}$$

16. Show that the work done in moving a object along a straight line from the point $(2,5)$ to the point $(10,5)$ by the constant force $\vec{f} = [4,3]$ is 32 ft-lb, where force is in pounds and distance is in feet. HINT. The displacement from $(2,5)$ to $(10,5)$ is the vector $[8,0]$. Use the result of Prob. 15.

17. Show that the work done in moving an object along a straight line from the point $(2,3)$ to the point $(12,8)$ by the constant force $\vec{f} = [4,6]$ is 70 ft-lb, where force is in pounds and distance is in feet.

11

TRIGONOMETRIC FUNCTIONS OF SUMS AND DIFFERENCES

In this chapter, we shall consider the basic relations used to express trigonometric functions of sums, products, and multiples of real numbers in terms of functions of the separate numbers.

11.1 ADDITION RELATIONS

The first of the relations to be considered is the identity for $\cos (u - v)$. Let u and v be real numbers such that $\pi/2 < v < u < \pi$ and let $P(u)$ and $Q(v)$ be the trigonometric points on the unit circle corresponding to u and v, respectively, as in Fig. 11.1. We will show that for these particular values of u and v, $\cos (u - v) = \cos u \cos v + \sin u \sin v$.

The coordinates of $P(u)$ are $(\cos u, \sin u)$ and the coordinates of $Q(v)$ are $(\cos v, \sin v)$ by the definitions of the sine and cosine functions (Sec. 5.2). Now let the point R on the unit circle be determined so that arc AR has measure $u - v$ units. Then R is the trigonometric point $R(u - v)$ and has coordinates $[\cos (u - v), \sin (u - v)]$. Since the arc AR has the same length as the arc PQ, the chords AR and PQ are equal. By the distance formula,

$$|AR|^2 = [\cos (u - v) - 1]^2 + [\sin (u - v) - 0]^2$$

$$= \cos^2 (u - v) - 2 \cos (u - v) + 1 + \sin^2 (u - v)$$

$$= [\cos^2 (u - v) + \sin^2 (u - v)] - 2 \cos (u - v) + 1$$

$$= 2 - 2 \cos (u - v)$$

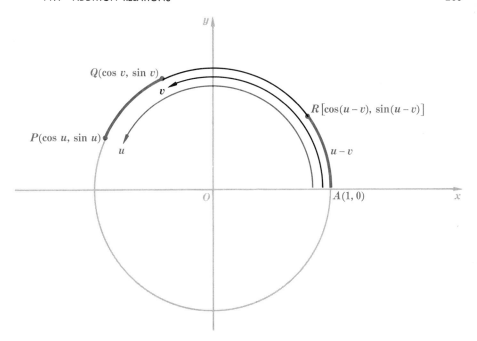

FIGURE 11.1 $\cos{(u - v)} = \cos u \cos v + \sin u \sin v$

and

$$|PQ|^2 = (\cos u - \cos v)^2 + (\sin u - \sin v)^2$$
$$= \cos^2 u - 2 \cos u \cos v + \cos^2 v + \sin^2 u - 2 \sin u \sin v + \sin^2 v$$
$$= (\cos^2 u + \sin^2 u) + (\cos^2 v + \sin^2 v) - 2 \cos u \cos v - 2 \sin u \sin v$$
$$= 2 - 2 \cos u \cos v - 2 \sin u \sin v$$

Since chord AR = chord PQ,

$$2 - 2 \cos{(u - v)} = 2 - 2 \cos u \cos v - 2 \sin u \sin v$$

Hence, $\qquad\qquad \cos{(u - v)} = \cos u \cos v + \sin u \sin v \qquad\qquad (11.1)$

To show that relation (11.1) holds for any two real numbers u and v, we reason thus: By the distance formula,

$$|PQ|^2 = 2 - 2 \cos u \cos v - 2 \sin u \sin v$$

as before. We now rotate the coordinate system so that the positive x axis passes through the terminal point of the arc v at Q. After this rotation, the new coordinates of Q are $(1,0)$ and the new coordinates of P are $[\cos{(u - v)}, \sin{(u - v)}]$. In this position

$$|PQ|^2 = [1 - \cos{(u - v)}]^2 + [0 - \sin{(u - v)}]^2$$
$$= 2 - 2 \cos{(u - v)}$$

Because the distance between P and Q is not changed when the co-ordinate system is rotated, it follows that the two expressions for $|PQ|^2$ are equal. This equality establishes relation (11.1).

EXAMPLE 1. Find $\cos(\pi/12)$.

Solution. Since $\pi/12 = \pi/3 - \pi/4$, we have

$$\cos\frac{\pi}{12} = \cos\left(\frac{\pi}{3} - \frac{\pi}{4}\right) = \cos\frac{\pi}{3}\cos\frac{\pi}{4} + \sin\frac{\pi}{3}\sin\frac{\pi}{4}$$

$$= \frac{1}{2}\left(\frac{\sqrt{2}}{2}\right) + \frac{\sqrt{3}}{2}\left(\frac{\sqrt{2}}{2}\right) = \frac{\sqrt{2} + \sqrt{6}}{4}$$

The relation (11.1) can now be used to develop additional relations. For example, if we set $u = \pi/2$ in (11.1), we immediately get

$$\cos\left(\frac{\pi}{2} - v\right) = \cos\frac{\pi}{2}\cos v + \sin\frac{\pi}{2}\sin v$$

$$= 0 \cdot \cos v + 1 \cdot \sin v$$

Hence, $$\cos\left(\frac{\pi}{2} - v\right) = \sin v \qquad (11.2)$$

Now, let $v = (\pi/2) - u$ in (11.2), so that

$$u = \frac{\pi}{2} - v$$

Then $$\cos u = \cos\left(\frac{\pi}{2} - v\right) = \sin v = \sin\left(\frac{\pi}{2} - u\right)$$

Therefore, $$\sin\left(\frac{\pi}{2} - u\right) = \cos u \qquad (11.3)$$

Since $$\tan\left(\frac{\pi}{2} - u\right) = \frac{\sin\left(\frac{\pi}{2} - u\right)}{\cos\left(\frac{\pi}{2} - u\right)} = \frac{\cos u}{\sin u}$$

we have $$\tan\left(\frac{\pi}{2} - u\right) = \cot u \qquad (11.4)$$

Similar relations can be derived for $\cot(\pi/2 - u)$, $\sec(\pi/2 - u)$, $\csc(\pi/2 - u)$. They are left as exercises.

Let us consider what happens if we take $u = 0$ in relation (11.1). In that case,

$$\cos(0 - v) = \cos 0 \cos v + \sin 0 \sin v$$

Thus, $\cos{(-v)} = \cos{v}$ (11.5)

It follows from (11.2) that

$$\sin{(-v)} = \cos{\left[\frac{\pi}{2} - (-v)\right]} = \cos{\left(\frac{\pi}{2} + v\right)}$$

$$= \cos{\left[v - \left(-\frac{\pi}{2}\right)\right]}$$

$$= \cos{v}\cos{\left(-\frac{\pi}{2}\right)} + \sin{v}\sin{\left(-\frac{\pi}{2}\right)}$$

Hence, $\sin{(-v)} = -\sin{v}$ (11.6)

We can now prove that

$$\tan{(-v)} = -\tan{v}$$ (11.7)

The proof of (11.7) is left as an exercise.

EXAMPLE 2.   Express cos 1.40 as a function of a number less than

$$\frac{\pi}{4} = 0.785$$

and express sin 60°20′ as a function of an angle less than 45°.

Solution. $\cos{1.40} = \sin{\left(\frac{\pi}{2} - 1.40\right)} = \sin{(1.57 - 1.40)} = \sin{0.17}$

and $\sin{60°20′} = \cos{(90° - 60°20′)} = \cos{29°40′}$

Functions of the sum of two numbers will be considered next.   Since

$$\cos{(u + v)} = \cos{[u - (-v)]}$$
$$= \cos{u}\cos{(-v)} + \sin{u}\sin{(-v)}$$

by (11.1), and since $\cos{(-v)} = \cos{v}$ and $\sin{(-v)} = -\sin{v}$, we have

$$\cos{(u + v)} = \cos{u}\cos{v} - \sin{u}\sin{v}$$ (11.8)

Furthermore,

$$\sin{(u + v)} = \cos{\left[\frac{\pi}{2} - (u + v)\right]}$$ by Eq. (11.2)

$$= \cos{\left[\left(\frac{\pi}{2} - u\right) - v\right]}$$

$$= \cos{\left(\frac{\pi}{2} - u\right)}\cos{v} + \sin{\left(\frac{\pi}{2} - u\right)}\sin{v}$$

by Eq. (11.1)

Consequently, by (11.2) and (11.3),

$$\sin (u + v) = \sin u \cos v + \cos u \sin v \tag{11.9}$$

Since $\sin (u - v) = \sin [u + (-v)]$
$$= \sin u \cos (-v) + \cos u \sin (-v)$$

we have $\sin (u - v) = \sin u \cos v - \cos u \sin v \tag{11.10}$

Finally $\tan (u + v) = \dfrac{\sin (u + v)}{\cos (u + v)}$

$$= \frac{\sin u \cos v + \cos u \sin v}{\cos u \cos v - \sin u \sin v}$$

Now, if $\cos u \neq 0$ and $\cos v \neq 0$, we can divide numerator and denominator of the right member by $\cos u \cos v$ and simplify. We thus obtain

$$\tan (u + v) = \frac{\tan u + \tan v}{1 - \tan u \tan v} \tag{11.11}$$

which holds for all cases except when $\tan u \tan v = 1$. In that case, $\tan (u + v)$ is undefined. The formula for $\tan (u - v)$ is left as an exercise.

EXAMPLE 3. Find $\cos (\pi/12)$ by using the fact that $\pi/12 = \pi/4 - \pi/6$.

Solution. $\cos \dfrac{\pi}{12} = \cos \left(\dfrac{\pi}{4} - \dfrac{\pi}{6} \right)$

$$= \cos \frac{\pi}{4} \cos \frac{\pi}{6} + \sin \frac{\pi}{4} \sin \frac{\pi}{6}$$

$$= \frac{\sqrt{2}}{2} \left(\frac{\sqrt{3}}{2} \right) + \frac{\sqrt{2}}{2} \left(\frac{1}{2} \right) = \frac{\sqrt{6} + \sqrt{2}}{4}$$

EXAMPLE 4. Show that $\sin (\pi - u) = \sin u$.

Solution. $\sin (\pi - u) = \sin \pi \cos u - \cos \pi \sin u$
$$= 0 \cdot \cos u - (-1) \sin u = \sin u$$

EXAMPLE 5. If $\sin u = \frac{4}{5}$, with u in the second quadrant, and $\cos v = -\frac{3}{5}$, with v in the third quadrant, find $\sin (u + v)$.

Solution. Since $\pi/2 < u < \pi$, $\cos u = -\sqrt{1 - \sin^2 u} = -\frac{3}{5}$, and since $\pi < v < 3\pi/2$, $\sin v = -\sqrt{1 - \cos^2 v} = -\frac{4}{5}$. Hence,

$$\sin (u + v) = \sin u \cos v + \cos u \sin v$$

$$= \frac{4}{5} \left(-\frac{4}{5} \right) + \left(-\frac{3}{5} \right) \left(-\frac{3}{5} \right) = -\frac{7}{25}$$

EXERCISE SET 11.1

1. Given that $\sin u = 0.3$ and $\cos v = 0.5$, u, v in first quadrant, find:

    (a) $\cos (u - v)$ (b) $\sin (u - v)$ (c) $\sin (u + v)$
    (d) $\cos (u + v)$ (e) $\cos (\pi/2 - u)$ (f) $\sin (\pi/2 - v)$
    (g) $\sin (\pi/2 + u)$ (h) $\cos (\pi/2 + v)$

2. Show that for any real value of θ,

    (a) $\cos (\pi - \theta) = -\cos \theta$ (b) $\cos (\theta - \pi) = -\cos \theta$
    (c) $\sin (\pi - \theta) = \sin \theta$ (d) $\sin (\theta - \pi) = -\sin \theta$
    (e) $\cos (\pi + \theta) = -\cos \theta$ (f) $\sin (\pi + \theta) = -\sin \theta$

3. Find the exact value of $\sin (5\pi/12)$.
4. Find the exact value of $\cos (7\pi/12)$.
5. Express each of the following as a function of a number less than $\pi/4$:

    (a) $\sin (5\pi/2)$ (b) $\cos (\pi/3)$ (c) $\tan (7\pi/12)$
    (d) $\sin 0.885$ (e) $\tan 0.92$ (f) $\cos 1$

6. Express each of the following as a function of an angle less than $45°$:

    (a) $\cos 75°$ (b) $\sin 54°$ (c) $\tan 78°$
    (d) $\sin 105°$ (e) $\cot 128°$ (f) $\cos 84°23'$

7. Use Table I and evaluate

    (a) $\cos 0.4 \cos 0.2 + \sin 0.4 \sin 0.2$ (b) $\sin .5 \cos 1 + \cos .5 \sin 1$
    (c) $\cos 0.2 \cos 0.1 - \sin 0.2 \sin 0.1$ (d) $\sin 2 \cos 1 - \cos 2 \sin 1$

8. Find the exact value of

    (a) $\sin 75°$ (b) $\sin 105°$ (c) $\cos 195°$
    (d) $\tan 15°$ (e) $\sin 15°$ (f) $\tan (-\pi/12)$

9. If $\sin u = -\frac{3}{5}$, $\pi < u < 3\pi/2$, and $\cos v = -\frac{4}{5}$, $\pi/2 < v < \pi$, find

    (a) $\sin (u + v)$ (b) $\sin (u - v)$ (c) $\cos (u + v)$ (d) $\cos (u - v)$

SOLVE the following problems, where $0 < v < u < \pi/2$:

10. Show that
$$\tan (u - v) = \frac{\tan u - \tan v}{1 + \tan u \tan v}$$
    $\tan u \tan v \neq -1$.
11. If $\sin u = \frac{4}{5}$ and $\cos v = \frac{15}{17}$, find the value of $\sin (u + v)$.
12. If $\cos u = \frac{12}{13}$ and $\sin v = \frac{7}{25}$, find the value of $\cos (u + v)$.
13. If $\tan u = \frac{1}{2}$ and $\tan v = \frac{1}{3}$, find the value of $\tan (u + v)$.
14. If $\sin u = \frac{3}{5}$ and $\sin v = \frac{5}{13}$, evaluate $\tan (u + v)$.
15. If $\cos u = \frac{3}{5}$ and $\sin v = \frac{5}{13}$, evaluate $\cos (u - v)$.
16. Given that $\tan u = 2$ and $\tan v = \frac{1}{3}$, find the value of $\tan (u - v)$. Use the result of Prob. 10.

EXPRESS each of the following as a function of one number:

17. $\cos (\pi/3) \cos (\pi/6) + \sin (\pi/3) \sin (\pi/6)$ 18. $\sin (\pi/3) \cos (\pi/4) + \cos (\pi/3) \sin (\pi/4)$
19. $\sin \frac{1}{2} \cos 1 + \cos \frac{1}{2} \sin 1$ 20. $\cos 2 \cos 0.5 - \sin 2 \sin 0.5$
21. $\frac{1}{2} \cos (\pi/12) - (\sqrt{3}/2) \sin (\pi/12)$

SIMPLIFY each of the following expressions:

22. $\sin(u - v) \cos v + \cos(u - v) \sin v$

23. $\cos(u + v) \cos v + \sin(u + v) \sin v$

24. Show that $\sin u \cos(u - \pi/6) - \cos u \sin(u - \pi/6) = \frac{1}{2}$.

25. If $0 < u < \pi/2$, and if $\sin 2u \cos 3u + \cos 2u \sin 3u = \frac{1}{2}$, find u. Give answer in terms of π.

DERIVE the following formulas:

26. $\cot\left(\dfrac{\pi}{2} - t\right) = \tan t$ **27.** $\sec\left(\dfrac{\pi}{2} - t\right) = \csc t$

28. $\csc\left(\dfrac{\pi}{2} - t\right) = \sec t$ **29.** $\tan(-t) = -\tan t$

30. $\tan(2\pi - v) = -\tan v$ **31.** $\sin(2\pi + t) = \sin t$

32. $\cos(2\pi + t) = \cos t$

PROVE the following identities:

33. $\dfrac{\sin(u + v)}{\cos u \cos v} = \tan u + \tan v$ **34.** $\dfrac{\cos(\alpha - \beta)}{\cos \alpha \sin \beta} = \tan \alpha + \cot \beta$

35. $\sin(u + v) \sin(u - v) = \sin^2 u - \sin^2 v$

36. $\cos(u + v) \cos(u - v) = \cos^2 u - \sin^2 v$

37. $\tan\left(\dfrac{\pi}{4} + t\right) = \dfrac{1 + \tan t}{1 - \tan t}$

38. $\sin 2t = 2 \sin t \cos t$ (HINT: Take $t = u = v$ in Eq. (11.9)

39. $\tan \alpha = \dfrac{\sin(\alpha + \beta) + \sin(\alpha - \beta)}{\cos(\alpha + \beta) + \cos(\alpha - \beta)}$ **40.** $\dfrac{\tan u + \tan v}{\tan u - \tan v} = \dfrac{\sin(u + v)}{\sin(u - v)}$

41. $\dfrac{1 + \tan u \tan v}{\tan u + \tan v} = \dfrac{\cos(u - v)}{\sin(u + v)}$ **42.** $\dfrac{\sin(u + v) \sin(u - v)}{\cos^2 u \cos^2 v} = \sec^2 u - \sec^2 v$

43. Prove that the relation (11.7) is an identity.

44. Prove that $\sin 3t = \sin 5t \cos 2t - \cos 5t \sin 2t$

11.2   TRIGONOMETRIC FUNCTIONS OF MULTIPLES OF t

The formulas which give the values of the sine, cosine, and tangent functions at $2t$ in terms of the values of these functions at t are direct consequences of the addition formulas.

Let $u = v = t$ in relation (11.9). Then

$$\sin(t + t) = \sin t \cos t + \cos t \sin t$$

Hence, $$\sin 2t = 2 \sin t \cos t \tag{11.12}$$

Similarly, since from the relation (11.8)

$$\cos(t + t) = \cos t \cos t - \sin t \sin t$$

$$\cos 2t = \cos^2 t - \sin^2 t \tag{11.13}$$

Two other forms for $\cos 2t$ can be obtained by using the identity $\sin^2 t + \cos^2 t = 1$. Thus, since

$$\cos 2t = \cos^2 t - \sin^2 t = \cos^2 t - (1 - \cos^2 t)$$

$$\cos 2t = 2 \cos^2 t - 1 \tag{11.14}$$

Also $\qquad \cos 2t = \cos^2 t - \sin^2 t = (1 - \sin^2 t) - \sin^2 t$

Hence, $\qquad \cos 2t = 1 - 2 \sin^2 t \tag{11.15}$

EXAMPLE 1. $\quad \sin \dfrac{2\pi}{3} = \sin 2 \left(\dfrac{\pi}{3} \right) = 2 \sin \dfrac{\pi}{3} \cos \dfrac{\pi}{3}$

$$= 2 \left(\frac{\sqrt{3}}{2} \right) \left(\frac{1}{2} \right) = \frac{\sqrt{3}}{2}$$

Since $\tan (t + t) = (\tan t + \tan t)/(1 - \tan t \tan t)$, $\tan t \tan t \neq 1$, we have the relation

$$\tan 2t = \frac{2 \tan t}{1 - \tan^2 t} \qquad \tan^2 t \neq 1 \tag{11.16}$$

If $\tan^2 t = 1$, $\tan 2t$ is undefined.

Two relations that are very useful in calculus courses can be derived by solving (11.14) for $\cos^2 t$ and (11.15) for $\sin^2 t$:

$$\sin^2 t = \frac{1 - \cos 2t}{2} \tag{11.17}$$

$$\cos^2 t = \frac{1 + \cos 2t}{2} \tag{11.18}$$

The values of the functions at $t/2$ in terms of their values at t can be found from (11.17) and (11.18). From (11.17),

$$\sin^2 u = \frac{1 - \cos 2u}{2} \qquad \text{for every } u \in R$$

Let $u = t/2$. Then

$$\sin^2 \frac{t}{2} = \frac{1 - \cos t}{2}$$

and $\quad \sin \dfrac{t}{2} = \sqrt{\dfrac{1 - \cos t}{2}} \qquad \dfrac{t}{2}$ in first or second quadrant

$$= -\sqrt{\frac{1 - \cos t}{2}} \qquad \frac{t}{2} \text{ in third or fourth quadrant} \tag{11.19}$$

Similarly,

$$\cos \frac{t}{2} = \sqrt{\frac{1 + \cos t}{2}} \qquad \frac{t}{2} \text{ in first or fourth quadrant}$$

$$= -\sqrt{\frac{1 + \cos t}{2}} \qquad \frac{t}{2} \text{ in second or third quadrant}$$

(11.20)

EXAMPLE 2. $\sin \dfrac{\pi}{8} = \sqrt{\dfrac{1 - \cos \dfrac{\pi}{4}}{2}} = \sqrt{\dfrac{1 - \dfrac{\sqrt{2}}{2}}{2}}$

$$= \frac{1}{2}\sqrt{2 - \sqrt{2}}$$

Since $\tan u = \sin u / \cos u$,

$$\tan \frac{t}{2} = \frac{\sin (t/2)}{\cos (t/2)} = \frac{2 \sin (t/2) \cos (t/2)}{2 \cos^2 (t/2)}$$

Then, from (11.12) and (11.20),

$$\tan \frac{t}{2} = \frac{\sin t}{1 + \cos t} \qquad \cos t \neq -1 \qquad (11.21)$$

An alternative form of this relation is

$$\tan \frac{t}{2} = \frac{1 - \cos t}{\sin t} \qquad \sin t \neq 0 \qquad (11.22)$$

The derivation is left as an exercise.

EXAMPLE 3. $\tan \left(-\dfrac{\pi}{8}\right) = \dfrac{\sin \left(-\dfrac{\pi}{4}\right)}{1 + \cos \left(-\dfrac{\pi}{4}\right)}$

$$= \frac{-(\sqrt{2}/2)}{1 + (\sqrt{2}/2)} = \frac{-\sqrt{2}}{2 + \sqrt{2}}$$

$$= 1 - \sqrt{2}$$

EXAMPLE 4. Find $\sin (-22.5°)$.

Solution. $\sin (-22.5°) = \sin \left(-\dfrac{45°}{2}\right)$

$$= -\sqrt{\frac{1 - \cos (-45°)}{2}}$$

$$= -\sqrt{\frac{1 - \dfrac{\sqrt{2}}{2}}{2}}$$

$$= -\frac{1}{2}\sqrt{2 - \sqrt{2}}$$

EXAMPLE 5.   Derive a relation for $\sin 3t$ in terms of $\sin t$.

Solution. $\sin 3t = \sin (2t + t)$

$$= \sin 2t \cos t + \cos 2t \sin t$$
$$= (2 \sin t \cos t) \cos t + (1 - 2 \sin^2 t) \sin t$$
$$= 2 \sin t \cos^2 t + \sin t - 2 \sin^3 t$$
$$= 2 \sin t (1 - \sin^2 t) + \sin t - 2 \sin^3 t$$
$$= 3 \sin t - 4 \sin^3 t$$

EXERCISE SET 11.2

FIND the exact value of the given function in each of the following by using the relations (11.12) through (11.22):

1. $\cos 15°$ 2. $\sin 22.5°$ 3. $\cos 67.5°$

4. $\tan 67.5°$ 5. $\sin 105°$ 6. $\cos 105°$

7. $\sin 15°$ 8. $\tan 15°$ 9. $\tan 22.5°$

10. $\tan 150°$ 11. $\cos 150°$ 12. $\sin 75°$

13. $\sec (3\pi/8)$ 14. $\cos (5\pi/16)$ 15. $\sin (\pi/24)$

16.   Given that $\pi/2 < t < \pi$ and $\cos t = -\frac{5}{13}$, find the value of

(a) $\sin 2t$ (b) $\cos 2t$ (c) $\tan 2t$

(d) $\sin (t/2)$ (e) $\cos (t/2)$ (f) $\tan (t/2)$

(g) $\sec 2t$ (h) $\csc 2t$ (i) $\cot 2t$

17.   Given that $\pi < t < 3\pi/2$ and $\sin t = -\frac{9}{11}$, find

(a) $\sin 2t$ (b) $\sin (t/2)$ (c) $\cos 2t$

(d) $\cos (t/2)$ (e) $\tan 2t$ (f) $\tan (t/2)$

18.   Given that $3\pi/2 < t < 2\pi$ and $\tan t = -3$, find the value of

(a) $\tan 2t$ (b) $\tan (t/2)$ (c) $\sin 2t$

(d) $\sin (t/2)$ (e) $\cos 2t$ (f) $\cos (t/2)$

19.   If $0 < u < \pi/2$ and $\cos u = \frac{12}{13}$, find the value of

(a) $\sin 3u$ (b) $\sin 4u$ (c) $\cos 3u$

(d) $\cos 4u$ (e) $\tan 3u$ (f) $\tan 4u$

PROVE the following identities:

20. $\cot t + \tan t = 2 \csc 2t$ 21. $\cot t - \tan t = 2 \cot 2t$

22. $\dfrac{\cot^2 t - 1}{\cot^2 t + 1} = \cos 2t$ 23. $\dfrac{1 - \sin 2x}{\sin x - \cos x} = \sin x - \cos x$

24. $\cot 2x = \dfrac{\cot^2 x - 1}{2 \cot x}$ 25. $\tan 2x - \sec 2x = \dfrac{\tan x - 1}{\tan x + 1}$

26. $\dfrac{2 \sin u \cos u}{\cos^2 u - \sin^2 u} = \tan 2u$ 27. $\dfrac{\sin^3 t - \cos^3 t}{\sin t - \cos t} = \dfrac{2 + \sin 2t}{2}$

28. $\dfrac{\tan t/2 + \cot t/2}{\cot t/2 - \tan t/2} = \sec t$ 29. $\dfrac{1 - \tan t/2}{1 + \tan t/2} = \sec t - \tan t$

30. $\dfrac{\sin 2t - 2 \sin t}{2 \sin t + \sin 2t} + \tan^2 \dfrac{t}{2} = 0$ **31.** $\dfrac{\tan t/2 + \cot t/2}{\cot t/2 - \tan t/2} = \dfrac{1}{\cos t}$

FIND the values of t, $0 \leq t \leq \pi/2$, which satisfy the given equation in each of the following:

32. $\cos t - \cos 2t = 0$ **33.** $\sin t - \sin 2t = 0$
34. $\sin t - \cos 2t = 0$ **35.** $\cos t - \sin 2t = 0$

FIND the values of t, $0 \leq t < 2\pi$, which make the given equation in each of the following a true statement:

36. $\tan t + \sin 2t = 0$ **37.** $\tan t - \tan 2t = 0$
38. $\cos 2t + \tan t = 1$ **39.** $4 \sin^2 (t/2) - \cos^2 t = 3$
40. $\tan (t/2) - 2 \sin t = 0$ **41.** Prove that the relation (11.22) is an identity.

11.3   FACTORING IDENTITIES

The following relations are called the *factoring identities*:

$$\sin (u + v) + \sin (u - v) = 2 \sin u \cos v \qquad (11.23)$$

Proof. $\sin (u + v) + \sin (u - v)$
$$= \sin u \cos v + \cos u \sin v + \sin u \cos v - \cos u \sin v$$
$$= 2 \sin u \cos v$$

$$\sin (u + v) - \sin (u - v) = 2 \cos u \sin v \qquad (11.24)$$
$$\cos (u + v) + \cos (u - v) = 2 \cos u \cos v \qquad (11.25)$$
$$\cos (u + v) - \cos (u - v) = -2 \sin u \sin v \qquad (11.26)$$

Proofs of (11.24), (11.25), and (11.26) are left as exercises.

EXAMPLE 1.   Express $\sin 2t \cos 5t$ as a sum of sines.

Solution.   From the relation (11.23), we have

$$\sin u \cos v = \frac{1}{2} [\sin (u + v) + \sin (u - v)]$$

Hence, $\sin 2t \cos 5t = \dfrac{1}{2} [\sin 7t + \sin (-3t)]$

$$= \frac{1}{2} (\sin 7t - \sin 3t)$$

Alternative forms of the preceding relations may be derived as follows:
Let $u + v = x$ and $u - v = y$, then

$$u = \frac{x + y}{2} \qquad \text{and} \qquad v = \frac{x - y}{2}$$

Substituting these values of u and v into (11.23),

$$\sin x + \sin y = 2 \sin \frac{x+y}{2} \cos \frac{x-y}{2} \qquad (11.27)$$

Similarly, by substituting in (11.24), (11.25), and (11.26), we obtain

$$\sin x - \sin y = 2 \cos \frac{x+y}{2} \sin \frac{x-y}{2} \qquad (11.28)$$

$$\cos x + \cos y = 2 \cos \frac{x+y}{2} \cos \frac{x-y}{2} \qquad (11.29)$$

$$\cos x - \cos y = -2 \sin \frac{x+y}{2} \sin \frac{x-y}{2} \qquad (11.30)$$

EXAMPLE 2. Express $\sin 6t + \sin 2t$ as a product of functions.

Solution. From the relation (11.27), we have

$$\sin x + \sin y = 2 \sin \frac{x+y}{2} \cos \frac{x-y}{2}$$

Hence $\qquad \sin 6t + \sin 2t = 2 \sin \dfrac{6t+2t}{2} \cos \dfrac{6t-2t}{2}$

$$= 2 \sin 4t \cos 2t$$

EXERCISE SET 11.3

EXPRESS each of the following products as sums:

1. $\sin 3t \cos 5t$
2. $\cos 6t \sin 2t$
3. $2 \sin 4t \cos 2t$
4. $2 \sin 8t \cos 6t$
5. $2 \sin 3 \cos 6$
6. $2 \cos 2 \sin 0.8$
7. $\cos 1 \cos 3$
8. $\sin 3 \sin 0.2$

EVALUATE without using tables:

9. $\sin 15° \cos 45°$
10. $\cos 127°30' \sin 7°30'$
11. $\sin 22.5° \sin 67.5°$
12. $\cos 15° \cos 165°$
13. $\sin 105° + \sin 15°$
14. $\cos 255° + \cos 165°$
15. $\sin 75° - \sin 15°$
16. $\cos 165° - \cos 75°$

PROVE each of the following identities:

17. $(\sin 4t - \sin 2t)/(\cos 4t + \cos 2t) = \tan t$
18. $\sin (\pi/6 + t) + \sin (\pi/6 - t) = \cos t$
19. $\tan u + \cot v = \cos (u - v)/\cos u \sin v$
20. $\tan (u + v)/2 = (\sin u + \sin v)/(\cos u + \cos v)$
21. $(\sin t/\sec 3t) + (\cos t/\csc 3t) = \sin 4t$ 22. $(\cos 6t + \cos 4t)/(\sin 6t - \sin 4t) = \cot t$

FIND the values of t, $0 \leq t \leq \pi/2$, which satisfy the given equation:

23. $\sin t - \sin 3t = 0$ **24.** $\cos t - \cos 3t = 0$

25. $\sin t + \sin 3t = 0$ **26.** $\cos 3t - \cos 5t = 0$

27. Find the values of t, $0 \leq t \leq 2\pi$, which satisfy the equation

$$\sin t - \sin 2t + \sin 3t = 0$$

28. Prove that the relation (11.24) is an identity.

29. Prove that the relation (11.25) is an identity.

30. Prove that the relation (11.26) is an identity.

12

INVERSE TRIGONOMETRIC FUNCTIONS

Certain functions and their inverses were discussed in Chap. 3. We shall now review very briefly the concepts developed in Secs. 3.7 and 3.8 and then extend the discussion to the trigonometric functions and their inverses. The inverse trigonometric functions are important in more advanced courses.

12.1 INVERSE FUNCTIONS

The function f defined by the equation $y = f(x)$ is a set of ordered pairs, that is,

$$f = \{(x,y) \mid y = f(x)\}$$

The inverse relation of f, designated by f^{-1}, is constructed by interchanging the first and second coordinates of each ordered pair of f. Hence,

$$f^{-1} = \{(x,y) \mid x = f(y)\}$$

The domain and range of f^{-1} are the range and domain, respectively, of the function f.

Now f^{-1} may or may not be a function. If no two distinct ordered pairs of f have the same *second* coordinate, it follows that no two distinct ordered pairs of f^{-1} have the same *first* coordinate. In this case, f^{-1} is a function and is called the inverse function of the function f.

If the equation $x = f(y)$ is equivalent to an equation $y = g(x)$, with y a function of x, we denote this function by

$$y = f^{-1}(x)$$

Furthermore, $f^{-1} = \{(x,y) \mid y = f^{-1}(x)\}$

Recall that the range of a function is also called the image of the function. The terms *range* and *image* will be used interchangeably.

EXAMPLE 1.   Determine whether or not the square-root function defined by the equation

$$y = f(x) = \sqrt{x} \qquad x \geq 0$$

has an inverse function. If it does, write the equation which defines the inverse function.

Solution.   The ordered pairs of f are $(x, \sqrt{x})$. If $x_1 > x_2$, then $\sqrt{x_1} > \sqrt{x_2}$. Thus, no two distinct ordered pairs of f can have the same second coordinate. Hence, f has an inverse function.

To find the inverse function, interchange x and y in the equation $y = \sqrt{x}$ to obtain

$$x = \sqrt{y}$$

Since $x \geq 0$, we can solve for y by squaring each side of the preceding equation. Hence

$$y = x^2$$

defines the inverse function. The graphs of the function and of the inverse function are shown in Fig. 12.1a and b respectively. Note that a horizontal line can intersect the graph of f in at most one point.

EXAMPLE 2.   Given the function f defined by the equation

$$y = f(x) = x^2 - 4x$$

Determine whether or not the inverse relation is an inverse function.

FIGURE 12.1   (*a*) $f(x) = \sqrt{x}, x \geq 0$; (*b*) $f^{-1}(x) = x^2, x \geq 0$

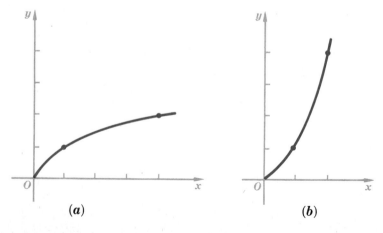

(a) (b)

Solution. Interchange x and y so that

$$x = y^2 - 4y$$

Solve for y: $\qquad y = 2 \pm \sqrt{4 + x}$

Hence, $\qquad f^{-1} = \{(x,y) \mid y = 2 \pm \sqrt{4 + x}\}$

To verify that f^{-1} is not a function, we note that $(x, 2 + \sqrt{4 + x})$ and $(x, 2 - \sqrt{4 + x})$ are distinct ordered pairs of f^{-1} and have the same first coordinate. The graphs of the function and the inverse relation are shown in Fig. 12.2*a* and *b*. Note that any horizontal line above the line $y = -4$ cuts the graph of the function in at least two points.

To determine from the graph of a function whether or not the inverse relation is a function, we need only consider a horizontal line through each point on the y axis whose ordinate is in the image set of the function. If each such line cuts the graph of f in at most one point, then f^{-1} is the inverse function. If any such line cuts the graph of f in more than one point, then f^{-1} is not a function, for if a horizontal line cuts the graph of f in more than one point, it means that the second coordinates of at least two distinct ordered pairs of f are equal. Hence, when the coordinates of all ordered pairs of f are interchanged, at least two distinct ordered pairs of f^{-1} will have the same first coordinate, and f^{-1} is not a function.

Restricting the domain

We can sometimes restrict the domain of a function so that the inverse relation is the inverse function of the function thus restricted. For example,

FIGURE 12.2 (*a*) $f(x) = x^2 - 4x$; (*b*) $f^{-1}(x) = 2 \pm \sqrt{x + 4}$

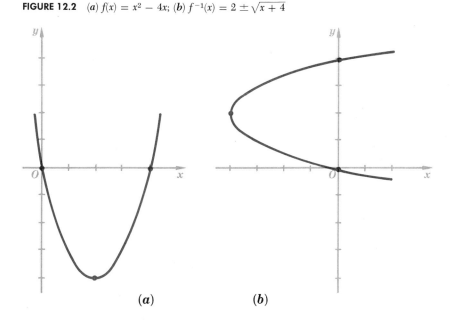

(*a*)   (*b*)

consider the function $f(x) = x^2 - 4x$, $x \in R$, which does not have an inverse function. The graph of f, Fig. 12.2a, indicates that for all $x \geq 2$, a horizontal line intersects the graph in at most one point. Let us now define a new function F whose domain is $\{x \mid x \geq 2\}$ and such that $F(x) = x^2 - 4x$, thus:

$$F(x) = x^2 - 4x \qquad x \geq 2$$

Since the graph of $F(x)$ is also the graph of $f(x)$ for all $x \geq 2$, it follows that the image of F is the set $\{y \mid y \geq -4\}$.

We may now be tempted to say that $F^{-1} = \{(x,y) \mid y = 2 \pm \sqrt{4 + x}\}$, but this is not so. Since the image of F must be the numbers in the domain of F^{-1}, however, it follows that y must be greater than or equal to 2. Hence, $y = 2 - \sqrt{4 + x}$ is not included in the image set of F^{-1}. Therefore,

$$F^{-1} = \{(x,y) \mid y = 2 + \sqrt{4 + x}\}$$

Thus, F^{-1} is the inverse function of F. The graphs of both F and F^{-1} are shown in Fig. 12.3. The heavy part of the graph of $F(x)$ represents that part of the graph of $f(x)$ for which $F(x) = f(x)$.

EXAMPLE 3.   Determine whether the inverse relation of

$$y = f(x) = \frac{3}{2} \sqrt{4 - x^2}$$

FIGURE 12.3   (a) $F(x) = x^2 - 4x$, $x \geq 2$; (b) $F^{-1}(x) = 2 + \sqrt{x + 4}$, $x \geq -4$

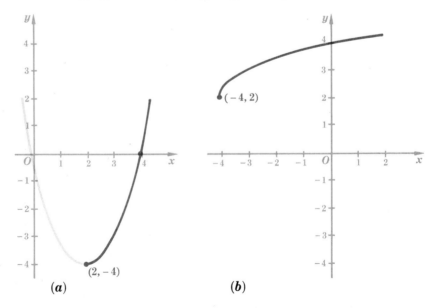

(a) (b)

is a function or not. If not, restrict the domain to the longest interval for which the inverse relation is a function.

Solution. The domain of f is $\{x \mid -2 \leq x \leq 2\}$, and the range of f is $\{y \mid 0 \leq y \leq 3\}$. Solving the inverse relation $x = \frac{3}{2}\sqrt{4 - y^2}$ for y, we obtain

$$f^{-1} = \left\{ (x,y) \mid y = \pm \frac{2}{3} \sqrt{9 - x^2} \right\}$$

Hence, f^{-1} is not a function. (Why?)

Let F be a new function such that

$$F(x) = \frac{3}{2} \sqrt{4 - x^2} \qquad 0 \leq x \leq 2$$

Since the domain of F is $\{x \mid 0 \leq x \leq 2\}$, the range of F^{-1} is $\{y \mid 0 \leq y \leq 2\}$. Therefore, since y is not negative,

$$F^{-1} = \left\{ (x,y) \mid y = \frac{2}{3} \sqrt{9 - x^2} \right\}$$

Hence, $y = \frac{2}{3}\sqrt{9 - x^2}$ is the inverse function of F.

EXERCISE SET 12.1

DETERMINE whether the inverse relation is a function or not. If it is not, restrict the domain of the function to the longest possible interval for which the inverse relation is a function. Specify the domain and range (image set) of the inverse function. Sketch the graphs of both the function and the inverse function.

1. $f(x) = \frac{1}{3}(5x - 1)$ $-1 \leq x \leq 5$ 2. $f(x) = -4 - \frac{2}{3}x$ $-3 \leq x \leq 9$

3. $y = 25 - x^2$ $-5 \leq x \leq 0$ 4. $f(x) = 3^x$ $x \in R$

5. $f(x) = \dfrac{1}{x} + x$ $x \geq 1$ 6. $f(x) = \dfrac{x^2 - 1}{x}$ $x \neq 0$

7. $f(x) = \dfrac{x^2 + 1}{x^2 - 1}$ $x \neq \pm 1$ 8. $y = \pm \dfrac{2x}{\sqrt{x^2 - 4}}$ $x \neq \pm 2$

12.2 INVERSE TRIGONOMETRIC FUNCTIONS

The function

$$f = \{(x,y) \mid y = \sin x\}$$

is called the *sine function*. The domain of f is $\{x \mid x \in R\}$ and the image of f is $\{y \mid -1 \leq y \leq 1\}$. We can show that the inverse relation of the sine function is not a function by showing that at least two distinct ordered pairs (x,y) of f have the same second coordinate. It is not difficult to verify that

$$(0,0),\ (\pi,0),\ (2\pi,0),\ \left(\frac{\pi}{6},\frac{1}{2}\right),\ \left(\frac{5\pi}{6},\frac{1}{2}\right)$$

are some of the ordered pairs of f. Hence, some of the ordered pairs of f^{-1} must be

$$(0,0),\ (0,\pi),\ (0,2\pi),\ \left(\frac{1}{2},\frac{\pi}{6}\right),\ \left(\frac{1}{2},\frac{5\pi}{6}\right)$$

Thus, $f(x) = \sin x$ does not have an inverse function. However, by restricting the domain of f it is possible to define a new function which does have an inverse function, and for which $f(x) = \sin x$.

To guide us in this matter, we consider the graph of $f(x) = \sin x$ and note that the longest interval on the x axis for which a horizontal line can intersect the curve in at most one point is of length π. For convenience, we select the interval from $-\pi/2$ to $\pi/2$ as the domain of our new function. We define this new function as follows:

$$\text{Sin } x = \sin x \qquad -\frac{\pi}{2} \leq x \leq \frac{\pi}{2} \qquad (12.1)$$

Principal sine function

We call Sin x the *principal* sine function, and we use the capital letter S to distinguish the principal sine function from the sine function. Note that the principal sine function is not defined outside the interval from $-\pi/2$ to $\pi/2$.

The solid part of the curve in Fig. 12.4a is the graph of $F(x) = \text{Sin } x$. The dotted part of the curve is the part of the graph of $f(x) = \sin x$ which lies outside the interval from $-\pi/2$ to $\pi/2$.

The domain of F is $\{x \mid -\pi/2 \leq x \leq \pi/2\}$ and the image of F is $\{y \mid -1 \leq y \leq 1\}$. Hence, the domain of F^{-1} is $\{x \mid -1 \leq x \leq 1\}$ and the image of F^{-1} is $\{y \mid -\pi/2 \leq y \leq \pi/2\}$.

The second coordinate of any ordered pair (x,y) of F^{-1} will be denoted by Sin^{-1} x (or Arcsin x). Thus,

$$F^{-1} = \{(x,y) \mid y = \text{Sin}^{-1} x\}$$

FIGURE 12.4 (a) $y = \text{Sin } x$; (b) $y = \text{Sin}^{-1} x$

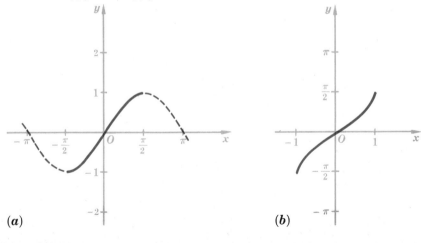

(a) (b)

We read $\text{Sin}^{-1} x$ as "the principal inverse sine of x," or as "the principal number whose sine is x." Thus, the equation $y = \text{Sin}^{-1} x$ (or $y = \text{Arcsin } x$) means that y is a real number whose sine is x, and such that

$$- \frac{\pi}{2} \le y \le \frac{\pi}{2}$$

DEFINITION. $y = \text{Sin}^{-1} x$ if and only if

$$\text{Sin } y = x$$

(12.2)

The graph of $y = \text{Sin}^{-1} x$ is shown in Fig. 12.4*b*.

EXAMPLE 1. Find $\text{Sin}^{-1} \frac{1}{2}$.

Solution. Let $y = \text{Sin}^{-1} \frac{1}{2}$; then $\text{Sin } y = \frac{1}{2}$. Now there is only one number y between $-\pi/2$ and $\pi/2$ such that $\text{Sin } y = \frac{1}{2}$, and that number is $\pi/6$. Hence,

$$y = \frac{\pi}{6}$$

EXAMPLE 2. Evaluate $\text{Sin}^{-1} (-0.5729)$.

Solution. Let $y = \text{Sin}^{-1} (-0.5729)$; then $\text{Sin } y = -0.5729$. From our previous work with trigonometric functions, and with the aid of Table I, we conclude that the only number y between $-\pi/2$ and $\pi/2$ such that

$$\text{Sin } y = -0.5729$$

is the number -0.61 (approx.). Hence,

$$y = -0.61 \text{ (approx.)}$$

Principal cosine function

The cosine function $f(x) = \cos x$, like the sine function, does not have an inverse function. By restricting the domain, however, we can define a principal cosine function which will have an inverse function. This time we select for the domain of the new function the interval from 0 to π and define the new function as follows:

$$\text{Cos } x = \cos x \qquad 0 \le x \le \pi \qquad (12.3)$$

The solid part of the curve in Fig. 12.5*a* represents the graph of

$$F(x) = \text{Cos } x$$

and the dotted part of the curve represents the part of $f(x) = \cos x$ which is outside the interval we have chosen for the domain of the principal cosine function. Note that the capital letter C is used to distinguish the principal cosine function $F(x) = \text{Cos } x$ from the cosine function $f(x) = \cos x$.

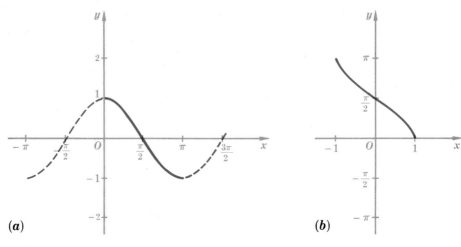

FIGURE 12.5   (a) $y = \text{Cos } x$; (b) $y = \text{Cos}^{-1} x$

The domain of $F(x) = \text{Cos } x$ is $\{x \mid 0 \leq x \leq \pi\}$ and the image of F is $\{y \mid -1 \leq y \leq 1\}$. The following remarks then apply to the inverse function.

The domain of F^{-1} is $\{x \mid -1 \leq x \leq 1\}$ and the image of F^{-1} is $\{y \mid 0 \leq y \leq \pi\}$. The second coordinate of the ordered pair (x,y) of F^{-1} is denoted by $\text{Cos}^{-1} x$ (or Arccos x). We read $\text{Cos}^{-1} x$ as "the principal inverse cosine of x" or as "the principal number whose cosine is x."

DEFINITION. $y = \text{Cos}^{-1} x$ if and only if

$$\text{Cos } y = x$$

(12.4)

The graph of $y = \text{Cos}^{-1} x$ is shown in Fig. 12.5b.

EXAMPLE 3.   Find the value of $\text{Cos}^{-1}(-\frac{1}{2})$.

Solution.   Let $y = \text{Cos}^{-1}(-\frac{1}{2})$; then $\text{Cos } y = -\frac{1}{2}$. There is only one number y between 0 and π such that $\text{Cos } y = -\frac{1}{2}$. That number is $2\pi/3$. Hence,

$$y = \frac{2\pi}{3}$$

Principal tangent function   In a similar manner, we restrict the domain of the tangent function $f(x) = \tan x$. We select the interval on the x axis from $-\pi/2$ to $\pi/2$ as the domain of a new function and define the new function as follows:

$$\text{Tan } x = \tan x \qquad -\frac{\pi}{2} < x < \frac{\pi}{2}$$

This new function $F(x) = \text{Tan } x$ has an inverse function. Elements in the image of F^{-1} are denoted by $\text{Tan}^{-1} x$ (or $\text{Arctan } x$) and are read "the principal inverse tangent of x" or as "the principal number whose tangent is x."

DEFINITION. $y = \text{Tan}^{-1} x$ if and only if

$$\text{Tan } y = x$$

$$(12.5)$$

The graphs of $y = \text{Tan } x$ and $y = \text{Tan}^{-1} x$ are shown in Fig. 12.6a and b.

EXAMPLE 4. Find the value of $\text{Tan}^{-1} 1$.

Solution. Let $y = \text{Tan}^{-1} 1$; then $\text{Tan } y = 1$. Hence, $y = \pi/4$, since $\pi/4$ is the only value of y between $-\pi/2$ and $\pi/2$ for which $\text{Tan } y = 1$.

We leave as an exercise the task of defining the principal cotangent function and its inverse function. We shall not discuss the principal secant and cosecant functions and their inverses in this book.

The following arrangement summarizes the definitions of the principal inverse trigonometric functions.

Function	Notation for Principal Inverse Function	Defining Equation	Domain	Image
sine	Sin^{-1}	$y = \text{Sin}^{-1} x$	$-1 \leq x \leq 1$	$-\frac{\pi}{2} \leq y \leq \frac{\pi}{2}$
cosine	Cos^{-1}	$y = \text{Cos}^{-1} x$	$-1 \leq x \leq 1$	$0 \leq y \leq \pi$
tangent	Tan^{-1}	$y = \text{Tan}^{-1} x$	$-\infty < x < \infty$	$-\frac{\pi}{2} < y < \frac{\pi}{2}$
cotangent	Cot^{-1}	$y = \text{Cot}^{-1} x$	$-\infty < x < \infty$	$0 < y < \pi$

The general procedure for evaluating an expression (or solving an equation) which involves inverse trigonometric functions is illustrated by the following examples.

EXAMPLE 5. Find the value of $\cos (\text{Sin}^{-1} \frac{1}{4} + \text{Cos}^{-1} \frac{1}{2})$.

Solution. Let $\alpha = \text{Sin}^{-1} \frac{1}{4}$ and $\beta = \text{Cos}^{-1} \frac{1}{2}$; then

$$\text{Sin } \alpha = \frac{1}{4} \qquad \text{Cos } \beta = \frac{1}{2}$$

Hence, $\mathrm{Cos}\ \alpha = \sqrt{1 - \dfrac{1}{16}}$ $\mathrm{Sin}\ \beta = \sqrt{1 - \dfrac{1}{4}}$

$$= \frac{\sqrt{15}}{4} \qquad\qquad\qquad = \frac{\sqrt{3}}{2}$$

from the fundamental identity $\sin^2 t + \cos^2 t = 1$. Therefore

$$\cos\left(\mathrm{Sin}^{-1}\frac{1}{4} + \mathrm{Cos}^{-1}\frac{1}{2}\right) = \cos(\alpha + \beta)$$

$$= \cos \alpha \cos \beta - \sin \alpha \sin \beta$$

$$= \frac{\sqrt{15}}{4} \cdot \frac{1}{2} - \frac{1}{4} \cdot \frac{\sqrt{3}}{2}$$

$$= \frac{\sqrt{15} - \sqrt{3}}{8}$$

EXAMPLE 6. Solve for x: $\mathrm{Cos}^{-1} x + \mathrm{Cos}^{-1}(2x) = \pi/2$.

Solution. Let $\alpha = \mathrm{Cos}^{-1} x$ and $\beta = \mathrm{Cos}^{-1}(2x)$. Then

$$\mathrm{Cos}\ \alpha = x \qquad \mathrm{Cos}\ \beta = 2x$$

Hence, $\mathrm{Sin}\ \alpha = \sqrt{1 - \mathrm{Cos}^2 \alpha} = \sqrt{1 - x^2}$

and $\mathrm{Sin}\ \beta = \sqrt{1 - \mathrm{Cos}^2 \beta} = \sqrt{1 - 4x^2}$

The given equation can now be written

$$\alpha + \beta = \frac{\pi}{2}$$

FIGURE 12.6   (a) $y = \mathrm{Tan}\ x$; (b) $y = \mathrm{Tan}^{-1} x$

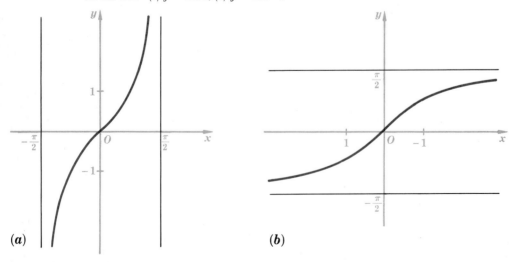

(a)

(b)

Hence, $$\text{Cos}\,(\alpha + \beta) = \cos\frac{\pi}{2} = 0$$

and $$\text{Cos}\,\alpha\,\text{Cos}\,\beta - \text{Sin}\,\alpha\,\text{Sin}\,\beta = 0$$

By substitution, we have

$$x(2x) - \sqrt{1 - x^2} \cdot \sqrt{1 - 4x^2} = 0$$

Then $$2x^2 = \sqrt{(1 - x^2)(1 - 4x^2)}$$

and $$4x^4 = 1 - 5x^2 + 4x^4$$

From this equation, we obtain the solution set $\{-\sqrt{5}/5, \sqrt{5}/5\}$.

EXERCISE SET 12.2

1. Define the principal cotangent function so that it will have an inverse function: HINT: Restrict the domain of the cotangent function.
2. Sketch the graph of the principal inverse cotangent function.
3. Evaluate the following without using tables:

    (a) $\text{Sin}^{-1} 1$

    (b) $\text{Sin}^{-1}(-\frac{1}{2})$

    (c) $\text{Tan}^{-1}(-1)$

    (d) $\text{Sin}^{-1}(-1)$

    (e) $\text{Cot}^{-1} 1$

    (f) $\text{Cos}^{-1}\left(-\dfrac{\sqrt{3}}{2}\right)$

    (g) $\text{Arcsin}\left(-\dfrac{\sqrt{3}}{2}\right)$

    (h) $\text{Arctan}\,\sqrt{3}$

    (i) $\text{Arccos}(-1)$

    (j) $\text{Arccos}\,\dfrac{\sqrt{3}}{2}$

    (k) $\text{Arcsin}\left(-\dfrac{\sqrt{2}}{2}\right)$

    (l) $\text{Arccot}(-1)$

    (m) $\text{Sin}^{-1}\,\dfrac{\sqrt{2}}{2}$

    (n) $\text{Cos}^{-1} 0$

    (o) $\text{Cos}^{-1}\left(-\dfrac{\sqrt{2}}{2}\right)$

4. Evaluate the following using Table I:

    (a) $\text{Sin}^{-1} 0.6442$

    (b) $\text{Arcsin}\ 0.5227$

    (c) $\text{Arccos}\ 0.8678$

    (d) $\text{Cos}^{-1} 0.8949$

    (e) $\text{Tan}^{-1} 1.592$

    (f) $\text{Cot}^{-1} 1.668$

    (g) $\text{Arccot}\ 1.431$

    (h) $\text{Arctan}\ 2.498$

    (i) $\text{Sin}^{-1}(-0.7833)$

    (j) $\text{Arcsin}(-0.9608)$

    (k) $\text{Arccos}(-0.1896)$

    (l) $\text{Cos}^{-1}(-0.0707)$

    (m) $\text{Tan}^{-1}(-5.798)$

    (n) $\text{Arctan}(-3.602)$

5. Express each of the following in terms of x or t:

    (a) $\cos(\text{Cos}^{-1} x)$

    (b) $\sin(\text{Sin}^{-1} x)$

    (c) $\text{Sin}^{-1}(\sin x)$

    (d) $\text{Cos}^{-1}(\cos x)$

    (e) $\tan(\text{Tan}^{-1} t)$

    (f) $\cot(\text{Cot}^{-1} t)$

    (g) $\text{Tan}^{-1}(\tan x)$

    (h) $\text{Cot}^{-1}(\cot x)$

6. Express each of the following in terms of x or t:

    (a) $\cos(\text{Tan}^{-1} x)$

    (b) $\sin(\text{Cot}^{-1} x)$

    (c) $\cos(\text{Cot}^{-1} t)$

    (d) $\tan(\text{Cot}^{-1} t)$

    (e) $\tan(\text{Cot}^{-1} 1/t)$

    (f) $\cos(\text{Sin}^{-1} t)$

7. Evaluate each of the following:

    (a) $\sin(\text{Cos}^{-1}\,\frac{3}{5})$

    (b) $\cos(\text{Sin}^{-1}\,\frac{3}{5})$

    (c) $\cos(\text{Sin}^{-1} 0.8)$

    (d) $\sin(\text{Cos}^{-1} 0.6967)$

    (e) $\sin(\text{Tan}^{-1}\,\frac{3}{4})$

    (f) $\cos(\text{Cot}^{-1}\,\frac{7}{24})$

    (g) $\cos[\text{Tan}^{-1}(-\frac{4}{3})]$

    (h) $\sin[\text{Cot}^{-1}(-\frac{24}{7})]$

8. Evaluate each of the following:

(a) $\sin (2 \, \text{Cos}^{-1} \, \tfrac{1}{2})$ (b) $\text{Cos}^{-1} (2 \sin \pi/3)$

(c) $\text{Sin}^{-1} (2 \cos \pi/3)$ (d) $\cos (2 \, \text{Arcsin} \, \tfrac{1}{2})$

9. Express in terms of t:

(a) $\tan (2 \, \text{Tan}^{-1} \, t)$ (b) $\sin (2 \, \text{Tan}^{-1} \, t)$

(c) $\cos (2 \, \text{Cot}^{-1} \, t)$ (d) $\cot (2 \, \text{Arccot} \, t)$

PROVE each of the following:

10. $\cos (\text{Sin}^{-1} w + \text{Cos}^{-1} z) = z\sqrt{1 - w^2} - w\sqrt{1 - z^2}$

11. $\sin (2 \, \text{Sin}^{-1} x) = 2x\sqrt{1 - x^2}$ **12.** $\cos (2 \, \text{Arccos} \, t) = 2t^2 - 1$

13. $\sin (\text{Arccos} \, t) = \cos (\text{Arcsin} \, t)$ **14.** $\tan \left(\tfrac{1}{2} \text{Arccos} \, u \right) = \sqrt{\dfrac{1 - u}{1 + u}}$

15. $\tan (\text{Arctan} \, u - \text{Arctan} \, v) = \dfrac{u - v}{1 + uv}$ **16.** $\tan (\text{Tan}^{-1} t + \text{Tan}^{-1} 1) = \dfrac{1 + t}{1 - t}$

17. $\text{Cos}^{-1} 2t \neq 2 \, \text{Cos}^{-1} t$ **18.** $\text{Sin}^{-1} 2t \neq 2 \, \text{Sin}^{-1} t$

19. $\text{Sin}^{-1} u + \text{Sin}^{-1} v \neq \text{Sin}^{-1} (u + v)$ **20.** $\text{Cos}^{-1} (u + v) \neq \text{Cos}^{-1} u + \text{Cos}^{-1} v$

SOLVE each of the following equations for x or t:

21. $\text{Arcsin} (-\tfrac{5}{13}) + \text{Arccos} \, \tfrac{4}{5} = \text{Arcsin} \, t$ **22.** $\text{Arccos} \, \tfrac{3}{5} - \text{Arcsin} \, \tfrac{4}{5} = \text{Arccos} \, t$

23. $\text{Tan}^{-1} 2x = 2 \, \text{Tan}^{-1} x$ **24.** Show that $\text{Cot}^{-1} 2t \neq 2 \, \text{Cot}^{-1} t$.

25. Prove that $\text{Sin}^{-1} (\sin u \sin v + \cos u \cos v) = \pi/2 - u + v$.

26. Show that $\text{Tan}^{-1} 3 + \text{Tan}^{-1} \tfrac{1}{3} = \pi/2$.

SHOW that the following relations hold:

27. $\text{Sin}^{-1} t = \text{Tan}^{-1} \dfrac{t}{\sqrt{1 - t^2}}$ **28.** $\sin (\text{Cos}^{-1} t) = \sqrt{1 - t^2}$

29. $\text{Tan}^{-1} t = \text{Sin}^{-1} \dfrac{t}{\sqrt{1 + t^2}}$ **30.** $\cos (\text{Arctan} \, t) = \dfrac{1}{\sqrt{1 + t^2}}$

SOLVE each of the following equations:

31. $2 \, \text{Sin}^{-1} t + \text{Cos}^{-1} 0 = \text{Cos}^{-1} (-1)$ **32.** $\text{Arccos} \, t + \text{Arccos} (1 - t) = \pi/2$

33. $\text{Arcsin} \, t + \text{Arccos} (1 - t) = \pi/2$ **34.** $2 \, \text{Tan}^{-1} x - \text{Sin}^{-1} x = 0$

35. $\text{Tan}^{-1} \sqrt{x} + 2 \, \text{Tan}^{-1} \sqrt{1 - x} = \pi/2$

13

SYSTEMS OF EQUATIONS. MATRICES

In preceding chapters we have discussed procedures for finding the solution sets of equations in one variable. We shall now consider methods for determining the solution sets of systems of equations in more than one variable.

13.1 LINEAR EQUATIONS IN TWO VARIABLES

An equation of the form

$$ax + by + c = 0 \tag{13.1}$$

where a, b, c are constants and a and b are not both zero, is a *linear equation in x and y*. If $b \neq 0$, we can solve this equation for y to obtain

$$y = -\frac{a}{b}x - \frac{c}{b}$$

The graph of the solution set of this equation is a straight line, as we pointed out in Sec. 3.10. In particular, if $a = 0$, then $b \neq 0$, and the graph of the solution set is the line $y = -(c/b)$. If $b = 0$, then $a \neq 0$, and the graph of the solution set is the line $x = -(c/a)$.

We turn our attention now to the solution of the system of linear equations in two variables,

$$\begin{aligned} a_1x + b_1y + c_1 &= 0 \\ a_2x + b_2y + c_2 &= 0 \end{aligned} \tag{13.2}$$

Let $L_1 = \{(x,y) \mid a_1x + b_1y + c_1 = 0\}$ be the solution set of the first equation and $L_2 = \{(x,y) \mid a_2x + b_2y + c_2 = 0\}$ the solution set of the second equation.   We define the solution set of the system (13.2) to be the set of all ordered pairs (x,y) that satisfy both equations.   Thus, the solution set of the system is the set

$$L_1 \cap L_2$$

If the graph of each equation is drawn on the same rectangular coordinate system, then the graph of the solution set of the system is the intersection of the graphs of L_1 and L_2.

If the graphs of L_1 and L_2 are distinct and not parallel, they will intersect in exactly one point (x_0, y_0).   Since this point lies on each line, its coordinates satisfy both equations.   The ordered pair (x_0, y_0) is called the *unique solution* of the system (13.2).

Determining the unique solution graphically

When the system (13.2) has a unique solution, the approximate values of x_0 and y_0 can be determined graphically, as follows: Use the same coordinate system and draw the graph of the solution set of each equation. Estimate the coordinates of the point of intersection.   The ordered pair whose elements are the coordinates of this point is the solution of the system.

EXAMPLE 1.   Solve the system graphically:

$$2x + 4y - 4 = 0$$
$$x - y - 5 = 0$$

Solution.   The graphs of the equations are shown in Fig. 13.1.   We estimate the coordinates of the point of intersection to be $(4, -1)$.   Hence, the ordered pair of real numbers $(4, -1)$ is the required solution.   The first number of

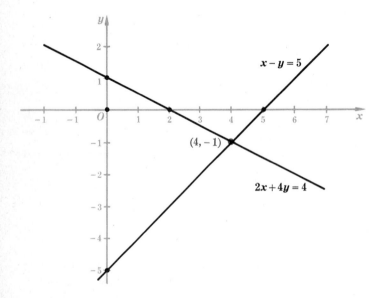

FIGURE 13.1
See Example 1.

the ordered pair is the value of x and the second is the value of y. We denote this solution as the set

$$\{(4, -1)\}$$

We shall frequently use the notation $L_1(x,y) = 0$ to denote the equation $a_1x + b_1y + c_1 = 0$ and $L_2(x,y) = 0$ to denote $a_2x + b_2y + c_2 = 0$. The system (13.2) can then be written

$$L_1(x,y) = 0$$
$$L_2(x,y) = 0$$

A system of equations that has at least one element in its solution set is called a *consistent* system. Not all systems are consistent, as we shall discover in the next section.

Consistent systems

13.2 EQUIVALENT SYSTEMS

If the system of Eqs. (13.2) has a unique solution, it may be found exactly by either of two algebraic methods, the method of *substitution* and the method of *elimination*. Each method is based upon the principle of replacing the given system of equations by an equivalent system. It was pointed out in Sec. 3.2 that equations are equivalent if they have identical solution sets. We define equivalent systems in a similar manner.

DEFINITION. Two systems of equations are equivalent if and only if they have identical solution sets.   (13.3)

For example, since the solution set of each of the following systems

$$5x - 3y = 9 \qquad\qquad 5x - 3y = 9$$
$$x + 2y = 7 \qquad \text{and} \qquad -13y = -26$$

is the ordered pair (3,2), the two systems are equivalent.

The general procedure in finding the solution set of a system of equations consists of replacing the given system by an equivalent system whose solution set is readily obtained. To find an equivalent system by the method of substitution, we use the fact that the system

Method of substitution

$$ax + by + c = 0$$
$$y = mx + b$$

and the system

$$ax + b(mx + b) + c = 0$$
$$y = mx + b$$

are equivalent systems. To show that they are equivalent, we reason as follows:

Since the second equation in each system is $y = mx + b$, the two systems are equivalent if the first equations are equivalent, that is, if $ax + by + c = 0$ is equivalent to $ax + b(mx + b) + c = 0$. By the substitution principle, these equations are equivalent.

The method of substitution consists of solving one of the equations for one variable, say y, in terms of the other, x. This expression for y is then substituted for y in the remaining equation. The result is an equation in the single variable, x. The given system can be replaced by the equations obtained in the first and second steps.

The new system is equivalent to the original system, that is, it has the same solution set. The new system is readily solved, as the following example will show.

EXAMPLE 1. Use the method of substitution to solve the system

$$3x - 5y = 11$$
$$x + 8y = -6$$

Solution. To avoid using fractions, we select the second equation and solve it for x in terms of y:

$$x = -6 - 8y$$

We substitute this value of x into the first equation to obtain

$$3(-6 - 8y) - 5y = 11$$

from which

$$y = -1$$

The given system can now be replaced by the equivalent system

$$x = -6 - 8y$$
$$y = -1$$

Any solution of this system must be of the form $(x, -1)$. If we substitute $y = -1$ into $x = -6 - 8y$, we obtain $x = 2$. Thus, the only ordered pair that satisfies both equations is $(2, -1)$. It now follows that the solution set of the original system is

$$\{(2, -1)\}$$

The solution should be checked by determining whether or not the ordered pair $(2, -1)$ makes *each* of the original equations a true statement.

Method of
elimination

The method of elimination for determining the solution set of the system (13.2) is also based on the principle of finding an equivalent system. We use the following theorems to find equivalent systems:

THEOREM. The position of any two equations of a given system can be interchanged. **(13.4)**

Proof. The solution set of a system is the intersection of the solution sets of the equations of the system. Since intersection of sets is associative and commutative, the theorem follows.

For example, if L_1 and L_2 are the solution sets of the two equations of a system, the solution set of the system is $L_1 \cap L_2$. If the equations are interchanged, the solution set is $L_2 \cap L_1$. But $L_1 \cap L_2 = L_2 \cap L_1$.

THEOREM. Any equation of a given system can be replaced by an equivalent equation. (13.5)

For example, if $L_1(x,y) = 0$ and $L_2(x,y) = 0$ are the equations of a system, then for every solution (x_1,y_1) of the system, $L_1(x_1,y_1) = 0$ and $L_2(x_1,y_1) = 0$. Now suppose that $L_3(x,y) = 0$ is equivalent to $L_2(x,y) = 0$, then $L_3(x_1,y_1) = 0$. Hence, the system

$$L_1(x,y) = 0$$
$$L_2(x,y) = 0$$

has the same solution set as the system

$$L_1(x,y) = 0$$
$$L_3(x,y) = 0$$

THEOREM. Either equation of the system

$$L_1(x,y) = 0$$
$$L_2(x,y) = 0$$

(13.6)

can be replaced by $L_1(x,y) + L_2(x,y) = 0$.

Proof. For every ordered pair (x_1,y_1) in the solution set of the system, it must be true that

$$L_1(x_1,y_1) = 0$$
and $$L_2(x_1,y_1) = 0$$
Therefore, $$L_1(x_1,y_1) + L_2(x_1,y_1) = 0$$

Thus, (x_1,y_1) satisfies each equation in the system

$$L_1(x,y) = 0$$
$$L_1(x,y) + L_2(x,y) = 0$$

and also satisfies each equation in the system

$$L_1(x,y) + L_2(x,y) = 0$$
$$L_2(x,y) = 0$$

It is left to the student to prove that every solution of either of the two preceding systems is also a solution of the system (13.6). To do this, let (x_2,y_2) be any solution of one of the systems, then show that (x_2,y_2) is also a solution of the system (13.6).

We frequently combine (13.5) and (13.6).

THEOREM. Either equation of the system

$$L_1(x,y) = 0$$
$$L_2(x,y) = 0 \qquad\qquad\qquad (13.7)$$

can be replaced by $aL_1(x,y) + bL_2(x,y) = 0$ provided a and b are nonzero real numbers.

The proof is similar to the proof of (13.6) and is left to the student. The following examples illustrate the use of these theorems.

EXAMPLE 2. Solve the system

$$2x - 7y - 5 = 0$$
$$3x - 9y - 8 = 0$$

Solution. The given system is equivalent to the system

$$a(2x - 7y - 5) + b(3x - 9y - 8) = 0$$
$$3x - 9y - 8 = 0$$

for any nonzero constants a and b. By a proper choice of a and b we can eliminate one of the variables from the first equation. To eliminate y, we take the smallest integral values of a and b that will do the job. Hence, we take $a = 9$ and $b = -7$. Then

$$9(2x - 7y - 5) - 7(3x - 9y - 8) = 0$$

or $\qquad\qquad\qquad -3x + 11 = 0$

We now replace the first equation of the given system by this new equation to obtain the equivalent system

$$-3x + 11 = 0$$
$$3x - 9y - 8 = 0$$

This new system is equivalent to the system

$$-3x + 11 = 0$$
$$c(-3x + 11) + d(3x - 9y - 8) = 0$$

To eliminate x from the second equation of this system, take $c = 1$ and $d = 1$, so that $1(-3x + 11) + 1(3x - 9y - 8) = 0$, or $-9y + 3 = 0$. Replace the second equation of the system with this equation to obtain the equivalent system

$$-3x + 11 = 0$$
$$-9y + 3 = 0$$

The unique solution is readily seen to be $(1\frac{1}{3}, \frac{1}{3})$. Hence, the solution set of the given system is the set $\{(1\frac{1}{3}, \frac{1}{3})\}$.

In practice the procedure is often abbreviated, as in the following example:

EXAMPLE 3. Solve the system

(1) $\hspace{3cm} 5x + 3y = 3$
(2) $\hspace{3cm} 9x + 5y = 7$

Solution.

(3) $\hspace{2.5cm} 45x + 27y = 27 \hspace{1.5cm}$ (1) multiplied by 9
(4) $\hspace{2.5cm} 45x + 25y = 35 \hspace{1.5cm}$ (2) multiplied by 5
(5) $\hspace{3.5cm} 2y = -8 \hspace{1cm}$ (4) subtracted from (3)
(6) $\hspace{3.7cm} y = -4 \hspace{1.5cm}$ (5) divided by 2

Since any solution of the system must be of the form $(x, -4)$ and must satisfy the equation $5x + 3y = 3$, we have

(7) $\hspace{2.8cm} 5x + 3(-4) = 3 \hspace{1.3cm}$ (6) substituted in (1)

Hence, $x = 3$, and the solution set is $\{(3, -4)\}$.

EXERCISE SET 13.1

SOLVE each of the following systems by drawing the graph of each equation and estimating the coordinates of their intersection:

1. $x - y = -2$
    $4x + 3y = 12$

2. $4y - x = -3$
    $8y - 3x = -1$

3. $2x + y = 4$
    $3x - 7y = -11$

4. $4x - 2y = -1$
    $2x + 6y = 10$

5. $3x - 2y = 11$
    $x - y = 7$

6. $x - 2y = -9$
    $3x + 4y = 8$

SOLVE each of the following systems by the method of substitution:

7. $2x - 3y = 4$
    $2x + y = 8$

8. $y = 2x + 8$
    $2x + y = 3$

9. $4x + y = 3$
    $8x + 26 = y - 1$

10. $1\frac{1}{2}x - 5y = 10$
     $5x + 10y = 0$

11. $x + 2y + 13 = 0$
     $2x - 3y + 5 = 0$

12. $x - y - 7 = 0$
     $3x - 2y - 11 = 0$

REPLACE each of the following systems with an equivalent system in which one equation involves x only:

13. $3x - y = 12$
     $2x + y = 8$

14. $5x + 2y = 11$
     $2x + 2y = 25$

15. $7x - 2y = 5$
     $3x + 3y = 7$

16. $3x - 4y = 10$
     $2x + 3y = 7$

REPLACE each of the following systems with an equivalent system in which one equation will involve y only:

17. $2x + 3y = 8$
     $2x - 5y = 16$

18. $5x - y = 7$
     $5x + 3y = 18$

19. $3x - y = 7$
     $2x + 3y = 9$

20. $5x + 3y = 6$
     $2x + y = 2$

REPLACE each of the following systems with an equivalent system in which one equation involves only one variable. Find the solution set of the given system from the solution set of the equivalent system.

21. $3x + y = 8$
     $2x - y = 7$

22. $4x - 8y = 4$
     $3x - 2y = 1$

23. $2x - \frac{5}{3}y = \frac{22}{3}$
     $4x - y = 6$

24. $\frac{9}{20}x + \frac{7}{20}y = 1$
     $6x - 5y = 23$

25. $3x + 8y = 6$
     $4x + 7y = 5$

26. $2x - y - 6 = 6$
     $3x - 6y + 6 = 0$

27. $3x = 5y - 4$
     $4y = -6x - 1\frac{2}{5}$

28. $x + 3y = 1$
     $4x + 10y = 26$

SOLVE each of the following systems by any method:

29. $3x - 2y = -16$
     $4x + 5y = -6$

30. $\sqrt{3}x - \sqrt{2}y = 3$
     $\sqrt{2}x + \sqrt{3}y = \sqrt{6}$

31. $(1 - i)x + 2y = -i$
     $(1 - i)x - 2y = i$

32. $3x + 2iy = 4$
     $ix - 2y = 3$

SOLVE each of the following systems. To do this, make a substitution that will yield a linear system.

33. $\dfrac{1}{x} + \dfrac{1}{y} = 3$

     $\dfrac{1}{x} - \dfrac{1}{y} = 1$ $(x \neq 0, y \neq 0)$

     HINT: Let $u = \dfrac{1}{x}$, $v = \dfrac{1}{y}$.

34. $\log x + 2 \log y = \log 2$ $(x \neq 0, y \neq 0)$
     $\log x + \log y = \log (0.1)$
     HINT: let $u = \log x$ and $v = \log y$.

35. $3^x + 5^y = 32$
     $2(3^x) - 3(5^y) = 39$
     HINT: Let $u = 3^x$ and $v = 5^y$.

36. $\dfrac{10}{x} + \dfrac{9}{y} = 17$

     $\dfrac{2}{x} - \dfrac{3}{y} = -3$

37. $\dfrac{25}{x} + \dfrac{4}{y} = 4$

     $\dfrac{10}{x} + \dfrac{6}{y} = -5$

38. $\dfrac{3}{x} + \dfrac{1}{2y} = 17$

     $\dfrac{1}{5x} - \dfrac{1}{4y} = 0$

39. $\dfrac{1}{x} - \dfrac{3}{y} = \dfrac{-5}{12}$

     $\dfrac{2}{x} + \dfrac{3}{y} = \dfrac{17}{12}$

40. $\dfrac{3}{x} + \dfrac{6}{y} = \dfrac{81}{2}$

     $\dfrac{4}{3x} + \dfrac{3}{2y} = -\dfrac{36}{5}$

41. $\log x + 3 \log y = 7$
     $5 \log x = 3 + \log y$

42. $Sin\ x - Cos\ y = 1$
     $Sin\ x + Cos\ y = 1$

43. The sum of the multiplicative inverses of two numbers is $18/77$, and the difference of the multiplicative inverses is $4/77$. Find the numbers.

44. If the numerator of a fraction is decreased by 2 and the denominator increased by 7, the resulting fraction equals $3/10$. If the numerator is increased by 5 and the denominator decreased by 4, the resulting fraction equals $13/9$. Find the original fraction.

45. At the beginning of a class period there are $5/8$ as many men as women in the room. One man and one woman come in late and there are then $7/11$ as many men as women in the class. Find the number of men and the number of women in the class at the beginning of the period.

46. One solution containing 20 percent alcohol is mixed with another solution containing 60 percent alcohol to make 10 gal of a solution that is 32 percent alcohol. How much of each solution is used?

47. Two cyclists who are 15 mi apart start at the same time traveling in the same direction. After 6 hr, the faster cyclist overtakes the slower one. If they had traveled toward each other they would have met after 2 hr. Find the average rate of travel of each.

48. A jet aircraft cruising with the wind took $17/11$ hr to make a 900-mi flight. The return trip against the wind required 1 hr 48 min. Find the speed of the aircraft in still air and the speed of the wind.

49. The sum of the digits of a two-digit number is three times the tens digit. If the number is multiplied by 4 and the product decreased by 54, the result is equal to the number obtained by reversing the digits of the original number. Find the original number.

50. Four times Mark's present age is 4 more than eight times what Matt's age was 2 years ago. Three times Matt's present age is 3 more than twice what Mark's age was 4 years ago. Find their present ages.

51. If the length of a certain rectangle is increased by 10 ft and the width decreased by 4 ft, the area is increased by 72 sq ft. If the length is decreased by 5 ft and the width increased by 6 ft, the area is increased by 42 sq ft. Find the dimensions of the rectangle.

52. The circumference of the rear wheel of a scooter is $7/8$ the circumference of the front wheel. In traveling a distance of $933\frac{1}{3}$ ft along a level street, the rear wheel makes 50 more revolutions than the front wheel. Find the circumference of each wheel.

13.3 INCONSISTENT AND DEPENDENT SYSTEMS

Let the straight lines L_1 and L_2 be the graphs of the solution sets of $a_1x + b_1y + c_1 = 0$ and $a_2x + b_2y + c_2 = 0$ respectively, drawn on the same coordinate system. If these lines are distinct and parallel, there is no point (x_0, y_0) at which they intersect. Thus, $L_1 \cap L_2 = \varnothing$. To describe such systems, we make the following definition.

> **DEFINITION.** A system of linear equations is inconsistent if it does not have a solution. $\qquad$ (13.8)

Inconsistent system An inconsistent system of equations results from imposing contradictory conditions on the variables. A simple example of an inconsistent system is the following:

$$x + y = 3$$
$$x + y = 5$$

The first equation imposes on x and y the condition that their sum is 3, whereas the second equation imposes the condition that their sum is 5.

Let us replace the given system by the equivalent system

$$x + y - 3 = 0$$
$$a(x + y - 3) + b(x + y - 5) = 0$$

for any nonzero constants a and b. Taking $a = 1$ and $b = -1$, we have the equivalent system

$$x + y - 3 = 0$$
$$2 = 0$$

We now reason as follows: If the original system has a solution, then 2 must equal 0. Since $2 \neq 0$, we conclude that the system does not have a solution and is therefore inconsistent. The graphs of the solution sets of the equations of the system are shown in Fig. 13.2a. Note that the graphs are parallel lines.

The following statement about inconsistent systems can be proved. We leave the proof to you.

> Two or more linear equations are inconsistent if and only if the graphs of any two of them are parallel.

If the graphs of the solution sets of the two equations of the system coincide, then every point of L_1 is also a point on L_2. Hence, any ordered pair (x,y) that satisfies one of the equations also satisfies the other. This means that the two equations are equivalent and $L_1 \cap L_2 = L_1 = L_2$. In this case, if one of the equations is multiplied by a suitable constant, the resulting equation is identical to the other. We then say that the equations are *dependent* equations.

Dependent equations

DEFINITION. A consistent system of equations is said to be dependent if there is one of them such that every solution of all the (13.9) other equations of the system is a solution of that one also.

For example, the system

(1) $2x - 7y = 21$
(2) $6x - 21y = 63$

is a dependent system. To show that this system is dependent, replace the system by the equivalent system

(3) $6x - 21y = 63$ (1) multiplied by 3
(4) $6x - 21y = 63$ (2) multiplied by 1

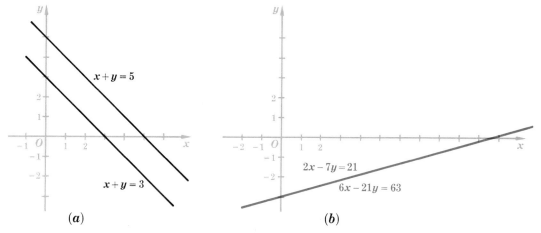

FIGURE 13.2   (*a*) Inconsistent system; (*b*) dependent system

It is now evident that every ordered pair (x,y) that satisfies one equation also satisfies the other.

The graph of the equations of the original system is the pair of coincident lines shown in Fig. 13.2*b*.

A system of equations of the form

$$a_1x + b_1y = 0$$
$$a_2x + b_2y = 0 \qquad (13.10)$$

Homogeneous systems

Trivial solution

is called a *homogeneous system*. By inspection, we see that $x = 0$, $y = 0$ satisfies each equation of the system, therefore the ordered pair $(0,0)$ is a solution. We call this solution the *trivial solution*.

The graph of each equation of the homogeneous system (13.10) is a straight line that passes through the origin. Hence, the lines are distinct and intersect at the origin, or they coincide. If they are distinct, the trivial solution is the only solution of the system. If they coincide, then there are infinitely many solutions. Therefore, a homogeneous system of linear equations is always consistent. It may or may not be a dependent system.

EXERCISE SET 13.2

DETERMINE which of the following systems are consistent, which are inconsistent, and which are dependent:

1. $9x - 12y = 36$
    $3x - 4y = 24$

2. $x + 2y = 3$
    $-5x - 10y = -15$

3. $-3x + 2y = 8$
    $6x - 4y = 16$

4. $3x - 5y = 10$
    $2x - 5y = 10$

5. $3y = 4x + 9$
$12 - 12x + 9y = 0$

6. $\frac{1}{2} x + \frac{3}{4} y = 2$
$2x - 3y = 8$

7. $y = 3x + 3$
$6x = 2y - 6$

8. $y = 2x$
$3y = 14 - 6x$

9. $6y = 12 - 2x$
$3y + x = 6$

10. $2x = 4y + 17\frac{1}{2}$
$3x = -3 - 4y$

FIND a value of k, if possible, such that each of the following systems has a unique solution:

11. $x + ky = 3$
$3x - 3y = 7$

12. $x - 3y = k$
$2x + 6y = 9$

13. $kx + 2y = 4$
$-5x + 6y = 11$

14. $2x - 5y = k$
$8x - ky = 2$

FIND a value of k, if possible, such that each of the following systems is consistent:

15. $3x - ky = 7$
$x + y = 9$

16. $kx + 5y = 10$
$x - 15y = 20$

17. $2x - y = 8$
$4x + ky = 8$

18. $4x + 2y = 6$
$2x + ky = 4$

FIND a value of k, if possible, such that each of the following systems is dependent:

19. $2x + y = 8$
$8x + 4y = k$

20. $x - 7y = k$
$-3x + 21y = 5$

21. $6x + y = k$
$kx - 3y = 20$

22. $5x - ky = 8$
$2x - 10y = 16\frac{2}{3}$

23. Show that the homogeneous system

$$2x + 3y = 0$$
$$6x + 9y = 0$$

is a dependent system.

24. Show that the system

$$3x + 4y = 0$$
$$6x - 8y = 0$$

is consistent.

In each of the following problems, the given system is

$$a_1x + b_1y = c_1$$
$$a_2x + b_2y = c_2$$

25. Show that an equivalent system is the system

$$(a_1b_2 - a_2b_1)x = b_2c_1 - b_1c_2$$
$$(a_1b_2 - a_2b_1)y = a_1c_2 - a_2c_1$$

HINT: Multiply each side of the first equation of the given system by b_2, each side of the second equation by b_1. Subtract.

26. Show that if $a_1b_2 - a_2b_1 \neq 0$, the system has a unique solution. HINT: Use the equivalent system of Prob. 25.

13.4 LINEAR EQUATIONS IN MORE THAN TWO VARIABLES

An equation of the form

$$ax + by + cz + d = 0 \qquad (13.11)$$

where a, b, c, d are constants and a, b, c are not all zero, is a linear equation in the three variables x, y, z. The solution set of the equation is the set of all ordered triples of numbers (x,y,z) that satisfy the equation. If L denotes the solution set, then

$$L = \{(x,y,z) \mid ax + by + cz + d = 0\}$$

We define the solution set of a system of three equations in the three variables x, y, z to be the set of all ordered triples (x,y,z) that satisfy *each* equation of the system. Thus, if L_1, L_2, L_3 are the solution sets of the first, second, and third equations of the system respectively, the solution set is the intersection

$$L_1 \cap L_2 \cap L_3$$

One of the methods for solving the system algebraically is called the method of *reducing the system to triangular form*.

Triangular form

DEFINITION. A system of three linear equations in three variables is in triangular form if the following conditions are satisfied:
    1. The first equation involves the first variable; it might or might not involve the second or third.
    2. The second equation involves the second variable but not the first; it might or might not involve the third.
    3. The third equation involves the third variable but not the first or second.

$$(13.12)$$

For example, the following system is in triangular form:

$$x + 3y - 5z - 14 = 0$$
$$2y + 3z - 1 = 0$$
$$6z + 6 = 0$$

This system is readily solved. From the third equation, $z = -1$. Substitution of $z = -1$ into the second equation yields $y = 2$. Substitution of $z = -1$ and $y = 2$ into the first equation yields $x = 3$. Thus, the ordered triple $(3,2,-1)$ is the solution of the system.
    To replace a given system by an equivalent system in triangular form, we use Theorems (13.4) to (13.6) and the following restatement of (13.7):

THEOREM. Any equation of the system

$$L_1(x,y,z) = 0$$
$$L_2(x,y,z) = 0$$
$$L_3(x,y,z) = 0$$

(13.7′)

can equivalently be replaced by a nonzero multiple of itself plus
a nonzero multiple of another equation.

The proof is similar to the proof for Theorem (13.7) and is left to the
student.

To apply these theorems, we write the equations so the variables are
in the same order in each of the equations.

EXAMPLE 1. Solve the system

(1) $x + 2y - 3z - 13 = 0$

(2) $3x - y + 4z + 8 = 0$

(3) $2x - 5y - 3z + 7 = 0$

Solution. To eliminate x from the second equation, we replace the second
equation by 1 times Eq. (2) minus 3 times Eq. (1):

$$3x - y + 4z + 8 - 3(x + 2y - 3z - 13) = 0$$

or

$$-7y + 13z + 47 = 0$$

To eliminate x from the third equation, replace the third equation by
1 times Eq. (3) minus 2 times Eq. (1):

$$2x - 5y - 3z + 7 - 2(x + 2y - 3z - 13) = 0$$

or

$$-9y + 3z + 33 = 0$$

The resulting equivalent system is now

(4) $x + 2y - 3z - 13 = 0$

(5) $-7y + 13z + 47 = 0$

(6) $-9y + 3z + 33 = 0$

To eliminate y from Eq. (6), we replace Eq. (6) by 7 times Eq. (6) minus 9
times Eq. (5):

$$7(-9y + 3z + 33) - 9(-7y + 13z + 47) = 0$$

or

$$-96z - 192 = 0$$

The resulting system

(7) $$x + 2y - 3z - 13 = 0$$
(8) $$-7y + 13z + 47 = 0$$
(9) $$-96z - 192 = 0$$

is equivalent to the original system. We see that $z = -2$ is the unique solution of Eq. (9). Then since $z = -2$, Eq. (8) yields a unique value for y, namely, $y = 3$. From Eq. (7) we obtain a unique value for x, $x = 1$. Hence, the solution set consists of the ordered triple

$$(1,3,-2)$$

This procedure is usually abbreviated as follows:

(4') $$x + 2y - 3z - 13 = 0$$
(5') $$-7y + 13z + 47 = 0 \qquad \text{1 times (2) minus 3 times (1)}$$
(6') $$-9y + 3z + 33 = 0 \qquad \text{1 times (3) minus two times (1)}$$

which reduces to

(7') $$x + 2y - 3z - 13 = 0$$
(8') $$-7y + 13z + 47 = 0$$
(9') $$-96z - 192 = 0 \qquad \text{7 times (6') minus 9 times (5')}$$

Then from (9'), $z = -2$; from (8'), $y = 3$; and from (7'), $x = 1$. Hence, the ordered triple $(x,y,z) = (1,3,-2)$ is the required solution.

EXERCISE SET 13.3

SOLVE the following systems by finding an equivalent system in triangular form:

1. $x + y - z = 2$
$x + 2y + z = 7$
$3x - y + 2z = 12$

2. $2x - z = 2$
$x - y = 5$
$y - z = -6$

3. $x - 2y + z = 7$
$y + 2z = 1$
$2x + 3z = 4$

4. $x - 2y + 4z = -3$
$2x + y - 3z = 11$
$3x + y - 2z = 12$

5. $2x + 4y + z = 0$
$5x + 3y - 2z = 1$
$4x - 7y - 7z = 6$

6. $x + 3y - 2z = 10$
$2x + y + 3z = 4$
$7x - 5y + 4z = -4$

7. $x + y + z = 6$
$2x + 2y + z = 11$
$3x - 4y - z = 0$

8. $x - 3y - 3z = 1$
$5x + 4y - 7z = 0$
$4x - 3y - 6z = 4$

9. $\begin{aligned} 3x - 2y &= 3 \\ 5x - 8z &= -19 \\ 5y - 7z &= 4 \end{aligned}$

10. $\begin{aligned} 2x - 3y + z &= 9 \\ x - 2y + 3z &= -2 \\ 3x - y + 2z &= 5 \end{aligned}$

11. $\begin{aligned} \frac{4}{x} - \frac{6}{y} + \frac{3}{z} &= -7 \\[4pt] \frac{3}{x} - \frac{5}{y} + \frac{2}{z} &= -5 \\[4pt] \frac{2}{x} + \frac{3}{y} + \frac{1}{z} &= 4 \end{aligned}$

12. $\begin{aligned} \frac{2}{x} + \frac{4}{z} &= 3 \\[4pt] \frac{2}{x} + \frac{3}{y} - \frac{2}{z} &= 8 \\[4pt] \frac{1}{x} + \frac{6}{y} + \frac{1}{z} &= 6 \end{aligned}$

HINT: Let $u = (1/x)$, $v = (1/y)$, $w = (1/z)$.

13. $\begin{aligned} x + y + 2z + 2w &= 9 \\ 4x - 3y + z - 5w &= 3 \\ 3x - 2y - 3z + 4w &= 6 \\ 2x + y - 2z - w &= -8 \end{aligned}$

14. $\begin{aligned} x + 4y + w &= 4 \\ x + 3z + 2w &= -11 \\ 2y + 3z + w &= -9 \\ x - y - 2z &= 7 \end{aligned}$

15. Part of an investment of \$20,000 is at 4 percent, part is at 5 percent, and the remainder is invested in a producing oil well. The first year, the oil well lost 2 percent, but the total net income on the three investments was \$560. The next year, the oil well paid a profit of 10 percent, and the total net income was \$1,160. How much was invested in the oil well?

16. The sum of the digits of a three-digit number is 11. If the order of the digits is reversed, the number is decreased by 396. The tens digit is one-half the hundreds digit. Find the number.

17. The sum of three integers is 84. The sum of the first two minus the third is 22, and the first minus the sum of the second and third is -34. Find the three integers.

18. Mark and Matt together can do a piece of work in 4 days. Mark and Mike can do the work in 6 days together. Matt and Mike together can do it in 5 days. How long would it take each alone to do the work?

19. In a three-digit number, the sum of the hundreds digit and the tens digit is 2 more than twice the units digit. The sum of the tens digit and the units digit equals the hundreds digit. The sum of the three digits is 14. Find the number.

20. The points $(1,1)$, $(2,4)$, $(3,9)$ lie on the curve whose equation is

$$y = ax^2 + bx + c$$

Find a, b, c.

13.5 SYSTEMS OF QUADRATIC EQUATIONS

Many problems require the solution of a system of equations in which at least one is a quadratic equation in two variables. In this section we shall consider methods for solving certain systems of this type.

Quadratic equation in two variables

An equation of the form

$$ax^2 + bxy + cy^2 + dx + ey + f = 0 \qquad (13.13)$$

where a, b, c are constants, not all zero, and d, e, f are any constants, is called a *quadratic equation in x and y*.

Solution by eliminating one variable

If a system consists of one linear and one quadratic equation, the method of substitution is effective. The steps used in this method are illustrated in the following example.

EXAMPLE 1.   Solve the system

$$x^2 + y^2 = 25$$
$$x + y = 7$$

Solution.   From the second equation, $y = 7 - x$.   Substitute this result into the first equation:

$$x^2 + (7 - x)^2 = 25$$

Hence,
$$x^2 - 7x + 12 = 0$$

and
$$x = 3 \quad \text{or} \quad x = 4$$

The corresponding values of y can be obtained from the equation $y = 7 - x$, as follows.   When $x = 3$, $y = 4$, and when $x = 4$, $y = 3$.   Hence, the solution set of the system is the set of ordered pairs (3,4) and (4,3).   The graphs of the equations, Fig. 13.3, can be used to check these solutions.

The method of substitution may be used to solve certain systems in which both equations are quadratic in x and y.   The following example illustrates the procedure.

EXAMPLE 2.   Solve the system

$$xy - 8 = 0$$
$$xy - 4x - 4y + 16 = 0$$

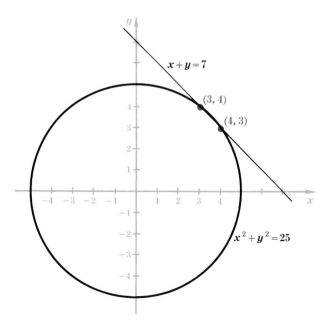

FIGURE 13.3

See Example 1.

Solution.   Solve the first equation for y, $y = 8/x$. Substitute $8/x$ for y in the second equation:

$$x\left(\frac{8}{x}\right) - 4x - 4\left(\frac{8}{x}\right) + 16 = 0$$

When simplified, this equation becomes

$$x^2 - 6x + 8 = 0$$

Hence, $x = 2$ or $x = 4$

If $x = 2$, the equation $y = 8/x$ yields $y = 4$. Hence, $(2,4)$ is a solution of the system. If $x = 4$, $y = 2$. Hence, $(4,2)$ is a solution. Thus, the solution set of the given system is the set $\{(2,4),(4,2)\}$.

Elimination by addition

When both equations of a system are of the form $ax^2 + cy^2 + f = 0$, the solution may be accomplished by regarding the system as linear in x^2 and y^2. It may be convenient at first to substitute u for x^2 and v for y^2, as in the following example.

EXAMPLE 3.   Solve the system

$$x^2 + y^2 = 13$$
$$2x^2 + 3y^2 = 30$$

Solution.   Let $u = x^2$ and $v = y^2$. The resulting system

$$u + v = 13$$
$$2u + 3v = 30$$

is linear in u and v. The solution of this system is

$$u = 9 \qquad\qquad v = 4$$

Hence, $x^2 = 9$ $y^2 = 4$

and $x = \pm 3$ $y = \pm 2$

Therefore, the solution of the given system is the set

$$\{(3,2),(3,-2),(-3,2),(-3,-2)\}$$

A system in which each equation is of the form

$$ax^2 + bxy + cy^2 + f = 0$$

Solution by substituting $y = vx$

can be solved by making the substitution $y = vx$ in each equation of the system. When this substitution has been made, the values of v can be found by solving each equation of the resulting system for x^2. The two expressions for x^2 are then set equal, and from the result v is determined. Once v becomes known, the values of x and y are easily obtained. An example will illustrate this method.

EXAMPLE 4.   Solve the system

$$2x^2 - 3xy + 4y^2 = 3$$
$$3x^2 - 4xy + 3y^2 = 2$$

Solution.   Let $y = vx$ in each equation.   Then

$$2x^2 - 3vx^2 + 4v^2x^2 = 3$$
$$3x^2 - 4vx^2 + 3v^2x^2 = 2$$

From the first equation,

$$x^2 = \frac{3}{2 - 3v + 4v^2}$$

From the second,

$$x^2 = \frac{2}{3 - 4v + 3v^2}$$

If we set one of these expressions for x^2 equal to the other and simplify, we obtain

$$v^2 - 6v + 5 = 0$$

Hence, $v = 1$ or $v = 5$

From the equations

$$x^2 = \frac{3}{2 - 3v + 4v^2} \qquad \text{and} \qquad y = vx$$

we see that if $v = 1$, $x^2 = 1$ and $y = x$.   Hence, $(1,1)$ and $(-1,-1)$ are solutions of the original system.   If $v = 5$, $x^2 = \frac{1}{29}$ and $y = 5x$.   Therefore

$$\left(\frac{\sqrt{29}}{29}, \frac{5\sqrt{29}}{29}\right) \qquad \text{and} \qquad \left(-\frac{\sqrt{29}}{29}, -\frac{5\sqrt{29}}{29}\right)$$

are solutions also.   Hence, the solution set of the system is the set

$$\left\{(1,1),(-1,-1),\left(\frac{\sqrt{29}}{29}, \frac{5\sqrt{29}}{29}\right),\left(-\frac{\sqrt{29}}{29}, -\frac{5\sqrt{29}}{29}\right)\right\}$$

EXERCISE SET 13.4

SOLVE the following systems (leave irrational numbers in radical form):

1. $x^2 - 3 = xy$
    $2x - y + 4 = 0$

2. $x^2 + y^2 = 5$
    $2x - 3y = 8$

3. $4x + 10y + 16 = 0$
    $3x - xy + 9 = 0$

4. $x + 2y = 4$
    $x^2 + 2y^2 = 9$

5. $y + (1/y) = x$
    $2x - 3y = 1$

6. $3x^2 - y^2 = 4$
    $3x - 2y = 1$

7. $x^2 + y^2 = 5$
    $2x^2 - 3y^2 = -10$

8. $y^2 - x^2 = 21$
    $3x^2 - y^2 = -13$

9. $8x^2 - 9y^2 = 7$
    $18y^2 - 4x^2 = 1$

10. $x^2 + 4y^2 = 61$
    $3x^2 + 2y^2 = 93$

11. $6x^2 - 7y^2 = 63$
    $2x^2 + 9y^2 = 13$

12. $15x^2 - 4y^2 = 8$
    $20x^2 + 12y^2 = 15$

13. $3x^2 + 2xy = 16$ **14.** $3x^2 - 2xy = 8$ **15.** $x^2 - xy + y^2 = 7$
      $4x^2 - 3xy = 10$ $4x^2 + 6xy = 28$ $3x^2 + 3xy - 3y^2 = 3$

16. $2xy - 2y^2 = 0$ **17.** $x^2 = y^2 - 27$ **18.** $2x^2 + y^2 = 44 - xy$
      $3x^2 = 6xy + 1$ $x^2 + xy + y^2 = 63$ $2y^2 = xy - x^2 + 16$

SOLVE the following problems, using two variables:

19. The sum of two numbers is 13 and the difference of their squares is 65. Find the numbers.
20. The diagonal of a rectangle is 13 ft. If its area is 8,640 sq in., find its dimensions.
21. The perimeter of a rectangle is 112 in. and its area is 768 sq in. Find its dimensions.
22. The area of a right triangle is 96 sq ft. If the hypotenuse is 20 ft, find the lengths of the two legs of the triangle.
23. The product of two numbers is 12. The sum of their reciprocals is the fraction $\frac{7}{12}$. Find the numbers.
24. In a two-digit number the sum of the squares of the digits is 13. The product of the digits is 6. Find the number.
25. The product of two integers exceeds their sum by 44. The quotient of the two integers is 8 less than their difference. Find the integers.
26. The sum of the squares of two numbers is 290, and the square of the larger is 48 more than the square of the smaller. Find the numbers.
27. The area of a rectangle is 300 sq ft. If the diagonal is 25 ft, find the dimensions of the rectangle.
28. The difference of the reciprocals of two numbers is 2. If the product of the numbers is $\frac{1}{35}$, find the numbers.
29. A two-digit number is three times the sum of its digits. The sum of the squares of the two digits is 53. Find the number.
30. The product of the two digits of a two-digit number is 18. If the digits are interchanged, the resulting number is 9 less than twice the original number. Find the original number.
31. A weight of x lb on one side of a teeter board balances a weight of 40 lb placed 6 ft from the fulcrum on the other side. If the unknown weight is moved 3 ft nearer the fulcrum, it balances a weight of 20 lb placed 7½ ft from the fulcrum. Find the unknown weight and the length of the teeter.
32. The hypotenuse of a right triangle is 30 in. The area is 225 sq in. Find the other two sides of the triangle.
33. Each of two rectangles has an area of 768 sq yd. The difference of their lengths is 16 yd, and the difference of their widths is 8 yd. Find the dimensions of each rectangle.

13.6   GRAPHS OF QUADRATIC EQUATIONS IN TWO VARIABLES

A knowledge of graphs of the simpler forms of quadratic equations in two variables is an important aid in understanding the solution of systems involving such equations. Suppose a system consists of two equations. If the graphs of the equations are sketched on the same coordinate system, the solution set of the system can be approximated by estimating the values of the coordinates of the points of intersection of the graphs. If the graphs do not intersect, the system has no real solution.

Certain forms of quadratic equations in x and y have graphs that are easily sketched, as indicated by the following discussion.

Circle. The graph of an equation that can be reduced to the form $(x - h)^2 + (y - k)^2 = r^2$ is a circle whose center is the point (h,k) and whose radius is r. We discussed this equation in Sec. 3.9.

Ellipse. The graph of an equation of the form

$$ax^2 + by^2 = c$$

where a, b, and c are positive constants and $a \neq b$ is an ellipse whose center is the origin and whose axes lie on the coordinate axes.

EXAMPLE 1. Sketch the graph of $9x^2 + 4y^2 = 36$.

Solution. When $x = 0$, $y = \pm 3$. Therefore, the y intercepts are $(0, \pm 3)$. When $y = 0$, $x = \pm 2$. Hence, the x intercepts are $(\pm 2, 0)$. Since x and y appear only as squares in this equation, it follows that if (x_1, y_1) is a point on the curve, then $(-x_1, y_1)$, $(-x_1, -y_1)$, and $(x_1, -y_1)$ are also points on the graph. We describe this property of the curve by saying that it is symmetric with respect to the y axis and to the x axis. Since

$$y = \frac{\pm 3\sqrt{4 - x^2}}{2}$$

we see that x cannot be greater than 2 nor less than -2. For each value of x between -2 and 2, there are two values of y that are real. The graph is shown in Fig. 13.4.

Hyperbola. Each of the equations

$$ax^2 - by^2 = c \qquad ay^2 - bx^2 = c$$

where a, b, and c are positive, represents a hyperbola whose center is at the origin and whose axes lie on the coordinate axes.

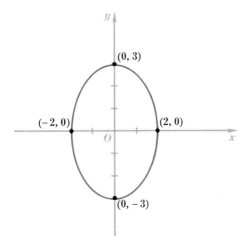

FIGURE 13.4
Graph of $9x^2 + 4y^2 = 36$

EXAMPLE 2.   Sketch the graph of $4x^2 - 9y^2 = 36$.

Solution.   When $y = 0$, $x = \pm 3$.   Hence, the x intercepts are $(\pm 3,0)$.
There are no y intercepts.   Since

$$y = \frac{\pm 2\sqrt{x^2 - 9}}{3}$$

we see that x cannot have values between -3 and 3.   The curve is symmetric
with respect to both axes.   Some of the ordered pairs of the relation
are shown in the following table.   Certain values of y are approximate; others
are exact.

x	-5	-4	-3	3	4	5
y	± 2.7	± 1.8	0	0	± 1.8	± 2.7

Note that the hyperbola has two branches.   The graph is shown in
Fig. 13.5.

Parabola.   An equation of the form

$$y = ax^2 + bx + c \qquad a \neq 0$$

represents a parabola whose axis is parallel to the y axis.   If $b = 0$ and $c = 0$,
then the axis of the parabola is the y axis.   An equation of the form

$$x = ay^2 + by + c \qquad a \neq 0$$

is a parabola whose axis is parallel to the x axis.   If $b = 0$ and $c = 0$, the axis
of the parabola is the x axis.   A discussion of the parabola whose axis is
parallel to the y axis was given in Sec. 8.3.

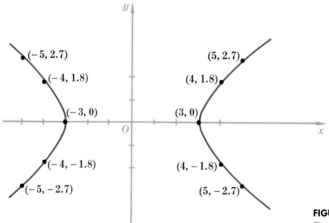

FIGURE 13.5
Graph of $4x^2 - 9y^2 = 36$

EXERCISE SET 13.5

SKETCH the graph of each of the following equations.

1. $x^2 + y^2 = 4$ 2. $16x^2 + 9y^2 = 288$
3. $4x^2 + 9y^2 = 144$ 4. $4x^2 - 25y^2 = 100$
5. $25x^2 - 16y^2 = 400$ 6. $x = y^2 - y - 5$
7. $y = x^2 - 2x + 1$ 8. $y^2 = 4x$
9. $x^2 = 9y$ 10. $9y^2 - x + 6y + 1 = 0$
11. $xy = 4$

SOLVE each of the following systems graphically. Sketch the graph of each equation on the same coordinate system and estimate the coordinates of the points of intersection.

12. $y = 9x - 6$ 13. $x = 10 + 2y$
    $y = 3x^2$ $y = x^2 + 2x - 15$

14. $2x - y = 1$ 15. $x^2 - y^2 = 6$
    $xy = 15$ $5x - 3y = 10$

13.7 MATRICES

In order to reduce the amount of labor involved in the process of solving systems of linear equations, we introduce the concept of a *matrix*. This concept is very useful in solving many of the problems confronting the modern mathematician.

To begin the discussion, we consider the system of linear equations

$$a_1x + b_1y = c_1$$
$$a_2x + b_2y = c_2$$

Array of coefficients   If we copy only the coefficients of the variables and the constants in the order in which they occur in the equations, we can abbreviate the writing of this system by the ordered array of numbers

$$\begin{pmatrix} a_1 & b_1 & c_1 \\ a_2 & b_2 & c_2 \end{pmatrix} \tag{13.14}$$

This array is usually enclosed within parentheses (or brackets, or double vertical bars). For example, the system

$$3x + 2y = 7$$
$$2x - 5y = 9$$

is represented by each of the following arrays:

$$\begin{pmatrix} 3 & 2 & 7 \\ 2 & -5 & 9 \end{pmatrix} \qquad \begin{bmatrix} 3 & 2 & 7 \\ 2 & -5 & 9 \end{bmatrix} \qquad \left\|\begin{matrix} 3 & 2 & 7 \\ 2 & -5 & 9 \end{matrix}\right\|$$

We shall use parentheses in this book.

The order of the numbers in the array

$$\begin{pmatrix} 3 & 2 & 7 \\ 2 & -5 & 9 \end{pmatrix}$$

is very important.   Any change in the order of elements would give an array which represents a different system of equations.

We note that this array is rectangular and consists of two *rows* and three *columns* of numbers.

DEFINITION.   A matrix is an ordered rectangular array of elements.   (**13.15**)

Dimensions of a matrix

If a matrix has m rows and n columns, we say that it has *dimensions m by n* and we call it an "*m-by-n* matrix."   A matrix consisting of only one row of elements,

$$(a_1 \quad a_2 \quad a_3 \quad \cdots \quad a_n)$$

Row matrix

is called a *row matrix*.   When a matrix consists of only one column of elements,

$$\begin{pmatrix} c_1 \\ c_2 \\ c_3 \\ \cdot \\ \cdot \\ \cdot \\ c_n \end{pmatrix}$$

Column matrix

Square matrix

it is called a *column matrix*.   A matrix which has the same number of columns as rows is called a *square matrix*.   Thus

$$\begin{pmatrix} 2 & 1 & 5 \\ 3 & 1 & 3 \\ 1 & 7 & 2 \end{pmatrix}$$

is a square matrix.

Let us consider the system of linear equations

$$\begin{matrix} a_1x + b_1y + c_1z = d_1 \\ a_2x + b_2y + c_2z = d_2 \\ a_3x + b_3y + c_3z = d_3 \end{matrix} \qquad (\textbf{13.16})$$

The coefficients of the variables can be exhibited as the square matrix

$$\begin{pmatrix} a_1 & b_1 & c_1 \\ a_2 & b_2 & c_2 \\ a_3 & b_3 & c_3 \end{pmatrix} \qquad (\textbf{13.17})$$

Coefficient matrix

called the *coefficient matrix* of the system (13.16). The constants on the right sides of the equations can be exhibited as the column matrix

$$\begin{pmatrix} d_1 \\ d_2 \\ d_3 \end{pmatrix} \tag{13.18}$$

The matrix

$$\begin{pmatrix} a_1 & b_1 & c_1 & d_1 \\ a_2 & b_2 & c_2 & d_2 \\ a_3 & b_3 & c_3 & d_3 \end{pmatrix} \tag{13.19}$$

obtained by adjoining to the coefficient matrix (13.17) the elements in the column matrix (13.18), is called the *augmented matrix* of the system (13.15).

Augmented matrix

EXAMPLE 1. Find the coefficient matrix and the augmented matrix of the system

$$2x + 3y - z = 5$$
$$3y - 2x + 2z = 7$$
$$3z + 5y - 6x = 1$$

Solution. Arrange the equations of the system so that each column of the coefficient matrix represents the coefficients of the same variable:

$$2x + 3y - z = 5$$
$$-2x + 3y + 2z = 7$$
$$-6x + 5y + 3z = 1$$

The coefficient matrix is

$$\begin{pmatrix} 2 & 3 & -1 \\ -2 & 3 & 2 \\ -6 & 5 & 3 \end{pmatrix}$$

and the augmented matrix is

$$\begin{pmatrix} 2 & 3 & -1 & 5 \\ -2 & 3 & 2 & 7 \\ -6 & 5 & 3 & 1 \end{pmatrix}$$

If the augmented matrix of a system of linear equations is given, the system can be found.

EXAMPLE 2. The augmented matrix of a system of linear equations in x, y, and z is

$$\begin{pmatrix} 3 & 2 & 1 & 3 \\ 2 & -3 & 2 & 2 \\ 1 & 5 & -4 & 1 \end{pmatrix}$$

Find the system of linear equations.

Solution. Let the elements in the first, second, and third columns be the coefficients of the variables x, y, and z respectively. Then the system of equations

$$3x + 2y + z = 3$$
$$2x - 3y + 2z = 2$$
$$x + 5y - 4z = 1$$

is the required system.

The operations on equations which we discussed in the theorems of Sec. 13.2 can be performed on the *rows* of numbers in the augmented matrix of the system to produce the augmented matrix of an equivalent system. These operations, called *elementary row operations*, are summarized in the following theorems. They always produce the augmented matrix of an equivalent system, and therefore are said to be *equivalent matrices*. The proof that such is the case follows from the fact that operating on the rows of the augmented matrix is the same thing as operating on the equations of the system. Review Theorems (13.4), (13.5), and (13.7).

Row operations

THEOREM. Any two rows of the augmented matrix can be interchanged to produce an equivalent matrix. **(13.20)**

THEOREM. Any row of the augmented matrix can be multiplied by any nonzero constant to produce an equivalent matrix. **(13.21)**

THEOREM. Any multiple of a row of the augmented matrix can be added to any different row to produce an equivalent matrix. **(13.22)**

The use of matrices to solve systems of linear equations is illustrated in the following examples.

EXAMPLE 3. Use matrix operations to solve the system

$$x + 3y = -1$$
$$5x - y = 11$$

Solution. Write the augmented matrix of the system:

$$\begin{pmatrix} 1 & 3 & -1 \\ 5 & -1 & 11 \end{pmatrix}$$

The objective is to get a zero in the first position of the second row. Hence, we replace the second row by itself minus 5 times the first row and write

$$\begin{pmatrix} 1 & 3 & -1 \\ 0 & -16 & 16 \end{pmatrix} \qquad \text{row 2 minus 5 times row 1}$$

This matrix represents the system

$$x + 3y = -1$$
$$-16y = 16$$

which is equivalent to the original system. From the second equation, $y = -1$. The first equation then yields $x = 2$. Hence, the solution of the given system is the set $\{(2, -1)\}$.

EXAMPLE 4. Solve the system

$$2x + y - z = 5$$
$$4x - 2y + z = -5$$
$$6x + 4y - 2z = 17$$

Solution. Write the augmented matrix of the system:

$$\begin{pmatrix} 2 & 1 & -1 & 5 \\ 4 & -2 & 1 & -5 \\ 6 & 4 & -2 & 17 \end{pmatrix}$$

We first transform this matrix to obtain a zero in the first position of row two and row three, as follows: Replace the second row by itself minus 2 times the first row. Then replace the third row by itself minus 3 times the first row. We write

$$\begin{pmatrix} 2 & 1 & -1 & 5 \\ 0 & -4 & 3 & -15 \\ 0 & 1 & 1 & 2 \end{pmatrix} \quad \begin{matrix} \text{row 2 minus 2 times row 1} \\ \text{row 3 minus 3 times row 1} \end{matrix}$$

We next work to obtain from this matrix a final matrix that will have zeros in the first two positions of the third row. Hence, we replace the third row by 4 times itself, plus the second row, and write

$$\begin{pmatrix} 2 & 1 & -1 & 5 \\ 0 & -4 & 3 & -15 \\ 0 & 0 & 7 & -7 \end{pmatrix} \quad \text{4 times row 3, plus row 2}$$

This is the augmented matrix of the system

$$2x + y - z = 5$$
$$-4y + 3z = -15$$
$$7z = -7$$

From the third equation $z = -1$; from the second, $y = 3$; and from the first, $x = \frac{1}{2}$. The solution is the ordered triple $(\frac{1}{2}, 3, -1)$.

In the preceding example, we reduced the given system of equations to an equivalent system in triangular form. This method imposes certain restrictions upon our choice as to which coefficient to use in eliminating the variables. The following method, called the "sweep out" method, allows more freedom of choice as to which coefficient to use. Let us begin with the augmented matrix of the system given in Example 4: Select any 1 or -1 (in a column of smallest numbers, generally) and sweep out the entire column. If a 1 or -1 does not occur, multiply by a constant to make one.

The augmented matrix is

$$\begin{pmatrix} 2 & 1 & -1 & 5 \\ 4 & -2 & 1 & -5 \\ 6 & 4 & -2 & 17 \end{pmatrix}$$

Taking the 1 in the first row and sweeping out the second column, we have

$$\begin{pmatrix} 2 & 1 & -1 & 5 \\ 8 & 0 & -1 & 5 \\ -2 & 0 & 2 & -3 \end{pmatrix} \qquad \begin{array}{l} \text{row 2 plus 2 times row 1} \\ \text{row 3 minus 4 times row 1} \end{array}$$

Next, taking the -1 in the second row and sweeping out the third column, we obtain

$$\begin{pmatrix} -6 & 1 & 0 & 0 \\ 8 & 0 & -1 & 5 \\ 14 & 0 & 0 & 7 \end{pmatrix} \qquad \begin{array}{l} \text{row 1 minus row 2} \\ \\ \text{row 3 plus 2 times row 2} \end{array}$$

Dividing the third row by 7 yields

$$\begin{pmatrix} -6 & 1 & 0 & 0 \\ 8 & 0 & -1 & 5 \\ 2 & 0 & 0 & 1 \end{pmatrix} \qquad \text{row 3 divided by 7}$$

Sweeping out the first column, we get a final matrix

$$\begin{pmatrix} 0 & 1 & 0 & 3 \\ 0 & 0 & -1 & 1 \\ 2 & 0 & 0 & 1 \end{pmatrix} \qquad \begin{array}{l} \text{row 1 plus 3 times row 3} \\ \text{row 2 minus 4 times row 3} \end{array}$$

This matrix represents the system

$$\begin{aligned} 2x &= 1 \\ y &= 3 \\ -z &= 1 \end{aligned}$$

Hence, the solution set is $\{(\frac{1}{2},3,-1)\}$ as before.

Any system of n linear equations involving n variables can be solved by the methods of this section.

EXERCISE SET 13.6

SOLVE each of the following systems, using matrix operations:

1. $3x + y - 3z = 2$
    $-x + 4y + 2z = 1$
    $x + 2y - z = 3$

2. $2x + 3y + z = 4$
    $3x - 2y - 2z = 4$
    $-3x + 8y + 9z = 3$

3. $2x + y = 1$
$3x - 4y + 2z = 3$
$5x - 3y + 4z = -5$

4. $4x - 2y - 3z + 1 = 0$
$x - 4z - 2 = 0$
$3x - y + 5z = 0$

5. $x - y - \frac{1}{2}z = \frac{3}{2}$
$2x + 3y - 5z = -2$
$5y + 3z = 2$

6. $2x + 3y - z - 3 = 0$
$y + z = x$
$y - x - z - 1 = 0$

7. $\dfrac{x - y}{3} - \dfrac{y - z}{4} = \dfrac{7}{3}$

   $\dfrac{y - z}{3} + \dfrac{x + z}{5} = -\dfrac{13}{15}$

   $\dfrac{x + z}{2} - \dfrac{x - y}{5} = \dfrac{43}{10}$

8. $\dfrac{3x - 2y}{5} - \dfrac{4z - 5y}{2} = \dfrac{19}{2}$

   $\dfrac{2x - 3z}{6} - \dfrac{x - 4y}{4} = \dfrac{7}{4}$

   $\dfrac{4x + z}{3} - \dfrac{3y + 5z}{2} = \dfrac{49}{3}$

9. $3x - 4y - 2z + w = 11$
$2x - 2y + 2z + w = 5$
$-2x + 2y + 3z - 2w = -13$
$2x + 4y - z + 3w = 14$

10. $u + 3x - 2y - z = -3$
$2u - x - y + 3z = 23$
$u + x + 3y - 2z = -12$
$3u - 2x + y + z = 22$

11. The middle digit of a three-digit number is half the sum of the other two digits. If the number is divided by the sum of its digits, the quotient is 20, and the remainder is 9. If 594 is added to the number, the digits will be reversed. Find the number.

12. The points $(1,4)$, $(-1,-2)$, $(2,4)$ are on the curve whose equation is $y = x^3 + ax^2 + bx + c$. Find the constants a, b, and c.

13.8 ALGEBRA OF MATRICES

We shall use capital letters to denote matrices. Thus,

$$A = \begin{pmatrix} 3 & -2 & 4 & 1 \\ 1 & 3 & 1 & 3 \end{pmatrix}$$

is a 2-by-4 matrix. Now consider the matrix

$$B = \begin{pmatrix} 3 & 4 & -2 & 1 \\ 1 & 1 & 3 & 3 \end{pmatrix}$$

which is also a 2-by-4 matrix having the same set of numbers for elements as the matrix A, but within which the elements are not in the same order as the elements of A. The system of equations represented by A is quite different from the system represented by B. In this case, we say that matrix A is not equal to matrix B. To determine when two matrices are equal, we make the following definition.

> **DEFINITION.** Two matrices A and B are equal if and only if they satisfy the following conditions:
> 1. They have the same dimensions. (13.23)
> 2. The elements in like positions of the two arrays are equal.

We use the notation $A = B$ to indicate that matrix A equals matrix B. If A does not equal B, we write $A \neq B$. For example, if

$$A = \begin{pmatrix} 2 & 3 & 6 \\ 2 & 1 & 5 \end{pmatrix} \qquad B = \begin{pmatrix} 3 & 2 & 6 \\ 1 & 2 & 5 \end{pmatrix} \qquad C = \begin{pmatrix} 2 & 3 & 6 \\ 2 & 1 & 5 \end{pmatrix}$$

then $A = C$, $A \neq B$, $B \neq C$.

The sum of two matrices is defined only if they have the same dimensions. To indicate this sum, we use the familiar $+$ sign with the understanding that addition of two matrices is the operation defined as follows.

> **DEFINITION.** If A and B are matrices having the same dimensions, then the sum of A and B, denoted by $A + B$, is a matrix such that each of its elements is the sum of the corresponding elements of A and B. $\qquad$ **(13.24)**

If $\qquad A = \begin{pmatrix} 3 & 1 & -2 \\ -2 & 0 & 5 \end{pmatrix} \qquad$ and $\qquad B = \begin{pmatrix} -5 & 2 & 3 \\ 6 & 1 & 8 \end{pmatrix}$

then $A + B$ is the matrix obtained as follows:

$$A + B = \begin{pmatrix} 3 + (-5) & 1 + 2 & -2 + 3 \\ -2 + 6 & 0 + 1 & 5 + 8 \end{pmatrix} = \begin{pmatrix} -2 & 3 & 1 \\ 4 & 1 & 13 \end{pmatrix}$$

Just as we defined the additive identity in the set of all real numbers and denoted it by 0, we define an additive identity in the set of all matrices and call it a *zero matrix*. We denote a zero matrix by 0, and we insist that it be a matrix such that

$$A + 0 = A$$

for any matrix A. This means that 0 must have the same dimensions as A and that its elements must be such that when each is added to the corresponding element of A, that element of A is unchanged.

> **DEFINITION.**   For the set of all m-by-n matrices, the zero matrix, denoted by 0, is an m-by-n matrix having all its elements equal to zero. $\qquad$ **(13.25)**

If A is the 3-by-2 matrix $\begin{pmatrix} 2 & 1 \\ 3 & 5 \\ 6 & 2 \end{pmatrix}$, then $0 = \begin{pmatrix} 0 & 0 \\ 0 & 0 \\ 0 & 0 \end{pmatrix}$. If $B = \begin{pmatrix} 3 & 7 & 5 \\ 1 & 1 & 8 \end{pmatrix}$,

then $0 = \begin{pmatrix} 0 & 0 & 0 \\ 0 & 0 & 0 \end{pmatrix}$.

The additive inverse of the matrix A, also called the negative of A, is denoted by $-A$ and is defined to be a matrix having the same dimensions as A and such that

$$A + (-A) = 0$$

This means that $-A$ is a matrix whose elements are the negatives of the corresponding elements of A. For example, if

$$A = \begin{pmatrix} 3 & -1 & 4 \\ 2 & 1 & -6 \end{pmatrix}$$

then

$$-A = \begin{pmatrix} -3 & 1 & -4 \\ -2 & -1 & 6 \end{pmatrix}$$

The following laws can be derived from the preceding definitions:

THE COMMUTATIVE LAW OF ADDITION. If A, B are matrices having the same dimensions, then

$$A + B = B + A$$

THE ASSOCIATIVE LAW OF ADDITION. If A, B, and C are matrices having the same dimensions, then

$$(A + B) + C = A + (B + C)$$

Scalar

Any number from the field of complex numbers is called a *scalar*. Thus, 2, $-\frac{5}{3}$, $3 + 2i$ are scalars. We define the product of a matrix A and a scalar k, denoted by kA, to be the matrix whose elements are k times the corresponding elements of A. For example, if

$$A = \begin{pmatrix} a & b & c \\ d & e & f \end{pmatrix}$$

then

$$kA = \begin{pmatrix} ka & kb & kc \\ kd & ke & kf \end{pmatrix}$$

To introduce the operation of matrix multiplication, we first define the product of a row matrix A of dimensions 1 by n, and a column matrix B of dimensions n by 1, as follows:

$$AB = (a_1 a_2 \cdots a_n) \begin{pmatrix} b_1 \\ b_2 \\ \cdot \\ \cdot \\ \cdot \\ b_n \end{pmatrix} = (a_1 b_1 + a_2 b_2 + \cdots + a_n b_n)$$

Note that the product AB of row matrix A and column matrix B is defined only if A has as many columns as B has rows. The product is a 1-by-1 matrix whose single element is the number $a_1b_1 + a_2b_2 + \cdots + a_nb_n$.

Scalar product This number is called the *scalar product* of A and B.

EXAMPLE 1. Find the product of A and B if $A = (2 \quad 3 \quad 5)$ and $B = \begin{pmatrix} 1 \\ 6 \\ 3 \end{pmatrix}$.

Solution. $AB = (2 \quad 3 \quad 5)\begin{pmatrix} 1 \\ 6 \\ 3 \end{pmatrix} = (2\cdot 1 + 3\cdot 6 + 5\cdot 3) = (35)$

We now define the product AB of any two matrices A and B where the number of columns of A is equal to the number of rows of B.

DEFINITION. If A has dimensions m by n and B has dimensions n by p, the product AB is a matrix having dimensions m by p, such that the element in the ith row and jth column is the scalar product of the ith row of A and the jth column of B.   **(13.26)**

EXAMPLE 2. If $A = \begin{pmatrix} 2 & 3 & 4 \\ 7 & 0 & 5 \end{pmatrix}$ and $B = \begin{pmatrix} 8 & 10 & 0 \\ 12 & 0 & 11 \\ 0 & 15 & 16 \end{pmatrix}$, find AB.

Solution.

$\begin{pmatrix} 2 & 3 & 4 \\ 7 & 0 & 5 \end{pmatrix}\begin{pmatrix} 8 & 10 & 0 \\ 12 & 0 & 11 \\ 0 & 15 & 16 \end{pmatrix}$

$= \begin{pmatrix} 2(8) + 3(12) + 4(0) & 2(10) + 3(0) + 4(15) & 2(0) + 3(11) + 4(16) \\ 7(8) + 0(12) + 5(0) & 7(10) + 0(0) + 5(15) & 7(0) + 0(11) + 5(16) \end{pmatrix}$

$= \begin{pmatrix} 52 & 80 & 97 \\ 56 & 145 & 80 \end{pmatrix}$

Note that the element in the first row and first column of AB is the scalar product of the first row of A and the first column of B. The element in the first row and second column of AB is the scalar product of the first row of A and the second column of B, and so on.

The product AB of any two matrices A and B is defined only if the number of columns of A is the same as the number of rows of B.

In general, the product of two matrices is not commutative. For example, if

$$A = \begin{pmatrix} 2 & 3 \\ 1 & 7 \end{pmatrix} \quad \text{and} \quad B = \begin{pmatrix} 5 & 2 \\ 3 & 6 \end{pmatrix}$$

then

$$AB = \begin{pmatrix} 2 & 3 \\ 1 & 7 \end{pmatrix}\begin{pmatrix} 5 & 2 \\ 3 & 6 \end{pmatrix} = \begin{pmatrix} 10+9 & 4+18 \\ 5+21 & 2+42 \end{pmatrix} = \begin{pmatrix} 19 & 22 \\ 26 & 44 \end{pmatrix}$$

$$BA = \begin{pmatrix} 5 & 2 \\ 3 & 6 \end{pmatrix}\begin{pmatrix} 2 & 3 \\ 1 & 7 \end{pmatrix} = \begin{pmatrix} 10+2 & 15+14 \\ 6+6 & 9+42 \end{pmatrix} = \begin{pmatrix} 12 & 29 \\ 12 & 51 \end{pmatrix}$$

Hence,

$$AB \neq BA$$

The set of all square matrices of the same dimensions has an identity matrix under the operation of matrix multiplication. The multiplicative identity matrix is denoted by I and must be such that

$$AI = A$$

for every matrix A. This means that if A is a square matrix of dimensions n by n, then I must be an n-by-n matrix such that each element on the main diagonal (the diagonal from upper left to lower right) is the number 1, and all other elements are zero.

Multiplicative identity matrix

The multiplicative identity matrix in the set of all 2-by-2 matrices is

$$I = \begin{pmatrix} 1 & 0 \\ 0 & 1 \end{pmatrix}$$

The multiplicative identity matrix in the set of all 3-by-3 matrices is

$$I = \begin{pmatrix} 1 & 0 & 0 \\ 0 & 1 & 0 \\ 0 & 0 & 1 \end{pmatrix}$$

For example, if the matrix A is equal to

$$\begin{pmatrix} 2 & 3 & 5 \\ 1 & 4 & 2 \\ 3 & 2 & 1 \end{pmatrix}$$

we have

$$AI = \begin{pmatrix} 2 & 3 & 5 \\ 1 & 4 & 2 \\ 3 & 2 & 1 \end{pmatrix}\begin{pmatrix} 1 & 0 & 0 \\ 0 & 1 & 0 \\ 0 & 0 & 1 \end{pmatrix} = \begin{pmatrix} 2 & 3 & 5 \\ 1 & 4 & 2 \\ 3 & 2 & 1 \end{pmatrix} = A$$

Also,

$$IA = \begin{pmatrix} 1 & 0 & 0 \\ 0 & 1 & 0 \\ 0 & 0 & 1 \end{pmatrix}\begin{pmatrix} 2 & 3 & 5 \\ 1 & 4 & 2 \\ 3 & 2 & 1 \end{pmatrix} = \begin{pmatrix} 2 & 3 & 5 \\ 1 & 4 & 2 \\ 3 & 2 & 1 \end{pmatrix} = A$$

It can be shown that if A is any square matrix, then

$$AI = IA$$

The multiplicative inverse of a matrix A is defined only if A is a square matrix. The multiplicative inverse of an n-by-n matrix A is denoted by A^{-1} and is an n-by-n matrix having the property that

$$AA^{-1} = I$$

For example, if the matrix $A = \begin{pmatrix} 1 & 3 \\ 2 & 5 \end{pmatrix}$, then $A^{-1} = \begin{pmatrix} -5 & 3 \\ 2 & -1 \end{pmatrix}$,

since $\quad \begin{pmatrix} 1 & 3 \\ 2 & 5 \end{pmatrix}\begin{pmatrix} -5 & 3 \\ 2 & -1 \end{pmatrix} = \begin{pmatrix} -5+6 & 3-3 \\ -10+10 & 6-5 \end{pmatrix} = \begin{pmatrix} 1 & 0 \\ 0 & 1 \end{pmatrix}$

It is not true that every square matrix has a multiplicative inverse. For example, the matrix

$$\begin{pmatrix} 2 & 3 \\ 4 & 6 \end{pmatrix}$$

does not have a multiplicative inverse.   To show that this is true, suppose that there is a multiplicative inverse.   Denote it by

$$\begin{pmatrix} x & y \\ z & w \end{pmatrix}$$

Then, $\qquad \begin{pmatrix} 2 & 3 \\ 4 & 6 \end{pmatrix}\begin{pmatrix} x & y \\ z & w \end{pmatrix} = \begin{pmatrix} 1 & 0 \\ 0 & 1 \end{pmatrix}$

and by Definition (13.26)

$$\begin{pmatrix} 2x + 3z & 2y + 3w \\ 4x + 6z & 4y + 6w \end{pmatrix} = \begin{pmatrix} 1 & 0 \\ 0 & 1 \end{pmatrix}$$

Therefore, by Definition (13.23), we have

(1) $\qquad\qquad\qquad \begin{aligned} 2x + 3z &= 1 \\ 4x + 6z &= 0 \end{aligned}$

and

(2) $\qquad\qquad\qquad \begin{aligned} 2y + 3w &= 0 \\ 4y + 6w &= 1 \end{aligned}$

Since each of the systems (1) and (2) is an inconsistent system, neither has a solution.   Our supposition is thus false and the given matrix has no multiplicative inverse.

A square matrix that does not have a multiplicative inverse is called a *singular matrix*.   A *nonsingular matrix* is a matrix that does have a multiplicative inverse.   This inverse is unique.   For a 2-by-2 matrix, the inverse may be found as in the following example.

EXAMPLE 3. Find A^{-1} if $A = \begin{pmatrix} a_1 & b_1 \\ a_2 & b_2 \end{pmatrix}$.

Solution. Let $A^{-1} = \begin{pmatrix} x & y \\ z & w \end{pmatrix}$. Then, since $AA^{-1} = I$,

$$\begin{pmatrix} a_1 & b_1 \\ a_2 & b_2 \end{pmatrix}\begin{pmatrix} x & y \\ z & w \end{pmatrix} = \begin{pmatrix} 1 & 0 \\ 0 & 1 \end{pmatrix}$$

$$\begin{pmatrix} a_1x + b_1z & a_1y + b_1w \\ a_2x + b_2z & a_2y + b_2w \end{pmatrix} = \begin{pmatrix} 1 & 0 \\ 0 & 1 \end{pmatrix}$$

Hence,

(1)
$$a_1x + b_1z = 1$$
$$a_2x + b_2z = 0$$

(2)
$$a_1y + b_1w = 0$$
$$a_2y + b_2w = 1$$

If $a_1b_2 - a_2b_1 \neq 0$, system (1) yields

$$x = \frac{b_2}{a_1b_2 - a_2b_1} \qquad z = \frac{-a_2}{a_1b_2 - a_2b_1}$$

From (2), $\qquad y = \dfrac{-b_1}{a_1b_2 - a_2b_1} \qquad w = \dfrac{a_1}{a_1b_2 - a_2b_1}$

If we let $\Delta = a_1b_2 - a_2b_1$, then

$$A^{-1} = \begin{pmatrix} \dfrac{b_2}{\Delta} & \dfrac{-b_1}{\Delta} \\ \dfrac{-a_2}{\Delta} & \dfrac{a_1}{\Delta} \end{pmatrix} = \begin{pmatrix} b_2 & -b_1 \\ -a_2 & a_1 \end{pmatrix}\frac{1}{\Delta}$$

We leave it to the student to prove that

$$AA^{-1} = I = A^{-1}A$$

This method can be used to find the inverse of any n-by-n nonsingular matrix. However, the task becomes quite arduous if n is large. There are other methods available, but all require a great amount of labor.

EXERCISE SET 13.7

FIND the values of x, y, z, w that satisfy the matrix equations:

1. $(x \quad y) = (3 \quad 5)$

2. $(z \quad w) = (5 \quad -3)$

3. $\begin{pmatrix} x & y \\ y & x \end{pmatrix} = \begin{pmatrix} y & 1 \\ x & y \end{pmatrix}$

4. $\begin{pmatrix} x + 5 & 3y + 2 \\ 2z + 4 & 3w + 2 \end{pmatrix} = \begin{pmatrix} 2x + 3 & y - 4 \\ 3 & 2w + 5 \end{pmatrix}$

5. $\begin{pmatrix} 2x+5 & y-1 & z+2 \\ y+3 & 2 & 2x \end{pmatrix} = \begin{pmatrix} 3x+2 & 0 & 0 \\ 2y+2 & -z & 6 \end{pmatrix}$

6. $\begin{pmatrix} x^2-1 & 3 \\ 2 & y^2-1 \end{pmatrix} = \begin{pmatrix} 0 & 3 \\ 2 & 3 \end{pmatrix}$

PERFORM the indicated operations:

7. $\begin{pmatrix} 3 & 1 & 2 \\ 4 & 2 & 5 \end{pmatrix} + \begin{pmatrix} 2 & 2 & 1 \\ 3 & 5 & 3 \end{pmatrix}$

8. $\begin{pmatrix} 2 & 7 & 1 \\ -3 & 4 & -6 \\ 1 & 2 & 5 \end{pmatrix} + \begin{pmatrix} 2 & 13 & 2 \\ 3 & -1 & 4 \\ -1 & -2 & 3 \end{pmatrix}$

9. $3\begin{pmatrix} x & -x & 4 \\ 0 & 5y & -3y \end{pmatrix} + 2\begin{pmatrix} -y & x & 1 \\ y & x & 3y \end{pmatrix}$

10. $5\begin{pmatrix} 3 & 4 \\ 1 & 2 \end{pmatrix} - 3\begin{pmatrix} 4 & 2 \\ 3 & 1 \end{pmatrix}$

11. $\begin{pmatrix} 2 & -2 & 3 \\ -1 & 0 & 7 \\ 5 & -1 & 4 \end{pmatrix} - \begin{pmatrix} -1 & 3 & -1 \\ 0 & 5 & -2 \\ 1 & 2 & 1 \end{pmatrix}$

12. $\begin{pmatrix} 2 & 5 & 4 \\ -3 & -2 & 1 \\ 2 & 4 & 0 \\ 1 & 3 & 5 \end{pmatrix} - \begin{pmatrix} 1 & 5 & 4 \\ -3 & -3 & 1 \\ 2 & 4 & 0 \\ 1 & 2 & 3 \end{pmatrix}$

13. $3\begin{pmatrix} 1 & 3 & 2 \\ 2 & 1 & 5 \\ 1 & 2 & 7 \end{pmatrix} - 5\begin{pmatrix} -2 & 1 & 3 \\ 4 & -3 & 6 \\ 1 & 5 & 5 \end{pmatrix}$

14. $(3 \quad -2 \quad 1 \quad -5)\begin{pmatrix} 3 \\ 2 \\ 1 \\ 5 \end{pmatrix}$

15. $(2 \quad 3 \quad 1 \quad -2)\begin{pmatrix} 3 \\ -2 \\ -1 \\ 2 \end{pmatrix}$

16. $\begin{pmatrix} 4 & -3 & 2 \\ 3 & 0 & 1 \\ 1 & -2 & 2 \end{pmatrix}\begin{pmatrix} 1 & 4 \\ -2 & 1 \\ 2 & 3 \end{pmatrix}$

17. $\begin{pmatrix} 2 & 1 & 3 \\ 5 & -1 & 0 \\ 4 & 1 & -2 \end{pmatrix}\begin{pmatrix} 3 & 1 & 2 \\ 1 & 2 & 1 \\ 3 & 0 & 1 \end{pmatrix}$

18. $\begin{pmatrix} 2 & -1 \\ 3 & 4 \\ 2 & 6 \end{pmatrix}\begin{pmatrix} 1 & 2 & 5 \\ 3 & 2 & 4 \end{pmatrix}$

19. $\begin{pmatrix} 2 & -1 \\ 0 & 3 \end{pmatrix}\begin{pmatrix} 1 & 0 & 2 & 3 \\ 0 & 0 & 1 & -5 \end{pmatrix}$

20. $\begin{pmatrix} x & 0 & 0 \\ 0 & y & 0 \\ 0 & 0 & z \end{pmatrix}\begin{pmatrix} a_1 & b_1 \\ a_2 & b_2 \\ a_3 & b_3 \end{pmatrix}$

21. If $X = \begin{pmatrix} 1 & 2 \\ 3 & 2 \\ 1 & 4 \end{pmatrix}$, $Y = \begin{pmatrix} 3 \\ 6 \end{pmatrix}$, $Z = (a \quad b \quad c)$, show that $(XY)Z = X(YZ)$.

22. For the matrices

$$A = \begin{pmatrix} a & b \\ c & d \end{pmatrix} \qquad B = \begin{pmatrix} e & f \\ g & h \end{pmatrix} \qquad C = \begin{pmatrix} i & j \\ k & l \end{pmatrix}$$

prove that (a) $(AB)C = A(BC)$; (b) $(A + B)C = AC + BC$.

23. The *transpose* of a matrix A, denoted by A', is the matrix obtained by interchanging the rows and columns of A. Thus, the first row of A' is the first column of A, the second row of A' is the second column of A, etc. Find A' if

$$A = \begin{pmatrix} 1 & 3 & 2 \\ 2 & 4 & 5 \\ 5 & 0 & 7 \end{pmatrix}$$

24. For the matrices of Prob. 22 show that

(a) $(A + B)' = A' + B'$
(b) $(kA)' = kA'$
(c) $(AB)' = B'A'$ (note the change in order)
(d) If $A + B = A + C$, then $B = C$

25. Given that $I = \begin{pmatrix} 1 & 0 \\ 0 & 1 \end{pmatrix}$, show that $I^{-1} = I$.

26. If $A = \begin{pmatrix} 3 & -4 & 2 \\ 1 & 5 & 9 \\ 2 & 6 & -7 \end{pmatrix}$, show that $AI = IA$.

27. The system of equations $a_1x + b_1y = c_1$, $a_2x + b_2y = c_2$ may be written in the compact matrix form

$$\begin{pmatrix} a_1 & b_1 \\ a_2 & b_2 \end{pmatrix}\begin{pmatrix} x \\ y \end{pmatrix} = \begin{pmatrix} c_1 \\ c_2 \end{pmatrix}$$

Write the system of equations which must be valid if

$$\begin{pmatrix} 2 & 3 \\ 5 & -2 \end{pmatrix}\begin{pmatrix} x \\ y \end{pmatrix} = \begin{pmatrix} 6 \\ -4 \end{pmatrix}$$

28. Express in matrix form the system $5x - 2y = 12$, $3x + 7y = -4$.

29. Find A^{-1} if $A = \begin{pmatrix} 2 & 5 \\ 3 & -7 \end{pmatrix}$.   **30.** Find A^{-1} if $A = \begin{pmatrix} 1 & 6 \\ 0 & 3 \end{pmatrix}$.

31. If $i^2 = -1$, the Pauli spin matrices are

$$I = \begin{pmatrix} 1 & 0 \\ 0 & 1 \end{pmatrix} \qquad A = \begin{pmatrix} -1 & 0 \\ 0 & -1 \end{pmatrix} \qquad B = \begin{pmatrix} 0 & 1 \\ -1 & 0 \end{pmatrix} \qquad C = \begin{pmatrix} 0 & -1 \\ 1 & 0 \end{pmatrix}$$

$$D = \begin{pmatrix} i & 0 \\ 0 & -i \end{pmatrix} \qquad E = \begin{pmatrix} -i & 0 \\ 0 & i \end{pmatrix} \qquad F = \begin{pmatrix} 0 & -i \\ -i & 0 \end{pmatrix} \qquad G = \begin{pmatrix} 0 & i \\ i & 0 \end{pmatrix}$$

Show that the set of Pauli spin matrices is closed under the operation of matrix multiplication.

32. For the square matrix A, we define $A^1 = A$, $A^2 = AA$, $A^3 = A^2A$. If $i^2 = -1$, and if

$A = \begin{pmatrix} 0 & -i \\ i & 0 \end{pmatrix}$, find A^3.

13.9 DETERMINANT OF A SQUARE MATRIX

A square matrix having dimensions n by n is said to be of *order n*. Thus, a 2-by-2 matrix is of order 2, a 3-by-3 matrix is of order 3, etc.

Determinant of a
matrix

Let A be a square matrix whose elements are real numbers. To the matrix A we assign a unique real number called the *determinant* of A and denoted by $d(A)$. We read $d(A)$ as "the determinant of A." We also refer to $d(A)$ as "the value of the determinant of A." Thus, d is a real-valued function whose ordered pairs are $[A,d(A)]$. The domain of d is the set of all square matrices having real-number elements, and the range is the set of all real numbers.

If the matrix is of order 1, that is, if $A = (a)$, then the real number a is defined to be the determinant of the matrix. Thus, if $A = (5)$, then

$$d(A) = d(5) = 5$$

For a 2-by-2 matrix

$$\begin{pmatrix} a_1 & b_1 \\ a_2 & b_2 \end{pmatrix}$$

the expression $a_1b_2 - a_2b_1$ is defined to be the determinant of the matrix, and this is denoted as follows:

$$d \begin{pmatrix} a_1 & b_1 \\ a_2 & b_2 \end{pmatrix} = \begin{vmatrix} a_1 & b_1 \\ a_2 & b_2 \end{vmatrix} = a_1b_2 - a_2b_1 \qquad (13.27)$$

Thus, $$d \begin{pmatrix} 2 & 3 \\ -1 & -5 \end{pmatrix} = \begin{vmatrix} 2 & 3 \\ -1 & -5 \end{vmatrix} = 2(-5) - (-1)(3) = -7$$

Note that the determinant of the matrix is indicated by enclosing the array of the matrix between vertical bars instead of parentheses. The determinant of a matrix of order n is said to be of order n.

Minor

The *minor* of any element of a determinant of order n is the determinant of order $n - 1$ obtained by deleting the row and column in which the element appears. Thus, in the third-order determinant.

$$\begin{vmatrix} a_1 & b_1 & c_1 \\ a_2 & b_2 & c_2 \\ a_3 & b_3 & c_3 \end{vmatrix}$$

the minor of the element a_1 is the second-order determinant

$$\begin{vmatrix} b_2 & c_2 \\ b_3 & c_3 \end{vmatrix}$$

the minor of b_2 is

$$\begin{vmatrix} a_1 & c_1 \\ a_3 & c_3 \end{vmatrix}$$

and the minor of the element a_3 is

$$\begin{vmatrix} b_1 & c_1 \\ b_2 & c_2 \end{vmatrix}$$

Cofactor

The *cofactor* of the element in the ith row and the jth column of a determinant is defined to be the minor of that element multiplied by $(-1)^{i+j}$. Hence, for the preceding third-order determinant, the cofactor of a_1 is

$$(-1)^{1+1} \begin{vmatrix} b_2 & c_2 \\ b_3 & c_3 \end{vmatrix}$$

the cofactor of b_1 is $$(-1)^{1+2} \begin{vmatrix} a_2 & c_2 \\ a_3 & c_3 \end{vmatrix}$$

the cofactor of c_1 is $\qquad (-1)^{1+3}\begin{vmatrix} a_2 & b_2 \\ a_3 & b_3 \end{vmatrix}$

and so on.

EXAMPLE 1. Find the cofactor of 3 in the determinant

$$\begin{vmatrix} 1 & 2 & 7 \\ 5 & 1 & 2 \\ 1 & 3 & 8 \end{vmatrix}$$

Solution. Since the element 3 is in the third row and second column, its cofactor is

$$(-1)^{3+2}\begin{vmatrix} 1 & 7 \\ 5 & 2 \end{vmatrix} = -\begin{vmatrix} 1 & 7 \\ 5 & 2 \end{vmatrix}$$

We now define the determinant of a matrix of order n as follows:

DEFINITION. The value of a determinant of order n, $n > 2$, is the sum of the products formed by multiplying each element of $\quad$ **(13.28)** the first row by its cofactor.

A third-order determinant is defined in terms of second-order determinants whose values can be found by Eq. (13.27). Thus,

$$\begin{vmatrix} a_1 & b_1 & c_1 \\ a_2 & b_2 & c_2 \\ a_3 & b_3 & c_3 \end{vmatrix} = a_1 \begin{vmatrix} b_2 & c_2 \\ b_3 & c_3 \end{vmatrix} - b_1 \begin{vmatrix} a_2 & c_2 \\ a_3 & c_3 \end{vmatrix} + c_1 \begin{vmatrix} a_2 & b_2 \\ a_3 & b_3 \end{vmatrix}$$

$$= a_1 b_2 c_3 - a_1 b_3 c_2 - a_2 b_1 c_3 + a_3 b_1 c_2$$
$$+ a_2 b_3 c_1 - a_3 b_2 c_1 \quad \textbf{(13.29)}$$

EXAMPLE 2.

$$\begin{vmatrix} 2 & 3 & -7 \\ 1 & 5 & 4 \\ 3 & 2 & 3 \end{vmatrix} = 2 \begin{vmatrix} 5 & 4 \\ 2 & 3 \end{vmatrix} - 3 \begin{vmatrix} 1 & 4 \\ 3 & 3 \end{vmatrix} - 7 \begin{vmatrix} 1 & 5 \\ 3 & 2 \end{vmatrix}$$

$$= 2(5 \cdot 3 - 2 \cdot 4) - 3(1 \cdot 3 - 3 \cdot 4) - 7(1 \cdot 2 - 3 \cdot 5)$$
$$= 132$$

Expansion of a determinant

Definition (13.28), applied to any determinant of order n, $n > 2$, is called the Laplace expansion of the determinant on elements of the first row. The definition reduces the problem of expanding a determinant of order n to that of expanding n determinants of order $n - 1$. The following theorem is fundamental in the study of determinants.

THEOREM.   A determinant may be evaluated by the Laplace expansion on elements of any row or column.   **(13.30)**

We will not give a proof of this theorem here, but the student should satisfy himself that it holds for any third-order determinant. If any of the elements of a row or column are zero, the determinant is usually expanded on that row or column which has the greatest number of zero elements.

EXAMPLE 3.   Evaluate the determinant

$$D = \begin{vmatrix} 1 & 2 & 1 & 5 \\ 2 & 4 & 0 & 1 \\ 0 & 0 & 1 & 0 \\ 5 & 2 & 7 & 3 \end{vmatrix}$$

Solution.   Let us expand the determinant on the elements of the third row. Then

$$D = 0\begin{vmatrix} 2 & 1 & 5 \\ 4 & 0 & 1 \\ 2 & 7 & 3 \end{vmatrix} - 0\begin{vmatrix} 1 & 1 & 5 \\ 2 & 0 & 1 \\ 5 & 7 & 3 \end{vmatrix} + \begin{vmatrix} 1 & 2 & 5 \\ 2 & 4 & 1 \\ 5 & 2 & 3 \end{vmatrix} - 0\begin{vmatrix} 1 & 2 & 1 \\ 2 & 4 & 0 \\ 5 & 2 & 7 \end{vmatrix}$$

$$= \begin{vmatrix} 1 & 2 & 5 \\ 2 & 4 & 1 \\ 5 & 2 & 3 \end{vmatrix} = \begin{vmatrix} 4 & 1 \\ 2 & 3 \end{vmatrix} - 2\begin{vmatrix} 2 & 1 \\ 5 & 3 \end{vmatrix} + 5\begin{vmatrix} 2 & 4 \\ 5 & 2 \end{vmatrix} = -72$$

EXERCISE SET 13.8

FIND the cofactor of each element in the second row of the following determinants:

1. $\begin{vmatrix} 2 & 3 \\ 1 & 5 \end{vmatrix}$ 2. $\begin{vmatrix} -2 & 6 \\ 3 & 5 \end{vmatrix}$ 3. $\begin{vmatrix} 1 & 3 & -5 \\ 2 & -6 & 0 \\ 3 & 1 & 7 \end{vmatrix}$ 4. $\begin{vmatrix} 1 & 2 & 4 \\ -3 & -8 & 6 \\ 5 & 1 & 3 \end{vmatrix}$

FIND the cofactor of each element in the second column of each of the following determinants:

5. $\begin{vmatrix} 1 & 2 \\ 4 & 3 \end{vmatrix}$ 6. $\begin{vmatrix} -2 & -3 \\ 0 & 6 \end{vmatrix}$ 7. $\begin{vmatrix} 0 & 3 & 1 \\ 1 & 0 & 2 \\ 2 & -1 & 5 \end{vmatrix}$ 8. $\begin{vmatrix} 2 & 0 & 1 \\ -3 & -1 & 5 \\ 7 & 2 & 3 \end{vmatrix}$

EVALUATE each of the following determinants:

9. $\begin{vmatrix} 4 & 3 \\ 2 & 7 \end{vmatrix}$ 10. $\begin{vmatrix} 6 & 3 \\ -5 & -4 \end{vmatrix}$ 11. $\begin{vmatrix} -6 & -5 \\ 7 & 9 \end{vmatrix}$

12. $\begin{vmatrix} 2x & y \\ -x & -y \end{vmatrix}$ 13. $\begin{vmatrix} 2x+1 & 3 \\ 4 & x \end{vmatrix}$ 14. $\begin{vmatrix} 9a & -4b \\ -5c & -d \end{vmatrix}$

15. $\begin{vmatrix} 1 & 3 & 2 \\ 4 & 5 & 3 \\ 7 & 2 & 1 \end{vmatrix}$

16. $\begin{vmatrix} 2 & 1 & -3 \\ -4 & -1 & 1 \\ 5 & 3 & 2 \end{vmatrix}$

17. $\begin{vmatrix} 6 & 7 & 8 \\ 4 & 0 & 3 \\ -1 & 3 & 2 \end{vmatrix}$

18. $\begin{vmatrix} 4 & 3 & 2 \\ 2 & 7 & 5 \\ 0 & 0 & 1 \end{vmatrix}$

19. $\begin{vmatrix} 3 & 2 & 6 \\ 0 & 1 & 0 \\ -4 & 7 & -5 \end{vmatrix}$

20. $\begin{vmatrix} 0 & -5 & -6 \\ 1 & 3 & 2 \\ 0 & 9 & 7 \end{vmatrix}$

21. $\begin{vmatrix} -4 & 3 & 1 & 2 \\ -6 & 1 & 4 & 0 \\ 5 & 2 & 0 & -7 \\ 1 & 3 & 0 & 0 \end{vmatrix}$

22. $\begin{vmatrix} 2 & 3 & 5 & 6 \\ 1 & 2 & 0 & 7 \\ 0 & 4 & 1 & 6 \\ -4 & 0 & 2 & 5 \end{vmatrix}$

23. $\begin{vmatrix} 1 & 1 & 0 & 0 & 0 \\ 1 & 0 & 1 & 0 & 0 \\ 1 & 0 & 0 & 1 & 0 \\ 1 & 1 & 1 & 1 & 1 \\ 1 & 0 & 0 & 0 & 1 \end{vmatrix}$

24. $\begin{vmatrix} 1 & 3 & 3 & 5 & 6 \\ 0 & 2 & 1 & 3 & 1 \\ 0 & 0 & 3 & 2 & 5 \\ 0 & 0 & 0 & 4 & 1 \\ 0 & 0 & 0 & 0 & 5 \end{vmatrix}$

25. $\begin{vmatrix} \sin x & \cos x \\ -\cos x & \sin x \end{vmatrix}$

26. $\begin{vmatrix} \cos x & \sin x \\ \sin x & \cos x \end{vmatrix}$

27. $\begin{vmatrix} \sec x & \tan x \\ \tan x & \sec x \end{vmatrix}$

28. $\begin{vmatrix} 1 & \cot x \\ -\cot x & 1 \end{vmatrix}$

13.10 PROPERTIES OF DETERMINANTS

The following theorems are basic to the study of determinants. Because the proofs can become involved and somewhat abstract, the proofs we shall present here are for second-order determinants only. We shall assume that the theorems can be proved for determinants of any order, and we encourage the student to formulate his own proofs of the theorems for determinants of the third order.

THEOREM. Interchanging two rows (or columns) of a determinant yields a determinant that is the negative of the original determinant.   **(13.31)**

Proof (for determinants of order 2).

$$\begin{vmatrix} a_1 & b_1 \\ a_2 & b_2 \end{vmatrix} = a_1b_2 - a_2b_1$$

If we interchange two rows, we have

$$\begin{vmatrix} a_2 & b_2 \\ a_1 & b_1 \end{vmatrix} = a_2b_1 - a_1b_2 = -(a_1b_2 - a_2b_1)$$

$$= -\begin{vmatrix} a_1 & b_1 \\ a_2 & b_2 \end{vmatrix}$$

It is left to the student to prove the corresponding result if the columns are interchanged.

COROLLARY. If two rows (or columns) of a determinant are alike, then the determinant is equal to zero. (13.32)

For example,

$$\begin{vmatrix} 3 & 5 \\ 3 & 5 \end{vmatrix} = 3 \cdot 5 - 3 \cdot 5 = 0$$

THEOREM. Interchanging the rows and columns of a determinant does not change its value. (13.33)

Proof (for determinants of order 2).

$$\begin{vmatrix} a_1 & b_1 \\ a_2 & b_2 \end{vmatrix} = a_1 b_2 - a_2 b_1$$

and

$$\begin{vmatrix} a_1 & a_2 \\ b_1 & b_2 \end{vmatrix} = a_1 b_2 - b_1 a_2 = a_1 b_2 - a_2 b_1$$

The conclusion follows.

THEOREM. If any row (or column) of a determinant is multiplied by a nonzero multiplier k, the resulting determinant is k times the original determinant. (13.34)

Proof (for determinants of order 2).

$$\begin{vmatrix} ka_1 & kb_1 \\ a_2 & b_2 \end{vmatrix} = ka_1(b_2) - a_2(kb_1) = k(a_1 b_2 - a_2 b_1)$$

$$= k \begin{vmatrix} a_1 & b_1 \\ a_2 & b_2 \end{vmatrix}$$

The cases in which the second row, the first column, and the second column are multiplied by k can be handled in a similar manner.

COROLLARY. A common factor of all the elements of the same row (or column) of a determinant may be removed and placed as a multiplier of the resulting determinant without changing the value of the determinant. (13.35)

For example,

$$\begin{vmatrix} 4 & 8 \\ 3 & 5 \end{vmatrix} = 4 \begin{vmatrix} 1 & 2 \\ 3 & 5 \end{vmatrix}$$

Another useful theorem in evaluating determinants of higher order is the following:

> **THEOREM.**   If a multiple of any row (or column) of a determinant is added to any other row (or column), the value of the resulting determinant is the same as that of the original determinant.   **(13.36)**

For example,

$$\begin{vmatrix} a_1 + ka_2 & b_1 + kb_2 \\ a_2 & b_2 \end{vmatrix} = \begin{vmatrix} a_1 & b_1 \\ a_2 & b_2 \end{vmatrix}$$

Proof. $\begin{vmatrix} a_1 + ka_2 & b_1 + kb_2 \\ a_2 & b_2 \end{vmatrix} = (a_1 + ka_2)b_2 - a_2(b_1 + kb_2)$

$$= a_1b_2 - a_2b_1 = \begin{vmatrix} a_1 & b_1 \\ a_2 & b_2 \end{vmatrix}$$

Theorem (13.36) is used to transform a given determinant into an equivalent determinant that will be easier to evaluate.   The following examples illustrate the principle.

EXAMPLE 1.   Evaluate the determinant

$$\begin{vmatrix} 1 & 3 & 1 \\ 4 & 2 & 2 \\ 5 & -1 & 3 \end{vmatrix}$$

Solution.   In applying Theorem (13.36), we look for the most efficient way to get zeros in a row or column.   Since there is a -1 in the third row, we see by inspection that the second column can be swept out by using multipliers 3 and 2.   Thus,

$$\begin{vmatrix} 1 & 3 & 1 \\ 4 & 2 & 2 \\ 5 & -1 & 3 \end{vmatrix} = \begin{vmatrix} 16 & 0 & 10 \\ 14 & 0 & 8 \\ 5 & -1 & 3 \end{vmatrix} \qquad \begin{array}{l} \text{row 1 plus 3 times row 3} \\ \text{row 2 plus 2 times row 3} \end{array}$$

$$= -(-1)\begin{vmatrix} 16 & 10 \\ 14 & 8 \end{vmatrix} = 2 \cdot 2 \begin{vmatrix} 8 & 5 \\ 7 & 4 \end{vmatrix} = -12$$

EXAMPLE 2.   Evaluate the determinant

$$\begin{vmatrix} -2 & 1 & -1 \\ 3 & 4 & 5 \\ 2 & 1 & -2 \end{vmatrix}$$

Solution.   Using the 1 in the first row, we can sweep out the first row by using multipliers 2 and 1, as follows:

$$\begin{vmatrix} -2 & 1 & -1 \\ 3 & 4 & 5 \\ 2 & 1 & -2 \end{vmatrix} = \begin{vmatrix} 0 & 1 & 0 \\ 11 & 4 & 9 \\ 4 & 1 & -1 \end{vmatrix} \quad \begin{array}{l} \text{column 1 plus 2 times column 2} \\ \text{column 3 plus 1 times column 2} \end{array}$$

$$= -1 \begin{vmatrix} 11 & 9 \\ 4 & -1 \end{vmatrix} = 47$$

Triangular
determinant

A determinant is called a *triangular determinant* if each element below the principal diagonal is zero.   The principal diagonal is the diagonal from upper left to lower right of the array.

THEOREM.   The value of a triangular determinant is the product of the elements of its principal diagonal.   **(13.37)**

Proof (for determinants of order 3).

$$\begin{vmatrix} a_1 & b_1 & c_1 \\ 0 & b_2 & c_2 \\ 0 & 0 & c_3 \end{vmatrix} = a_1 \begin{vmatrix} b_2 & c_2 \\ 0 & c_3 \end{vmatrix} = a_1 b_2 c_3$$

EXAMPLE 3.

$$\begin{vmatrix} 1 & 3 & 1 \\ 2 & 5 & 4 \\ 3 & 2 & 1 \end{vmatrix} = \begin{vmatrix} 1 & 3 & 1 \\ 0 & -1 & 2 \\ 0 & -7 & -2 \end{vmatrix} \quad \begin{array}{l} \text{row 2 minus 2 times row 1} \\ \text{row 3 minus 3 times row 1} \end{array}$$

$$= \begin{vmatrix} 1 & 3 & 1 \\ 0 & -1 & 2 \\ 0 & 0 & -16 \end{vmatrix} \quad \text{row 3 minus 7 times row 2}$$

$$= 1(-1)(-16) = 16$$

13.11   SOLUTION OF SYSTEMS BY DETERMINANTS

One of the applications of the theory of determinants is in the study of systems of linear equations.   We shall now demonstrate a method of solving such systems called *Cramer's rule*.   Consider the system of equations

$$\begin{aligned} a_1 x + b_1 y + c_1 z &= d_1 \\ a_2 x + b_2 y + c_2 z &= d_2 \\ a_3 x + b_3 y + c_3 z &= d_3 \end{aligned} \qquad (13.38)$$

Denote the determinant of the coefficient matrix by D.   Then

$$x \cdot D = x \begin{vmatrix} a_1 & b_1 & c_1 \\ a_2 & b_2 & c_2 \\ a_3 & b_3 & c_3 \end{vmatrix} = \begin{vmatrix} a_1x & b_1 & c_1 \\ a_2x & b_2 & c_2 \\ a_3x & b_3 & c_3 \end{vmatrix} \qquad \text{by Theorem (13.34)}$$

$$= \begin{vmatrix} a_1x + b_1y & b_1 & c_1 \\ a_2x + b_2y & b_2 & c_2 \\ a_3x + b_3y & b_3 & c_3 \end{vmatrix} \qquad \text{by Theorem (13.36)}$$

$$= \begin{vmatrix} a_1x + b_1y + c_1z & b_1 & c_1 \\ a_2x + b_2y + c_2z & b_2 & c_2 \\ a_3x + b_3y + c_3z & b_3 & c_3 \end{vmatrix} \qquad \text{by Theorem (13.36)}$$

For each element in the first column of this determinant, substitute its value from Eqs. (13.38). Then

$$x \cdot D = x \begin{vmatrix} a_1 & b_1 & c_1 \\ a_2 & b_2 & c_2 \\ a_3 & b_3 & c_3 \end{vmatrix} = \begin{vmatrix} d_1 & b_1 & c_1 \\ d_2 & b_2 & c_2 \\ d_3 & b_3 & c_3 \end{vmatrix}$$

Similarly,

$$y \cdot D = \begin{vmatrix} a_1 & d_1 & c_1 \\ a_2 & d_2 & c_2 \\ a_3 & d_3 & c_3 \end{vmatrix} \qquad z \cdot D = \begin{vmatrix} a_1 & b_1 & d_1 \\ a_2 & b_2 & d_2 \\ a_3 & b_3 & d_3 \end{vmatrix}$$

When $D \neq 0$, we can divide and thus obtain

$$x = \frac{\begin{vmatrix} d_1 & b_1 & c_1 \\ d_2 & b_2 & c_2 \\ d_3 & b_3 & c_3 \end{vmatrix}}{D} \qquad y = \frac{\begin{vmatrix} a_1 & d_1 & c_1 \\ a_2 & d_2 & c_2 \\ a_3 & d_3 & c_3 \end{vmatrix}}{D} \qquad z = \frac{\begin{vmatrix} a_1 & b_1 & d_1 \\ a_2 & b_2 & d_2 \\ a_3 & b_3 & d_3 \end{vmatrix}}{D} \qquad (13.39)$$

We observe that the value of each unknown is given by a fraction whose denominator is the coefficient determinant D and whose numerator is the determinant obtained from D by replacing the coefficients of the unknown with the corresponding constants d_1, d_2, d_3 from the equations.

This method, called Cramer's rule, can be applied to any system of n linear equations in n variables, provided $D \neq 0$.

EXAMPLE 1. Use Cramer's rule to solve the system

$$2x + 3y - z = -1$$
$$x - 6y - 5z = 4$$
$$3x + 4y + 2z = 14$$

Solution. $D = \begin{vmatrix} 2 & 3 & -1 \\ 1 & -6 & -5 \\ 3 & 4 & 2 \end{vmatrix} = -57 \neq 0$

Hence, by Eqs. (13.39),

$$x = \frac{\begin{vmatrix} -1 & 3 & -1 \\ 4 & -6 & -5 \\ 14 & 4 & 2 \end{vmatrix}}{-57} = 6 \qquad y = \frac{\begin{vmatrix} 2 & -1 & -1 \\ 1 & 4 & -5 \\ 3 & 14 & 2 \end{vmatrix}}{-57} = -3$$

$$z = \frac{\begin{vmatrix} 2 & 3 & -1 \\ 1 & -6 & 4 \\ 3 & 4 & 14 \end{vmatrix}}{-57} = 4$$

Thus, the solution of the system is the ordered triple $(6, -3, 4)$.   The check is left as an exercise.

Determinant equal to zero

If $D = 0$ and any one of the numerators of Eqs. (13.39) is different from zero, the given system has no solution, and the system is inconsistent.   If $D = 0$ and each numerator is zero, the system may or may not have a solution.   A discussion of this case will not be given here.

The values obtained for x, y, and z of Eqs. (13.39) are unique because $D \neq 0$ and the determinants involved are all unique numbers.   It can be shown that these values, and only these values, satisfy each equation of the system.   Hence, the solution set of the system is the ordered triple (x,y,z) determined by Eqs. (13.39).

EXERCISE SET 13.9

SOLVE each of the following systems, using determinants:

1. $3x + 4y = 1$
    $5x + 2y = 11$

2. $3x + 2y = 2$
    $2x - 3y = 36$

3. $10x + 9y = 12$
    $8x + 2y = 7$

4. $(x/5) + (y/4) = \frac{5}{2}$
    $(x/2) - (y/5) = \frac{23}{5}$

5. $2(x + y) - 7(x - y) = 5$
    $3(x + y) - 10(x - y) = 8$

6. $(2/x) + (8/y) + 7 = 0$
    $(1/x) - (2/y) - 4 = 0$

7. $x + y - z = 2$
    $x + 2y + z = 7$
    $3x - y + 2z = 12$

8. $2x - z - 2 = 0$
    $x - y - 5 = 0$
    $y - z + 6 = 0$

9. $x - 2y + z = 7$
    $y + 2z = 1$
    $2x + 3z = 4$

10. $x - 2y + 4z = -3$
     $2x + y - 3z = 11$
     $3x + y - 2z = 12$

11. $(2/x) + (3/y) + (1/z) = 4$
     $(4/x) - (6/y) + (3/z) = -7$
     $(3/x) - (5/y) + (2/z) = -5$

12. $8x + 3y - 18z = 1$
     $16x + 6y - 6z = 7$
     $4x + 9y + 12z = 9$

13. $2w - x + y + z = 2$
     $w + 5y - 4z = -2$
     $3x + 2y - 3z = 0$
     $w + 2x + 2y - z = 7$

14. $4r + 5s - 2t + 6u = 7$
     $3r - 4s + 8t + 3u = 8$
     $r + s + 2t + 3u = 4$
     $r + 2s - 4t - 3u = -2$

13.12   LINEAR PROGRAMMING

We have discussed previously the graphic solution of systems of inequalities (Sec. 8.8). Let us now consider the problem of finding a solution of a system of linear inequalities in the two variables x and y that will yield the maximum and minimum values of some given polynomial in x and y. This application of mathematics is called *linear programming* and is relatively new to mathematics. It is widely used in industry and government and usually requires the use of electronic computers. We can consider here only very simple examples that may seem quite artificial. They do, however, illustrate the method. We consider first an example that is a review of the procedure for finding graphically the solution set of a system of inequalities in x and y.

EXAMPLE 1.   Find the solution set of the system

$$3x - 2y + 6 \geq 0$$
$$x + y \leq 6$$
$$1 \leq x \leq 4$$
$$y \geq 1$$

Solution.   We draw the graphs of the equations (see Fig. 13.6)

$$3x - 2y + 6 = 0$$
$$x + y = 6$$
$$x = 1$$
$$x = 4$$
$$y = 1$$

and then shade the region indicated by the inequalities. We note that the points whose coordinates satisfy the inequality $3x - 2y + 6 \geq 0$ must lie on or below the line $3x - 2y + 6 = 0$, since for this inequality $y \leq 3x/2 + 3$. The points satisfying $x + y \leq 6$ must lie on or below the line $x + y = 6$, and points that satisfy $1 \leq x \leq 4$ lie on or between the lines $x = 1$ and $x = 4$. The inequality $y \geq 1$ is satisfied by the coordinates of all points on or above the line $y = 1$. The solution set of the system is then the intersection of all these half-planes and is the polygon $ABCDE$ of Fig. 13.6.

Suppose now that an expensive model of a manufactured item yields a net profit of $2 on each one sold, and a cheaper model brings a profit of only $1 each. If x and y denote the respective number of the expensive and cheaper models produced per unit of time, then the profit P per unit of time is given by

$$P = 2x + y$$

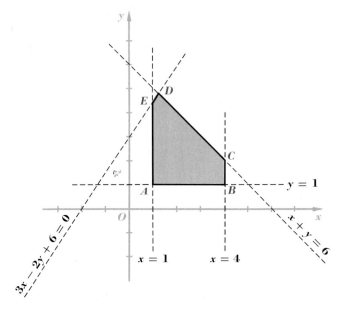

FIGURE 13.6 The shaded area represents the solution set.

There are, however, certain restrictions on x and y imposed by the manufacturing process. Let these constraints be given by

$$3x - 2y + 6 \geq 0$$
$$x + y \leq 6$$
$$1 \leq x \leq 4$$
$$y \geq 1$$

which are the inequalities considered in Example 1.

How can we determine x and y so that the profit P is a maximum if all items are sold? To answer this question, we must rely on a theorem that we state here without proof.

THEOREM. If the solution set of a polynomial in x and y is a convex polygon, then the polynomial has its greatest and least **(13.40)** values when x and y are the coordinates of a vertex of the polygon.

Therefore, to find the maximum profit per unit of time, we must find the vertexes of the polygon $ABCDE$ shown in Fig. 13.6 and then determine from the equation $P = 2x + y$ which vertex has coordinates that yield the greatest value of P.

The coordinates of the vertexes are found to be $A = (1,1)$, $B = (4,1)$, $C = (4,2)$, $D = (\%,2\%)$, $E = (1,\%)$ by solving the appropriate pairs of equations simultaneously. For example, to find $D(\%,2\%)$, we solve the system

$$3x - 2y + 6 = 0$$
$$x + y = 6$$

Since $P = 2x + y$

we have at A $P = 2(1) + 1 = 3$

at B $P = 2(4) + 1 = 9$

at C $P = 2(4) + 2 = 10$

at D $P = 2\left(\dfrac{6}{5}\right) + \dfrac{24}{5} = \dfrac{36}{5}$

at E $P = 2(1) + \dfrac{9}{2} = \dfrac{13}{2}$

It is now clear that the maximum profit is obtained at $x = 4$ and $y = 2$ and is \$10 per unit of time.

EXERCISE SET 13.10

1. Find the maximum and minimum values of the polynomial $f = 3x - 2y$ if the following restrictions must be satisfied: $3x - 2y + 6 \geq 0$, $x + y \leq 6$, $1 \leq x \leq 4$, $y \geq 1$.

2. Use the restrictions of Prob. 1 to find the maximum value of the polynomial $f = 5x - 3y$ subject to those restrictions.

3. Find the maximum value of $f = 20x + 40y$ where the following restrictions hold: $3x + 4y \leq 60$, $x + 3y \leq 30$, $x \geq 0$, $y \geq 0$.

4. Use the restrictions of Prob. 3 to find the maximum value of $P - 20x - 40y$.

5. The solution set of the system of inequalities $x - 2y \leq 0$, $2x - 3y + 6 \geq 0$, $x + 3y \leq 0$ defines a convex polygon. Find the maximum and minimum values of $g = 2x + y$, subject to the restraints indicated by the system of inequalities.

6. The profit per hour in producing x units of one product and y units of another is given by $P = 4x + 3y$. In the production process the following restrictions must prevail: $2x + y \leq 10$, $x - y \geq 2$, $x + 6y \geq 8$. Find the values of x and y that yield the maximum hourly profit.

7. Use the restrictions of Prob. 6 to find the approximate maximum profit per hour if the profit is given by $P = 6x + 2y - 12$.

8. A manufacturer uses two machines to produce two items A and B. Each product A requires 2 hr on the first machine and 5 hr on the second machine. Each product B requires 4 hr on the first machine and 2 hr on the second. The time available on each machine is 16 hr/day. If the profit on each A item is \$6 and on each B item \$8, find the number of articles A and B that should be produced each day to yield the maximum profit. What is the maximum profit?

14

SEQUENCE FUNCTIONS

In this chapter we shall consider the range of a special type of function because of its importance in mathematics. This function, whose domain is the set of positive integers, is the basis for the study of sequences and series. We shall also discuss the method of proof by mathematical induction, arithmetic and geometric series, and the expansion of a binomial having a rational exponent.

14.1 SEQUENCES AND SERIES

Sequence function

Let f be a function having as its domain the set of positive integers $P = \{1,2,3, \ldots ,n, \ldots\}$. We call f a *sequence function*. Denote $f(n)$ by a_n, so that $f(1) = a_1$, $f(2) = a_2$, $f(3) = a_3, \ldots , f(n) = a_n, \ldots$. Then

$$f = \{(1,a_1),(2,a_2),(3,a_3), \ldots ,(n,a_n), \ldots\}$$

Sequence

The elements in the range of f considered in the order $a_1, a_2, a_3, \ldots , a_n, \ldots$ are said to form a *sequence*. For example, if $f(x) = 2x + 3$, $x \in P$, then $a_1 = f(1) = 5$, $a_2 = f(2) = 7$, $a_3 = f(3) = 9$, $a_4 = f(4) = 11, \ldots$. Hence,

$$f = \{(1,5),(2,7),(3,9),(4,11), \ldots\}$$

and the range of f

$$\{5,7,9,11, \ldots\}$$

is the sequence associated with the function.

Finite sequence

If the domain of f is the set of positive integers $\{1,2,3,\ldots,n\}$ for some fixed n, then the range of f is called a *finite sequence*. Thus, if $f(x) = x/(x+1)$, $x \in \{1,2,3,4\}$, then

$$f = \left\{ \left(1,\tfrac{1}{2}\right), \left(2,\tfrac{2}{3}\right), \left(3,\tfrac{3}{4}\right), \left(4,\tfrac{4}{5}\right) \right\}$$

and the range

$$\left\{ \tfrac{1}{2}, \tfrac{2}{3}, \tfrac{3}{4}, \tfrac{4}{5} \right\}$$

is a finite sequence. If $f(x) = 2$, $x \in \{1,2,3,4\}$, then

$$f = \{(1,2),(2,2),(3,2),(4,2)\}$$

and the set $\{2,2,2,2\}$ is a finite sequence.

Terms of a sequence

The elements a_1, a_2, a_3, $\ldots$ of the range of a sequence function are the *terms* of the sequence. For convenience, we often use the symbol $\{a_n\}$ to denote the sequence.

Since there is a first positive integer, a second, a third, and so on, there is a first term of the sequence, a second term, a third term, and so on.

We may specify a sequence in either of two ways. We may give the first term, a_1, and then express the nth term, a_n, in terms of preceding terms, as in the following example.

EXAMPLE 1. Let the sequence a_n be defined by the following conditions:

$$a_1 = 5$$
$$a_n = 3a_{n-1} + 7 \qquad \text{for all } n > 1$$

Find the first four terms of the sequence.

Solution.
$a_1 = 5$
$a_2 = 3a_1 + 7 = 22$
$a_3 = 3a_2 + 7 = 73$
$a_4 = 3a_3 + 7 = 226$

Hence, the first four terms of the sequence are 5, 22, 73, 226.

Recursion formula

An equation like $a_n = 3a_{n-1} + 7$, which tells how to obtain any term (other than the first) from the preceding term, is called a *recursion formula*. A second way to specify a sequence is to express a_n in terms of n.

EXAMPLE 2. Let the sequence $\{a_n\}$ satisfy the condition that

$$a_n = \frac{3}{2^{n-1}} \qquad n \geq 1$$

Find the first five terms of the sequence.

Solution. $a_1 = 3/2^0 = 3$, $a_2 = 3/2^1$, $a_3 = 3/2^2$, $a_4 = 3/2^3$, and $a_5 = 3/2^4$. Hence, the first five terms of the sequence are 3, ³⁄₂, ³⁄₄, ³⁄₈, ³⁄₁₆.

Instead of writing the explicit equation $a_n = 3/2^{n-1}$, $n \geq 1$, of Example 2, we could just as well denote the sequence by

$$\left\{ \frac{3}{2^{n-1}} \right\}$$

EXAMPLE 3. Find the tenth term of

$$\left\{ \frac{1}{2^n + 1} \right\}$$

Solution. Since $n = 10$,

$$a_{10} = \frac{1}{2^{10} + 1} = \frac{1}{1{,}025}$$

Series

If a sequence consists of terms for which the operation of addition is defined, the indicated sum of the terms of the sequence is called a *series*. Thus, associated with the sequence a_1, a_2, a_3, ... is the series

$$a_1 + a_2 + a_3 + \cdots$$

If a sequence is finite, the series associated with it is a finite series. For example, if n is a fixed positive integer, the finite sequence

$$5, 7, 9, \ldots, (2n + 3)$$

leads to the finite series

$$S_n = 5 + 7 + 9 + \cdots + (2n + 3)$$

where S_n denotes the sum of the n terms of the sequence.

Summation symbol

We often represent a series with the aid of the symbol Σ. This symbol, the Greek letter sigma, is used to denote a sum. The preceding series can be written in the compact form

$$S_n = \sum_{k=1}^{n} (2k + 3)$$

where the term on the right represents the series obtained by successively replacing k in the expression $(2k + 3)$ with 1, 2, 3, ..., n.

EXAMPLE 4. If

$$S_n = \sum_{j=1}^{n} \frac{j}{2j + 1}$$

find S_5.

Solution.   Replacing j in the expression $j/(2j + 1)$ by 1, 2, 3, 4, 5 successively, we obtain the expanded form of the series

$$S_5 = \frac{1}{3} + \frac{2}{5} + \frac{3}{7} + \frac{4}{9} + \frac{5}{11}$$

To indicate that a series has an infinite number of terms, we use the notation

$$S_\infty = \sum_{k=1}^{\infty} a_k$$

where a_k is an expression for the kth term.

EXAMPLE 5.   Find an expression for a general term and write in sigma notation:

$$x^3 - x^6 + x^9 - x^{12} + \cdots$$

Solution.   An expression for the kth term is $(-1)^{k+1}x^{3k}$.   Hence, the series can be written

$$\sum_{k=1}^{\infty} (-1)^{k+1}x^{3k}$$

A finite series has a finite sum.   For example, $1 + \frac{1}{2} + \frac{1}{3} + \frac{1}{4}$ has the sum $\frac{25}{12}$.   An infinite series has no last term, therefore we cannot find the sum of all the terms by adding them.   We shall discuss in a later section the meaning, if any, of certain infinite sums.

EXERCISE SET 14.1

WRITE the first four terms of the sequence $\{a_n\}$ defined by each of the following:

1. $a_1 = 3,\ a_n = 3a_{n-1} + 5$
2. $a_1 = 2,\ a_n = \frac{1}{2}\,na_{n-1}$
3. $a_1 = 5,\ a_{n+1} = 1/-a_n$
4. $a_1 = 3,\ a_n = (a_{n-1})^2$
5. $a_1 = 2,\ a_2 = 3,\ a_n = \frac{1}{3}(a_{n-1} + a_{n-2})$
6. $a_1 = 5,\ a_2 = 7,\ a_n = \frac{1}{2}(a_{n-1} + a_{n-2})$

WRITE the first four terms of each of the following sequences:

7. $\left\{\dfrac{1}{2n}\right\}$
8. $\left\{\dfrac{n}{n+1}\right\}$
9. $\left\{\dfrac{1}{n(n+1)}\right\}$

10. $\left\{\dfrac{2^n}{n+1}\right\}$
11. $\{n^n\}$
12. $\{(-1)^n 2^n\}$

13. $\left\{\dfrac{(-1)^{n-1}}{n^3}\right\}$
14. $\left\{\left(\dfrac{1}{2}\right)^{2n+1} + \left(-\dfrac{1}{2}\right)^{2n}\right\}$
15. $\{2^n - 1\}$

16. $\left\{\dfrac{1 \times 3 \times 5 \times \cdots \times (2n-1)}{(2n)!}\right\}$
17. $\left\{\dfrac{1}{n}\log 10^n\right\}$

WRITE each of the following series using the summation symbol Σ:

18. $1 + 2 + 4 + 8 + \cdots + 64$

19. $1 + \frac{1}{2} + \frac{1}{4} + \frac{1}{8} + \cdots + \frac{1}{128}$

20. $1 + 3 + 5 + \cdots$

21. $\frac{1}{2} + \frac{3}{4} + \frac{5}{6} + \frac{7}{8} + \cdots$

22. $1 + \frac{1}{4} + \frac{1}{9} + \frac{1}{16} + \cdots$

23. $2 - 4 + 6 - 8 + \cdots$

24. $x + \dfrac{x^3}{1 \cdot 2} + \dfrac{x^5}{1 \cdot 2 \cdot 3} + \dfrac{x^7}{1 \cdot 2 \cdot 3 \cdot 4} + \cdots$

25. $x + \dfrac{x^2}{1 \cdot 2} + \dfrac{x^3}{1 \cdot 2 \cdot 3} + \dfrac{x^4}{1 \cdot 2 \cdot 3 \cdot 4} + \cdots$

WRITE in expanded form each of the following:

26. $\displaystyle\sum_{n=1}^{5} (-1)^n$

27. $\displaystyle\sum_{k=1}^{3} 1 - (-1)^k$

28. $\displaystyle\sum_{j=1}^{4} \frac{1}{j+1}$

29. $\displaystyle\sum_{s=1}^{5} x^{s-1}$

30. $\displaystyle\sum_{r=1}^{3} 3r^2$

31. $\displaystyle\sum_{j=1}^{\infty} \frac{j}{(j+1)^j}$

32. $\displaystyle\sum_{k=0}^{\infty} \frac{1}{3^k}$

14.2   MATHEMATICAL INDUCTION

We now consider a method of proof called *mathematical induction.* This method is one of the most useful tools at our disposal and is especially useful in proving propositions about series.   It is based on a fundamental property of the set of positive integers.   We call this property of the positive integers the *axiom of finite induction.*

Induction axiom

Every set M of positive integers contains all positive integers if

(1) $1 \in M$ (14.1)

(2) $k + 1 \in M$ whenever $k \in M$

Like any other axiom, the axiom of finite induction is not susceptible of proof in the system of which it is an axiom.   It seems a reasonable assumption, however.   For by (1), $1 \in M$; and by (2), $1 + 1 = 2 \in M$.   Applying (2) again, we get $2 + 1 = 3 \in M$.   Applying (2) once again, we get $3 + 1 = 4 \in M$, and so on indefinitely.   The following examples illustrate the use of the axiom of finite induction.

EXAMPLE 1.   Prove that if m and n are positive integers and a is any number, then

$$a^m \cdot a^n = a^{m+n}$$

Proof.

(1)   By our definition of positive-integer exponents,

$$a^1 = a$$
$$a^{k+1} = a^k \cdot a$$

(2)   Let m be any positive integer and let

$$M = \{n \mid n \in P \text{ and } a^m \cdot a^n = a^{m+n}\}$$

(3)   Since $a^m \cdot a^1 = a^m \cdot a = a^{m+1}$ by definition, we have $1 \in M$, and M is not the empty set.

(4)   Suppose $k \in M$; then k is a positive integer and $a^m \cdot a^k = a^{m+k}$ by the definition of the set M.

(5)   Now

$$
\begin{aligned}
a^m \cdot a^{k+1} &= a^m(a^k \cdot a) && \text{why?}\\
&= (a^m \cdot a^k)a && \text{associative law}\\
&= (a^{m+k})a && \text{from step 4}\\
&= a^{(m+k)+1} && \text{why?}\\
&= a^{m+(k+1)} && \text{associative law}
\end{aligned}
$$

(6)   Hence, $(k + 1) \in M$ by definition of the set M.

(7)   Since $1 \in M$ and $(k + 1) \in M$ whenever $k \in M$, the axiom of finite induction assures us that M contains all the positive integers.

(8)   Hence, $a^m \cdot a^n = a^{m+n}$ for all positive integers m and n.

EXAMPLE 2.   Prove the proposition

$$1 + 3 + 5 + \cdots + (2n - 1) = n^2$$

Proof.

(1)   Let $M = \{n \mid n \in P \text{ and } 1 + 3 + 5 + \cdots + (2n - 1) = n^2\}$.

(2)   Since $1 = 1^2$, we see that $1 \in M$.

(3)   Suppose $k \in M$. Then the sum of k terms of the series is given by

$$1 + 3 + 5 + \cdots + (2k - 1) = k^2$$

(4)   The sum of $k + 1$ terms is found by adding the $(k + 1)$th term to the left member. The $(k + 1)$th term is $(2k + 1)$. Then, from step 3,

$$
\begin{aligned}
[1 + 3 + 5 + \cdots + (2k - 1)] + (2k + 1) &= k^2 + (2k + 1)\\
&= k^2 + 2k + 1\\
&= (k + 1)^2
\end{aligned}
$$

(5)   Hence, $k + 1 \in M$ whenever $k \in M$.

(6)   From steps 2 and 5, we conclude that the proposition is true for all positive integers n.

These two examples illustrate an important consequence of the axiom of finite induction which we refer to as the *principle of mathematical induction* and state as a theorem.

Mathematical
induction

THEOREM. Let $P(n)$ be a statement about the positive integer n. If $P(1)$ is true and if $P(k + 1)$ is true whenever $P(k)$ is true, then $P(n)$ is true for every positive integer n.   **(14.2)**

Proof.

(1)   Let $M = \{n \mid n \in P \text{ and } P(n) \text{ is true}\}$.
(2)   Since $P(1)$ is true, by hypothesis, we have $1 \in M$.
(3)   Also, $P(k + 1)$ is true whenever $P(k)$ is true, by hypothesis.
(4)   Hence, $(k + 1) \in M$ whenever $k \in M$.
(5)   Therefore, M contains all the positive integers, by the axiom of finite induction.
(6)   This means that $P(n)$ is true for every positive integer n.

Proofs by the principle of mathematical induction require two things:
Part 1.   A proof that the proposition is true for $n = 1$.
Part 2.   A proof that if the proposition is true for $n = k$, then it is true for $n = k + 1$.

EXAMPLE 3.   Prove that for all positive integers n,

$$1^3 + 2^3 + 3^3 + \cdots + n^3 = \left[\frac{n(n + 1)}{2}\right]^2$$

Solution.   The proof consists of two parts:

Part 1.   We must first show that the assertion is true for $n = 1$.
When $n = 1$, we have on the left $1^3 = 1$; on the right we have

$$\left[\frac{1(1 + 1)}{2}\right]^2 = 1$$

Hence, the proposition is true for $n = 1$.

Part 2.   We must next show that if the proposition is true for $n = k$, then it is true for $n = k + 1$.
If the assertion is true for $n = k$, then we must have

$$1^3 + 2^3 + 3^3 + \cdots + k^3 = \left[\frac{k(k + 1)}{2}\right]^2$$

Since

$$(k + 1)^3 = (k + 1)^3$$

it follows that

$$1^3 + 2^3 + 3^3 + \cdots + k^3 + (k + 1)^3 = \left[\frac{k(k + 1)}{2}\right]^2 + (k + 1)^3$$

$$= \frac{k^2(k + 1)^2}{2^2} + (k + 1)^3$$

$$= (k + 1)^2 \left(\frac{k^2}{2^2} + k + 1 \right)$$

$$= (k + 1)^2 \left(\frac{k + 2}{2} \right)^2$$

$$= \left[\frac{(k + 1)(k + 2)}{2} \right]^2$$

Hence, if the proposition is true for $n = k$, it is true for $n = k + 1$. By the principle of mathematical induction, the proposition is true for every positive integer n.

Proof by induction
In the proof of a general proposition by the principle of mathematical induction, it is *absolutely essential* that the truth of each part be established.

EXERCISE SET 14.2

PROVE each of the following propositions, where n is a positive integer, using mathematical induction:

1. $1 + 2 + 3 + \cdots + n = \frac{1}{2} n(n + 1)$
2. $2 + 4 + 6 + \cdots + 2n = n(n + 1)$
3. $2 + 2^2 + 2^3 + \cdots + 2^n = 2(2^n - 1)$
4. $1^2 + 2^2 + 3^2 + \cdots + n^2 = \frac{1}{6} n(n + 1)(2n + 1)$
5. $\dfrac{1}{1 \cdot 2} + \dfrac{1}{2 \cdot 3} + \dfrac{1}{3 \cdot 4} + \cdots + \dfrac{1}{n(n + 1)} = \dfrac{n}{n + 1}$
6. $1 + 5 + 9 + \cdots + (4n - 3) = n(2n - 1)$
7. $1 + \dfrac{1}{2} + \dfrac{1}{2^2} + \cdots + \dfrac{1}{2^{n-1}} = \dfrac{2^n - 1}{2^{n-1}}$
8. $1^2 + 3^2 + 5^2 + \cdots + (2n - 1)^2 = \frac{1}{3} n(2n + 1)(2n - 1)$
9. $1^3 + 2^3 + 3^3 + \cdots + n^3 = \frac{1}{4} n^2(n + 1)^2$
10. $1^3 + 3^3 + 5^3 + \cdots + (2n - 1)^3 = n^2(2n^2 - 1)$
11. $\displaystyle\sum_{j=1}^{n} j(j + 1) = \frac{1}{3} n(n + 1)(n + 2)$ 12. $\displaystyle\sum_{j=1}^{n} j(j + 2) = \frac{1}{6} n(n + 1)(2n + 7)$

PROVE the following by mathematical induction, given that a and b are any numbers and m, n are positive integers:

13. $(ab)^n = a^n b^n$
14. $(a^m)^n = a^{mn}$
15. $1^n = 1$
16. (a) Show that the following proposition is true for $n = 1$: $840 - n^2$ is divisible by n.
    (b) Show that the proposition is true for $n = 2, 3, 4, \ldots, 8$. Try $n = 14$ and $n = 15$.
    (c) Is the proposition true for all n?
    (d) Can you prove this proposition by mathematical induction? Give reasons for your answer.
17. Show that if the proposition

$$2 + n = 3 + n$$

is true for $n = k$, then it is also true for $n = k + 1$. Does this prove the proposition? Give reasons for your answer.

18. Show that if the proposition

$$1 + 2 + 3 + \cdots + n = \frac{n^2 + n + 1}{2}$$

is true for $n = k$, then it is true for $n = k + 1$. Does this prove the proposition? Give reasons.

19. Show that the statement

$$n^2 - n + 11 = \text{a prime number}$$

is true for $n = 1, 2, 3, \ldots, 10$. Can you prove that if the proposition is true for $n = k$, then it is also true for $n = k + 1$? Is the statement true for all n?

20. Show that $8^n - 3^n$ is divisible by 5.

21. Show that $a^n - b^n$ is divisible by $a - b$.

PROVE the following by mathematical induction, where n is a positive integer:

22. De Moivre's theorem: $[r(\cos \theta + i \sin \theta]^n = r^n(\cos n\theta + i \sin n\theta)$

23. $\sin x + \sin 2x + \cdots + \sin nx = \dfrac{\sin \frac{1}{2}(n + 1)x \sin \frac{1}{2} nx}{\sin \frac{1}{2} x}$

24. $\sin t + \sin 3t + \cdots + \sin (2n - 1)t = \dfrac{\sin^2 nt}{\sin t}$

25. $\cos t + \cos 3t + \cdots + \cos (2n - 1)t = \dfrac{\sin 2nt}{2 \sin t}$

PROVE the following, if possible, by mathematical induction, where the brackets denote the greatest integer function (see Exercise Set 3.3):

26. $\left[\dfrac{x + 1}{2}\right] + \left[\dfrac{x + 2}{2}\right] = x + 1$

27. $\left[\dfrac{x - 3}{2}\right] + \left[\dfrac{x}{2}\right] = x - 2$

28. $x + \left[\dfrac{2x - 5}{3}\right] = \left[\dfrac{2x + 1}{3}\right] + \left[\dfrac{x}{2}\right] + \left[\dfrac{x - 3}{2}\right]$

14.3   PROGRESSIONS

Arithmetic sequence

The sequence $\{a_1, a_2, a_3, \ldots, a_n\}$ is called an *arithmetic sequence* if and only if there is a constant d such that

$$a_n - a_{n-1} = d \qquad \text{for every } n > 1 \tag{14.3}$$

The constant d is called the *common difference* of the arithmetic sequence. Equation (14.3) can be written as

$$a_n = a_{n-1} + d \tag{14.4}$$

Thus, each term of an arithmetic sequence (after the first) can be obtained by adding d to the preceding term.

From Eq. (14.4), we conclude that $a_2 = a_1 + d$, $a_3 = a_2 + d =$

$(a_1 + d) + d = a_1 + 2d$, $a_4 = a_1 + 3d$, $a_5 = a_1 + 4d$, and so on. Therefore, it seems a good guess that for any $n > 1$

$$a_n = a_1 + (n - 1)d \tag{14.5}$$

We leave it to the student to prove (by mathematical induction) that the proposition is indeed true.

EXAMPLE 1. Find the 37th term of the arithmetic sequence 8, 11, 14, ...

Solution. Here $n = 37$, $a_1 = 8$, and $d = a_2 - a_1 = 11 - 8 = 3$. Then, by Eq. (14.5),

$$a_{37} = 8 + (37 - 1)(3) = 116$$

Arithmetic series

An arithmetic sequence is also called an *arithmetic progression* (abbreviated AP). The indicated sum of the terms is an arithmetic series. To derive an equation that will give the sum of the n terms of a finite arithmetic series, we reason as follows: Let S_n denote the sum of the first n terms of the AP. Then

$$S_n = a_1 + (a_1 + d) + (a_1 + 2d) + \cdots + [a_1 + (n - 1)d]$$

Now, if we reverse the order of writing the same sum, we obtain

$$S_n = [a_1 + (n - 1)d] + [a_1 + (n - 2)d] + \cdots + (a_1 + d) + a_1$$

If we add corresponding sides of the two equations, we find that in each column of the right side the sum of the two terms is precisely $a_1 + [a_1 + (n - 1)d]$. Since there are n columns to add,

$$2S_n = n[2a_1 + (n - 1)d]$$
$$S_n = \frac{n}{2}[2a_1 + (n - 1)d] \tag{14.6}$$

Also,

$$S_n = \frac{n}{2}\{a_1 + [a_1 + (n - 1)d]\}$$
$$= \frac{n}{2}(a_1 + a_n) \qquad \text{by Eq. (14.5)} \tag{14.7}$$

EXAMPLE 2. Find the first term and the common difference of an AP if the sum of the first 17 terms is 187 and the 17th term is 27.

Solution. Since $n = 17$, $S_{17} = 187$, $a_{17} = 27$, we have, from Eq. (14.7),

$$187 = \frac{17}{2}(a_1 + 27)$$

Hence

$$a_1 = -5$$

Since
$$a_n = a_1 + (n - 1)d$$
$$27 = -5 + (17 - 1)d$$

and
$$d = 2$$

Geometric sequence   The sequence $\{a_1, a_2, a_3, \ldots, a_n\}$ is called a *geometric sequence* if and only if

(1) $a_n \neq 0$ for every n

(2)   There is a constant $r \neq 0$ such that

$$\frac{a_n}{a_{n-1}} = r \qquad \text{for every } n > 1 \tag{14.8}$$

Common ratio   The constant r is called the *common ratio* of the geometric sequence. We can write Eq. (14.8) in the equivalent form

$$a_n = ra_{n-1} \tag{14.9}$$

Thus, each term of a geometric sequence (after the first) can be obtained by multiplying the preceding term by r.

From Eq. (14.9), we have $a_2 = ra_1$, $a_3 = ra_2 = r(ra_1) = r^2 a_1$, $a_4 = r^3 a_1$, and so on. Hence, it seems a good guess that

$$a_n = a_1 r^{n-1} \qquad n > 1 \tag{14.10}$$

The proof that (14.10) is valid for all $n > 1$ can be obtained by mathematical induction. The details of the proof are left as an exercise.

Geometric series   A geometric sequence is also called a *geometric progression* (abbreviated GP). The indicated sum of the terms is a geometric series. To find the sum of the terms of a finite geometric progression, let S_n denote the sum of the first n terms. Then

$$S_n = a_1 + (a_1 r) + (a_1 r^2) + \cdots + (a_1 r^{n-1})$$

and since $r \neq 0$,

$$rS_n = a_1 r + (a_1 r^2) + (a_1 r^3) + \cdots + (a_1 r^{n-1}) + (a_1 r^n)$$

Now, subtract each side of the second equation from the corresponding sides of the first to obtain

$$S_n - rS_n = a_1 - a_1 r^n$$

or
$$S_n(1 - r) = a_1 - a_1 r^n$$

If $r \neq 1$, we can divide by $1 - r$ and get the equation

$$S_n = \frac{a_1 - a_1 r^n}{1 - r} = \frac{a_1 r^n - a_1}{r - 1} \qquad r \neq 1 \tag{14.11}$$

If $r = 1$, the given geometric series becomes

$$S_n = a_1 + a_1 + a_1 + \cdots + a_1 = na_1$$

We leave it to the student to find the sum S_n if $r = -1$.

The proof of Eq. (14.11) by mathematical induction is left to the student as an exercise.

Since $a_n = a_1 r^{n-1}$ and since $r \neq 0$, we have $ra_n = a_1 r^n$. The relations (14.11) then become

$$S_n = \frac{a_1 - ra_n}{1 - r} = \frac{ra_n - a_1}{r - 1} \qquad r \neq 1 \qquad (14.12)$$

EXAMPLE 3. Find the number of terms n and the common ratio r of the GP in which the first term is 256, the last term is 81, and the sum of the n terms is 781.

Solution. From (14.12),

$$781 = \frac{81r - 256}{r - 1}$$

Hence, $\qquad\qquad 781r - 781 = 81r - 256$

and $\qquad\qquad\qquad r = \frac{3}{4}$

Since $a_n = a_1 r^{n-1}$, we have

$$81 = 256 \left(\frac{3}{4}\right)^{n-1}$$

and $\qquad\qquad \frac{81}{256} = \left(\frac{3}{4}\right)^{n-1}$

We know that $\qquad\qquad \frac{81}{256} = \left(\frac{3}{4}\right)^{4}$

Therefore, $\qquad\qquad n - 1 = 4$

and $\qquad\qquad\qquad n = 5$

EXERCISE SET 14.3

FIND the indicated term in each of the following arithmetic sequences:

1. $13, 11, 9, \ldots$ (25th term)   2. $2\sqrt{2}, 4\sqrt{2}, 6\sqrt{2}, \ldots$ (7th term)
3. $2, \frac{2}{3}, -\frac{2}{3}, \ldots$ (11th term)
4. Find the 99th term of an arithmetic sequence whose seventh term is 29 and whose 44th term is 177.
5. Which term of the arithmetic sequence $-7, -5, -3, \ldots$ is 65?
6. A body falling freely in space (there is no resistance) falls approximately 16 ft the first second, 48 ft the next second, 80 ft the next, and so on. Find the number of feet (approximately) such a body will fall during the 15th second.

FIND the sum of the indicated terms of each of the following arithmetic series:

7. $5 + 3 + 1 + \cdots$ (40 terms)
8. $2 + \frac{5}{4} + \frac{1}{2} + \cdots$ (8 terms)
9. $3 + 7 + 11 + \cdots$ (25 terms)
10. $4 + 9 + 14 + \cdots$ (30 terms)
11. If $a_n = 109$, $d = 6$, $n = 18$, find S_n.
12. If $a_9 = 1.8$, $a_1 = 1$, find S_9.
13. The fourth term of an AP is 4, and the tenth term is 7. Find the sum of the first 47 terms.
14. The terms between two nonconsecutive terms of an arithmetic sequence are called *arithmetic means* between those two terms. Find five arithmetic means between 42 and -24.
15. Find seven arithmetic means between 3 and 15.
16. Insert eight arithmetic means between -0.7 and 11.
17. Find the arithmetic mean of 6 and 14.   HINT: Find one arithmetic mean between 6 and 14.
18. Find the arithmetic mean of -36 and 94.
19. The sum of three numbers in arithmetic progression is 42. The sum of their squares is 830. Find the three numbers.
20. The sequence $\{a_1, a_2, \ldots, a_n\}$ is called a *harmonic* sequence if and only if
$$\left\{ \frac{1}{a_1}, \frac{1}{a_2}, \frac{1}{a_3}, \ldots, \frac{1}{a_n} \right\}$$
is an arithmetic sequence. Insert three harmonic means between 5 and 9.

FIND the indicated term in each of the following geometric sequences:

21. 2, 4, 8, ... (9th term)
22. $\frac{4}{5}$, $-\frac{8}{15}$, $\frac{16}{45}$, ... (6th term)
23. 0.3, 0.03, 0.003, ... (7th term)
24. $\frac{2}{5}$, $\frac{1}{5}$, $\frac{1}{10}$, ... (7th term)
25. The nth term of the geometric sequence $\frac{5}{2}$, 5, 10, ... is 640. Find the number of terms of the sequence.
26. The common ratio of a geometric sequence is 2. The 13th term is 512. Write the first five terms of the sequence.
27. Write the first five terms of the geometric sequence in which the third term is $\sqrt{2}$ and the sixth term is $\sqrt{54}$.
28. An exhaust pump removes one-fourth of the air from a bell-jar with each stroke. What fractional part of the air originally in the bell-jar remains after five strokes of the pump?

FIND the sum of the indicated terms of each of the following geometric series:

29. $4 + 12 + 36 + \cdots$ (6 terms)
30. $12 - 18 + 27 - \cdots$ (5 terms)
31. $\frac{1}{8} + \frac{1}{4} + \frac{1}{2} + \cdots$ (8 terms)
32. $0.0018 + 0.018 + 0.18 + \cdots$ (7 terms)
33. Find the sum of the terms of the geometric progression $16 + 8 + 4 + \cdots + \frac{1}{16}$.
34. The sum of n terms of a GP is $\frac{728}{27}$, and the last term is $\frac{2}{27}$. If the common ratio is $\frac{1}{3}$, find the number of terms and the first term.
35. In a certain culture the number of bacteria doubles every 20 min. How many times the number of bacteria in the original culture will there be at the end of 2 hr? (Assume that no loss occurs.)
36. Three numbers are in arithmetic sequence. If the first is increased by 9, the second by 7, and the third by 9, the resulting numbers are in geometric sequence. Find the numbers.
37. The terms between two nonconsecutive terms of a geometric sequence are called the *geometric means* between those two terms. Find two geometric means between 40 and 5.
38. Insert three geometric means between 24 and $\frac{3}{2}$.
39. Find the geometric means of -3 and -48.   HINT: Find one geometric mean between -3 and -48.
40. The population of a certain city increases at the rate of 5 percent per year. If the present population is 300,000, find the expected population 6 years from now.

14.4 INFINITE GEOMETRIC SERIES

In this section, we shall consider some of the properties of the infinite geometric series

$$a_1 + a_1r + a_1r^2 + \cdots + a_1r^{n-1} + \cdots$$

If $|r| > 1$, then each term of the series is numerically larger than the preceding term. Hence, no definite value representing such an infinite sum can exist. If $|r| < 1$, we can assign a meaning to the "sum" of the infinitely many terms by considering what happens to r^n as n is allowed to increase without bound. For example, take $r = \frac{2}{3}$ and consider r^n as n increases from 4 to 12 to 50. When $n = 4$,

$$\left(\frac{2}{3}\right)^4 < \frac{1}{5}$$

when $n = 12$,

$$\left(\frac{2}{3}\right)^{12} < \frac{1}{100}$$

and when $n = 50$,

$$\left(\frac{2}{3}\right)^{50} < \frac{1}{100,000,000}$$

It now appears that by taking n sufficiently large, the difference between $(\frac{2}{3})^n$ and zero can be made as small as we please.

To describe this property of r^n, $|r| < 1$, more accurately, let ε be any positive real number. Then regardless of how small ε may be, we can always find a value of n, say N, such that

$$|r^n| < \varepsilon \qquad \text{for all } n \geq N$$

We summarize this statement by writing

Limit

$$\lim_{n \to \infty} r^n = 0 \qquad |r| < 1 \tag{14.13}$$

and read "the limit of r^n as n increases without bound is zero."

Now from Eq. (14.11), the sum of the first n terms of a geometric series can be written

$$S_n = \frac{a_1}{1 - r} - \frac{a_1}{1 - r}(r^n)$$

Hence, if $-1 < r < 1$, the difference between S_n and $a_1/(1 - r)$ can be made as small as we please simply by taking n sufficiently large. To indicate that such is the case, we write

$$\lim_{n \to \infty} S_n = \frac{a_1}{1 - r} \qquad |r| < 1 \tag{14.14}$$

If $r = 1$, $S_n = a_1 + a_1 + a_1 + \cdots$. Therefore, if $a_1 \neq 0$, S_n increases beyond bound in absolute value as n increases without bound.

If $r = -1$, $S_n = a_1 - a_1 + a_1 - \cdots$. Hence, $S_n = a_1$, or $S_n = 0$, according to whether n is odd or even.

We denote the limit of S_n in Eq. (14.14) by S and call it the "sum" of the infinite geometric series. Hence,

$$S = \frac{a_1}{1 - r} \qquad |r| < 1 \qquad\qquad (14.15)$$

Note that S is not an arithmetic sum but the limit of the sum of n terms as n is allowed to increase without bound.

EXAMPLE 1. Find the "sum" of the geometric series $3 + \tfrac{3}{2} + \tfrac{3}{4} + \cdots$.

Solution. Since $S = \dfrac{a_1}{1 - r}$

$$S = \frac{3}{1 - \tfrac{1}{2}} = 6$$

EXAMPLE 2. Find the rational number which is equivalent to the repeating decimal $0.363636\ldots$.

Solution. The decimal $0.363636\ldots$ can be written in the equivalent form

$$0.36 + 0.0036 + 0.000036 + \cdots$$

which can be considered as an infinite geometric series whose first term is 0.36 and whose common ratio is 0.01. Hence,

$$S = \frac{0.36}{1 - 0.01} = \frac{36}{99} = \frac{4}{11}$$

EXAMPLE 3. Find the rational number equivalent to the repeating decimal $2.35242424\ldots$.

Solution. $2.35242424\ldots = 2.35 + 0.0024 + 0.000024 + \cdots$

$$= 2.35 + \frac{0.0024}{1 - 0.01}$$

$$= \frac{235}{100} + \frac{24}{9,900}$$

$$= \frac{23,289}{9,900}$$

EXERCISE SET 14.4

FIND the "sum" of each of the following infinite geometric series:

1. $12 + 6 + 3 + \cdots$
2. $0.9 + 0.03 + 0.001 + \cdots$
3. $7 + \tfrac{7}{2} + \tfrac{7}{4} + \cdots$
4. $60 + 6 + 0.6 + \cdots$

5. $5 + 1 + \frac{1}{5} + \cdots$ **6.** $22 - 2 + \frac{2}{11} - \cdots$

7. $13/10^2 + 13/10^4 + 13/10^6 + \cdots$ **8.** $1 - \frac{1}{4} + \frac{1}{16} - \frac{1}{64} + \cdots$

FIND the rational number equivalent to each of the following repeating decimals. Check your answer by dividing.

9. $0.7777\ldots$ **10.** $1.32454545\ldots$ **11.** $0.124124124\ldots$

12. $0.696969\ldots$ **13.** $7.1272727\ldots$ **14.** $0.279279279\ldots$

15. A rubber ball drops from a height of 30 ft. It always rebounds to a height equal to one-third of the distance through which it falls. Find the total distance traveled by the ball before it comes to rest.

16. The sum of an infinite geometric progression is $\frac{9}{4}$, and the first term is $\frac{3}{2}$. Find the common ratio.

17. The length of the side of a square is 9 in. A second square is inscribed by joining the midpoints of the sides of the first square, a third by joining the midpoints of the sides of the second square, and so on. Find the sum of the areas of the infinite number of inscribed squares thus formed.

14.5 BINOMIAL THEOREM

If n is a positive integer, then $(a + b)^n$ is defined by

$$(a + b)^1 = a + b$$
$$(a + b)^{k+1} = (a + b)^k(a + b)$$

Expansion of a binomial

The result of performing all the operations indicated by $(a + b)^n$ is a series of terms called the *expansion* of the binomial. Now, by definition,

$$(a + b)^1 = a + b$$

and by actually performing the indicated multiplications, we have

$$(a + b)^2 = a^2 + 2ab + b^2$$
$$(a + b)^3 = a^3 + 3a^2b + 3ab^2 + b^3$$
$$(a + b)^4 = a^4 + 4a^3b + 6a^2b^2 + 4ab^3 + b^4$$
$$(a + b)^5 = a^5 + 5a^4b + 10a^3b^2 + 10a^2b^3 + 5ab^4 + b^5$$

From these identities, we observe that if $n = 1, 2, 3, 4, 5$, then the expansion of $(a + b)^n$ contains $n + 1$ terms, with the following properties:

1. The first term of the expansion is a^nb^0. The exponents of a then decrease by 1 in each successive term.

2. The second term is $na^{n-1}b$. The exponents of b then increase by 1 in each successive term.

3. The last term is a^0b^n.

4. If the coefficient of any term is multiplied by the exponent of a in the term and this product divided by the number of the term, the coefficient of the next term is obtained.

If we assume that these properties hold when n is any positive integer, then

$$(a + b)^n = a^n + \frac{n}{1} a^{n-1}b + \frac{n}{1}\frac{n-1}{2} a^{n-2}b^2$$

$$+ \frac{n}{1}\frac{n-1}{2}\frac{n-2}{3} a^{n-3}b^3 + \cdots + b^n \qquad (14.16)$$

Binomial formula

This identity is called the *binomial formula*. The statement that the binomial formula holds for every positive integer n is called the *binomial theorem*. The proof of the theorem can be established by mathematical induction, and the interested student is encouraged to work out the details of the proof. For the present, we assume that the theorem is valid and postpone the formal proof to the next chapter. At that time, we will develop a proof which may be more meaningful than a proof by mathematical induction.

EXAMPLE 1. Expand $(2x - y)^5$.

Solution. To apply the binomial formula, take $2x = a$ and $(-y) = b$ and write the binomial in the form $[(2x) + (-y)]^5$. Then

$$[(2x) + (-y)]^5 = (2x)^5 + \frac{5}{1} (2x)^4(-y) + \frac{5}{1}\frac{4}{2} (2x)^3(-y)^2$$

$$+ \frac{5}{1}\frac{4}{2}\frac{3}{3} (2x)^2(-y)^3 + \frac{5}{1}\frac{4}{2}\frac{3}{3}\frac{2}{4} (2x)(-y)^4$$

$$+ \frac{5}{1}\frac{4}{2}\frac{3}{3}\frac{2}{4}\frac{1}{5}(-y)^5$$

$$= (2x)^5 + 5(2x)^4(-y) + 10(2x)^3(-y)^2$$

$$+ 10(2x)^2(-y)^3 + 5(2x)(-y)^4 + (-y)^5$$

Now simplify each term and obtain

$$(2x - y)^5 = 32x^5 - 80x^4y + 80x^3y^2 - 40x^2y^3 + 10xy^4 - y^5$$

EXAMPLE 2. Write the first four terms and the last term of the expansion of $[(3/x) - (x/3)]^6$.

Solution. $$\left(\frac{3}{x} - \frac{x}{3}\right)^6 = \left(\frac{3}{x}\right)^6 + \frac{6}{1}\left(\frac{3}{x}\right)^5\left(-\frac{x}{3}\right) + \frac{6}{1}\frac{5}{2}\left(\frac{3}{x}\right)^4\left(-\frac{x}{3}\right)^2$$

$$+ \frac{6}{1}\frac{5}{2}\frac{4}{3}\left(\frac{3}{x}\right)^3\left(-\frac{x}{3}\right)^3 + \cdots + \left(-\frac{x}{3}\right)^6$$

$$= \frac{729}{x^6} - \frac{486}{x^4} + \frac{135}{x^2} - 20 + \cdots + \frac{x^6}{729}$$

To derive an expression for the $(r + 1)$th term of the binomial expansion, we first note the following facts.

1. The exponent of b in any term is 1 less than the number of the term. Hence, the exponent of b in the $(r + 1)$th term is r.

2. The sum of the exponents of a and b in any term is n. Thus the exponent of a in the $(r + 1)$th term is $n - r$.

3. The numerator of the coefficient of the $(r + 1)$th term contains the r factors

$$n(n - 1)(n - 2) \cdots n - (r - 1) = n(n - 1)(n - 2) \cdots (n - r + 1)$$

The denominator contains the r factors $1 \cdot 2 \cdot 3 \cdot \cdots \cdot r$. Hence, the $(r + 1)$th term is given by

$$\frac{n(n - 1)(n - 2) \cdots (n - r + 1)}{1 \cdot 2 \cdot 3 \cdot \cdots \cdot r} a^{n-r} b^r \qquad (14.17)$$

EXAMPLE 3.   Find the fifth term of $(2x^2 + 3y)^9$.

Solution.   Since the fifth term is desired, and our formula (14.17) is for the $(r + 1)$th term, we have $r + 1 = 5$ and $r = 4$. Therefore the exponent of $a = 2x^2$ is $n - 4 = 5$. The exponent of $b = 3y$ is 4. The fifth term is

$$\frac{9 \cdot 8 \cdot 7 \cdot 6}{1 \cdot 2 \cdot 3 \cdot 4}(2x^2)^5(3y)^4 = 126(2x^2)^5(3y)^4$$

$$= 326{,}592x^{10}y^4$$

EXAMPLE 4.   Find the term free of x in the expansion of $(x^3 + 1/x)^8$.

Solution.   The $(r + 1)$th term in the expansion will contain the factors $(x^3)^{8-r}(1/x)^r$ and therefore will involve x^{24-4r}. Then, if the term is to be free of x, we must have

$$24 - 4r = 0$$

Hence,

$$r = 6$$

The required term is the seventh term:

$$\frac{8 \cdot 7 \cdot 6 \cdot 5 \cdot 4 \cdot 3}{1 \cdot 2 \cdot 3 \cdot 4 \cdot 5 \cdot 6}(x^3)^2\left(\frac{1}{x}\right)^6 = 28$$

In order to make a complete proof of the binomial theorem by mathematical induction, we need an expression for a typical term of the expansion. The formula for the $(r + 1)$th term provides this needed ingredient. We now state the theorem formally.

THE BINOMIAL THEOREM.   If n is a positive integer,

$$(a + b)^n = a^n + na^{n-1}b + \frac{n(n-1)}{2!} a^{n-2}b^2 + \cdots$$

$$+ \frac{n(n-1)(n-2) \cdots (n-r+1)}{r!} a^{n-r}b^r + \cdots + b^n \quad (14.18)$$

The symbol $r!$, where r is a positive integer, denotes the product of the positive integers $1 \cdot 2 \cdot 3 \cdot 4 \cdot \cdots \cdot r$.   We read this symbol as "r factorial." Thus, $7! = 1 \cdot 2 \cdot 3 \cdot 4 \cdot 5 \cdot 6 \cdot 7$.

To prove the binomial theorem by the principle of mathematical induction, we first show that the proposition is true for $n = 1$.   Since

$$(a + b)^1 = a + b = a^1 + \frac{1}{1} a^0 b$$

we see that the assertion is true for $n = 1$.

It is now necessary to show that if the proposition is true for $n = k$, then it is true for $n = k + 1$.   The details of this part of the proof and the conclusion are left as an exercise.   We shall consider another proof of this theorem in the next chapter after we have studied combinations.

EXERCISE SET 14.5

EXPAND each of the following:

1. $(a + b)^6$
2. $(x + 2)^7$
3. $\left(x + \dfrac{1}{x}\right)^6$
4. $(x^3 - 2y^2)^6$
5. $(y^{2/3} - x^{2/3})^4$
6. $(x^{-1} - y^{-2})^5$
7. $(y^{1/2} + x^{1/2})^6$
8. $(1 + x)^6 - (1 - x)^6$
9. $(a + b)^7 + (a - b)^7$
10. $\left(\dfrac{x}{\sqrt{y}} - \dfrac{y}{\sqrt{x}}\right)^4$

11.   Write the first three terms of $(2x^2 + \frac{1}{2}x^{-1/2})^6$.

12.   Write the first three terms of $\left(\dfrac{a}{b^2} - \dfrac{b}{a^2}\right)^5$.

FIND the indicated term in each of the following:

13. $(2a - b)^7$; 5th term
14. $(2a^2 - 3b)^8$; 4th term
15. $(1 + xy)^9$; 6th term
16. $(x^2 - y)^{10}$; middle term
17. $\left(\dfrac{a}{b} - \dfrac{b}{a}\right)^{10}$; the term free of a and b
18. $\left(\dfrac{2x}{y} + \dfrac{y}{2x}\right)^8$; the term free of x and y
19. $(x^{3/2} + 2x^{1/2})^{10}$; the term involving x^{12}
20. $\left(x^2 - \dfrac{2}{x}\right)^8$; the term involving x^4

21. Write as a binomial and evaluate $(101)^3$. Try $(100 + 1)^3$.
22. Write as a binomial and evaluate $(99)^4$.

14.6 BINOMIAL SERIES

The substitution $1 = a$ and $x = b$ in the binomial formula yields

$$(1 + x)^n = 1 + nx + \frac{n(n-1)}{2!} x^2 + \frac{n(n-1)(n-2)}{3!} x^3 + \cdots$$

$$+ \frac{n(n-1)(n-2) \cdots (n-r+1)}{r!} x^r + \cdots \quad (14.19)$$

If n is a positive integer, the expansion contains a finite number, $n + 1$, of terms. If n is a real number, but not a positive integer or zero, the expansion will not terminate. The resulting infinite series is called the *binomial series*.

Binomial series

It is shown in more advanced works that if $|x| < 1$, the sum of the first k terms of the binomial series can be made to approximate $(1 + x)^n$ to any desired degree of accuracy by taking k sufficiently large.

EXAMPLE 1. If $x^2/2 < 1$, expand to three terms and simplify $(2 - x^2)^{-2}$.

Solution. $(2 - x^2)^{-2} = \left[2 \left(1 - \frac{x^2}{2} \right) \right]^{-2} = 2^{-2} \left(1 - \frac{x^2}{2} \right)^{-2}$

$$= 2^{-2} \left[1^{-2} + (-2)(1^{-3}) \left(-\frac{x^2}{2} \right) \right.$$

$$\left. + (-3)(1^{-4}) \left(-\frac{x^2}{2} \right)^2 + \cdots \right]$$

$$= \frac{1}{4} + \frac{x^2}{4} + \frac{3x^4}{16} + \cdots$$

EXAMPLE 2. If $|y| < 4x^2$, expand to three terms $(4x^2 - y)^{1/2}$.

Solution. $(4x^2 - y)^{1/2} = (4x^2)^{1/2} + \frac{1}{2} (4x^2)^{-1/2}(-y)$

$$+ \left(-\frac{1}{8} \right) (4x^2)^{-3/2}(-y)^2 + \cdots$$

$$= 2x - \frac{y}{4x} + \frac{y^2}{64x^3} - \cdots$$

EXAMPLE 3. Use the binomial series to expand $\sqrt[4]{17}$ to three terms, then evaluate to two decimal places.

Solution. To apply the binomial series, we must write $\sqrt[4]{17}$ as $(16 + 1)^{1/4}$, not as $(1 + 16)^{1/4}$. Then

$$(16 + 1)^{1/4} = (16)^{1/4} + \frac{1}{4}(16)^{-3/4}(1) + \left(-\frac{3}{32}\right)(16)^{-7/4}(1)^2 + \cdots$$

$$= 2 + \frac{1}{32} - \frac{3}{4096} + \cdots$$

$$= 2 + 0.03 - 0.00073 + \cdots$$

$$= 2.03 \text{ (approx.)}$$

EXERCISE SET 14.6

WRITE the first four terms of each expansion, assuming that each of the following binomials can be expanded by use of the binomial series:

1. $(2 - m)^{-1}$ 2. $(1 - 2y)^{-6}$ 3. $(y + 2)^{-2}$
4. $(2 - x^2)^{-3}$ 5. $(8 - x)^{2/3}$ 6. $(x + 5y)^{-1}$
7. $(1 - x)^{-1/2}$ 8. $(x^2 + y)^{1/2}$ 9. $(4 + y)^{1/2}$

WRITE each of the following as a binomial series and evaluate to three decimal places:

10. $(1.02)^4$ 11. $(1.01)^5$ 12. $(1.06)^{-3}$
13. $(1.08)^{3/4}$ 14. $(1.03)^{-3/5}$ 15. $(1.06)^{1/3}$
16. $(1.05)^{-5}$

FIND the principal root correct to three decimal places of each of the following (use a binomial series):

17. $\sqrt[5]{31}$ 18. $\sqrt[3]{9}$ 19. $\sqrt{1.02}$
20. $\sqrt[4]{14}$ 21. $\sqrt[3]{-28}$ 22. $\sqrt{0.98}$

15

PERMUTATIONS, COMBINATIONS, PROBABILITY

In this chapter we shall consider some of the concepts that are basic to the study of probability and statistics. To develop a good working knowledge of probability would require much more space than is available. There are some simple concepts, however, that do not require extensive discussion. We shall consider these notions before we define probability.

15.1 FUNDAMENTAL PRINCIPLES

We recall that the cartesian product of the finite sets A and B is the set of all ordered pairs (a,b) such that a is an element of A and b is an element of B. Thus, $A \times B = \{(a,b) \mid a \in A \text{ and } b \in B\}$. For example, if $A = \{a,b,c,d\}$ and $B = \{r,s\}$, then $A \times B = \{(a,r),(a,s),(b,r),(b,s),(c,r),(c,s),(d,r),(d,s)\}$.

Note here that there are four possible ways to select the first coordinates of the ordered pairs of $A \times B$. For the first coordinate, we may select a, b, c, or d. With each choice of a first coordinate, there are two ways to select the second coordinate. Thus, for each of the four elements of A taken as a first coordinate there are two elements of B to be taken as second coordinates. Hence, the total number of ordered pairs that can be formed is the number of elements in set A multiplied by the number of elements in set B. We recall (Exercise Set 1.3) that the symbol $n(A)$ denotes the number of elements in any finite set A and $n(B)$ denotes the number of elements in set B. Thus, we have

$$n(A \times B) = n(A)n(B) = (4)(2) = 8$$

The above illustrates the following theorem:

THEOREM.   If P and Q are finite sets of distinct elements, then

$$n(P \times Q) = n(P)n(Q)$$

(15.1)

This theorem can be generalized to apply to any finite number of sets. Thus, if $P, Q, R, \ldots$ are finite sets of distinct elements, then

$$n(P \times Q \times R \times \cdots) = n(P)n(Q)n(R) \cdots$$

The following examples illustrate the use of the theorem.

EXAMPLE 1.   We wish to pick a class president from a group of five boys, a secretary from a group of three girls, and a sponsor from a group of three faculty members.   In how many ways can this be done?

Solution.   Let $P = \{b_1, b_2, b_3, b_4, b_5\}$, $Q = \{g_1, g_2, g_3\}$, $R = \{f_1, f_2, f_3\}$ represent the candidates for president, secretary, and sponsor respectively. Consider each slate of candidates to be an element of $P \times Q \times R$. From the generalization of Theorem (15.1), we obtain

$$n(P \times Q \times R) = n(P)n(Q)n(R)$$
$$= 5(3)(3) = 45$$

EXAMPLE 2.   Find the number of three-letter code words that can be formed from the letters a, b, c, d if no letter is to be repeated in any code word.

Solution.   Let $P = \{a, b, c, d\}$, then $n(P) = 4$.   Now, any one of the four letters of P can be chosen for the first letter in a code word. After the first letter of a code word is chosen, there are three letters remaining from which the second letter can be selected. Call this set Q; then $n(Q) = 3$. After the first and second letters are chosen, there are two letters left from which the third letter can be taken. Call this set R; then $n(R) = 2$.

Consider each code word as an element of $P \times Q \times R$. Then

$$n(P \times Q \times R) = 4(3)(2) = 24$$

The preceding examples illustrate a principle which is important in problems dealing with enumeration.

THE FUNDAMENTAL PRINCIPLE.   Let the sets $S_1, S_2, \ldots, S_r$ have $n_1, n_2, \ldots, n_r$ elements respectively.   The number of ways of selecting first an element from S_1, then an element from $S_2, \ldots,$ finally an element from S_r is the number

(15.2)

$$n_1 n_2 \cdots n_r$$

EXAMPLE 3. How many three-digit integers (positive) less than 700 can be formed from the digits 1, 3, 5, 7, 9 if repetition of digits is permitted?

Solution. Since the required positive integer is to be less than 700, the hundreds digit can be selected from the set {1,3,5}. Thus, the hundreds digit can be selected in three different ways. Because repetitions are permitted, the tens digit can be selected from the set {1,3,5,7,9} in five different ways. Similarly, the units digit can be chosen in five different ways. Hence, by the Fundamental Principle, the total number of three-digit positive integers less than 700 is

$$3(5)(5) = 75$$

n factorial

As we noted in Sec. 14.5, the product of all the positive integers from 1 to n is denoted by $n!$ and read "n factorial." Thus,

$$1! = 1$$
$$2! = 1(2)$$
$$3! = 1(2)(3)$$
$$\cdots\cdots\cdots\cdots\cdots$$
$$n! = 1(2)(3) \cdots (n)$$

and
$$(n + 1)! = 1(2)(3) \cdots (n)(n + 1)$$

Hence, it follows that

$$(n + 1)! = n!(n + 1)$$

If this statement is to hold for $n = 0$, then we must have

$$(0 + 1)! = 0!(0 + 1)$$

The discussion suggests the following recursive definition:

DEFINITION. If n is a nonnegative integer,

$$0! = 1 \tag{15.3}$$
$$n! = (n - 1)!n \qquad n \geq 1$$

From the definition, we immediately have the following theorem:

THEOREM. If $n \in P$ and $r \in P$, and $r < n$, then

$$n! = r!(r + 1)(r + 2) \cdots (n) \tag{15.4}$$

EXAMPLE 4. Find the value of $7!/5!$.

Solution. By Theorem (15.4),

$$\frac{7!}{5!} = \frac{5!(6)(7)}{5!} = 6(7) = 42$$

EXAMPLE 5. Find the value of n if $[(n + 3)!]/[(n + 1)!] = 30$.

Solution. $30 = \dfrac{(n + 3)!}{(n + 1)!} = \dfrac{(n + 1)!(n + 2)(n + 3)}{(n + 1)!}$

$$= (n + 2)(n + 3)$$

Hence, $30 = n^2 + 5n + 6$ or $n^2 + 5n - 24 = 0$

Therefore, $n = 3$ or $n = -8$

We do not define the factorial of a negative integer. Consequently, the fraction $[(n + 3)!]/[(n + 1)!]$ has no meaning for $n = -8$. We conclude that $n = 3$ is the only possible solution. That 3 actually is a solution is easily seen. For if $n = 3$, then

$$\frac{(3 + 3)!}{(3 + 1)!} = \frac{6!}{4!} = 30$$

If n is any other integer, the given equation is not a true statement.

EXERCISE SET 15.1

1. $\dfrac{9!}{11!}$

2. $\dfrac{5! - 8!}{4! - 7!}$

3. $\dfrac{4!5!}{6!7!}$

4. $\dfrac{k!}{(k - 1)!}$

5. $\dfrac{(k + 1)!}{k!}$

6. $\dfrac{4! + 5(4!)}{3!}$

7. $\dfrac{(n - 1)!(n + 1)!}{(n!)(n!)}$

8. $\dfrac{(2n)!}{(2n - 1)!}$

9. $\dfrac{k!}{(k - 2)!}$

10. $\dfrac{(n + 1)!}{(n - 1)!}$

11. $\dfrac{n!(n + 1)!}{(n - 1)!(n + 2)!}$

12. $\dfrac{k!(k - 2)!}{[(k - 1)!]^2}$

SOLVE each of the following, using the Fundamental Principle:

13. How many four-digit numbers can be formed? The digits are 0, 1, 2, 3, 4, 5, 6, 7, 8, 9. The first digit cannot be zero.

14. An automobile agency offers a choice of 5 body styles, a choice of 3 motors, and a choice of 12 colors. In how many ways can a buyer choose a car?

15. If three fair dice are tossed, in how many ways can they fall?

16. Three students enter a stadium that has 12 gates. If no two of them choose the same gate, in how many ways can they enter?

17. Three fraternities have 30, 40, and 50 members respectively. How many committees consisting of three persons can be selected if all the fraternities are to be represented on each committee and no one belongs to two different fraternities?

18. How many possible license plates can be made consisting of a letter of the English alphabet followed by a number of four digits chosen from the set of digits? Here, the first digit may be a zero.

15.2 PERMUTATIONS

Each linear ordering of the elements of a finite set is called a *permutation* of the set. A finite set is linearly ordered if there is a first element, a second, a third, etc. For example, the elements of the set {3,4,5} can be arranged in the following six ways.

$$
\begin{array}{ccc}
3\ 4\ 5 & 4\ 3\ 5 & 5\ 3\ 4 \\
3\ 5\ 4 & 4\ 5\ 3 & 5\ 4\ 3
\end{array}
$$

Each of these arrangements is a linear ordering of the set, and therefore each is a permutation of the set.

We observe here that there are three ways of selecting the first digit of an arrangement. After the first digit has been selected, there are two possible ways to choose the second digit, and then there is only one way to select the third digit. By the Fundamental Principle, there are

$$3 \cdot 2 \cdot 1 = 3! = 6$$

ways of permuting the three digits.

Frequently, we wish to consider the selection and arrangement of a specified number of elements of a set. For example, we may wish to select r elements from a set of n elements and then arrange these r elements in order. We define the number of all such permutations as follows:

DEFINITION. The number of different arrangements, each consisting of r elements, that can be selected from a set of n distinct elements, is called the number of permutations of n elements taken r at a time. (15.5)

Since the number of permutations of n elements taken r at a time is a function of n and r, we denote it by the symbol $P(n,r)$.

THEOREM. $P(n,r) = n(n - 1)(n - 2) \cdots (n - r + 1)$ (15.6)

To see that the theorem is valid, we reason that the first of the r positions in a permutation can be filled in n different ways. Then the second position can be filled in $n - 1$ ways, the third in $n - 2$ ways, and so on. Thus, the number of ways of filling each position is n minus the number of positions already filled. When the rth element is to be chosen, $r - 1$ places have already been filled. Hence, the rth position can be filled in $n - (r - 1) = n - r + 1$ ways. By the Fundamental Principle, the theorem follows.

EXAMPLE 1. Evaluate $P(12,4)$.

Solution. $P(12,4) = 12(11)(10)(9) = 11,880$

EXAMPLE 2. A basketball coach has a squad of ten players. If each man can play at any position, how many different teams can the coach field?

Solution. Since there are five positions to be filled and there are ten men available, we have

$$P(10,5) = 10(9)(8)(7)(6) = 30,240$$

If $r = n$, then the last factor of $P(n,r)$ is $n - n + 1 = 1$, and therefore Theorem (15.6) becomes

$$P(n,n) = n(n - 1)(n - 2) \cdots (1) = n! \qquad (15.7)$$

EXAMPLE 3. The first six letters of the alphabet are to be arranged to form code words. How many such words can be obtained if no letter is repeated in any arrangement?

Solution. Here we require the number of permutations of six things taken six at a time. Hence,

$$P(6,6) = 6! = 720$$

Permutations with some like elements

A more difficult problem arises when some of the elements of a set are alike. For example, suppose there are four red balls, one white, one green, and one black ball to be drawn from a bag. In how many different orders can they be drawn? We reason as follows: If the red balls were distinguishable from each other, the number of permutations would be 7!. However, since we cannot distinguish one red ball from another, we see that the four red ones could be rearranged among themselves without altering the identity of the selection. Hence, if P denotes the number of distinguishable permutations, then

$$4!P = 7! \qquad \text{and} \qquad P = \frac{7!}{4!}$$

This reasoning can be generalized as follows:

THEOREM. If in a set of n elements, n_1 are alike, n_2 others are alike, and so on, then the number of distinct permutations P of the n elements is given by

$$(15.8)$$

$$P = \frac{n!}{n_1!n_2! \cdots}$$

EXAMPLE 4. How many distinct permutations can be formed from the letters of the word OKLAHOMA?

Solution. There are eight letters, of which two are A's and two are O's. Hence,

$$P = \frac{8!}{2!2!} = 10{,}080$$

EXERCISE SET 15.2

1. Evaluate (a) $P(7,5)$; (b) $P(10,6)$; (c) $P(18,3)$; (d) $P(40,4)$.

2. Prove that $P(n,r) = \dfrac{n!}{(n-r)!}$; $\dfrac{P(n,r)}{r!} = \dfrac{P(n,n-r)}{(n-r)!}$.

3. Prove that $P(n,n) = P(n,n-1)$.   4. If $P(n,3) = 60$, find n.

5. Find n if $P(n+2,2) = 5P(n-1,2)$.   6. If $5P(n,2) = 14P(n-2,2)$, find n.

7. Three men enter a bus that has twenty empty seats. In how many different ways can they seat themselves?

8. If there are ten men available for the four backfield positions of a football team, and if each man can play at any position, in how many ways can a backfield be selected?

9. How many permutations can be formed from the letters of the word BACKSPIN if four letters are taken at a time?

10. If ten chairs are placed in a row, in how many ways can five boys be seated in consecutive chairs?

11. A baseball coach has his best hitter bat in fourth place and his pitcher in last place. How many batting orders are then possible?

12. How many odd four-digit numbers can be formed from the digits 1, 2, 4, 5, 6, 8, 9 if no repetitions are allowed?

13. Find the number of four-letter code words that can be formed from the five vowels and twenty-one consonants if vowels and consonants are to alternate. Repetitions are allowed.

14. A club has a membership of twenty-five. A president, a vice president, a secretary, and a treasurer are to be elected from the membership. Only ten of the members are eligible for president and vice president. How many sets of officers are possible?

15. Find the number of different arrangements that can be made of the letters in the word COLLEGE.

16. In how many ways can five people be seated in a row if two particular ones are not permitted to sit next to each other?

17. In how many ways can five people be seated in a row if two particular ones must sit next to each other?

18. In how many ways can five persons be seated in consecutive chairs of a row that contains nine chairs?

19. On a library shelf there are two algebra books, three history books, three geometry books, and three English grammar books. Find the number of ways of arranging these books on the shelf if the books on each subject are to be together.

20. In how many ways can eight persons be seated at a round table? HINT: One person can be placed anywhere at the table; the remaining ones can then be permuted.

21. Twelve persons are to sit at a round table. Two particular people insist on sitting opposite each other. Find the number of ways the twelve can be seated.

22. Eight persons are to sit at a round table. Two particular people insist on sitting next to each other. Find the number of ways of seating the eight persons.

15.3   COMBINATIONS

If we select and arrange r elements from a set of n elements, we call the result a permutation of n things taken r at a time. If we select but *do not arrange* r elements from a set of n elements, we call the result a *combination of n elements taken r at a time.*

> **DEFINITION.** The number of different subsets of r elements that can be selected from a set of n elements, when the order of elements within each subset is disregarded, is called the *number of combinations of n things taken r at a time.*   **(15.9)**

We denote the number of combinations of n elements taken r at a time by the symbol $C(n,r)$.

Combination versus permutation

The essential difference between permutations and combinations is that of order, or arrangement. For example, $a\,b\,c$ and $c\,b\,a$ are different permutations but are identical combinations.

Let us consider the number of combinations of the letters a, b, c, d taken three at a time. The different sets of letters taken three at a time, disregarding the order in which the letters are arranged, are

$$a\,b\,c \qquad a\,b\,d \qquad a\,c\,d \qquad b\,c\,d$$

From each of these four combinations we can form $P(3,3) = 3!$ different permutations of the letters. Therefore, each of the four combinations furnishes 3! permutations to the total number of permutations. This means that there are $3!C(4,3)$ permutations in all. Hence,

$$3!C(4,3) = P(4,3)$$

and

$$C(4,3) = \frac{P(4,3)}{3!}$$

Since the symbol $C(n,r)$ represents the number of possible ways of selecting r elements from a set of n different elements, and since there are $P(r,r) = r!$ arrangements of the elements of any selection, there are $r!C(n,r)$ permutations of the n elements taken r at a time. Hence, $r!C(n,r) = P(n,r)$. Now, if we define $C(n,0) = 1$, we have the following theorem:

> **THEOREM.** $C(n,r) = \dfrac{P(n,r)}{r!}$ $0 \le r \le n$
>
> $$= \frac{n(n-1)\cdots(n-r+1)}{r!}$$ **(15.10)**

EXAMPLE 1. Find the number of different committees of five persons each that can be selected from a group of twelve persons.

Solution. If the order in which the members of a committee are chosen is disregarded, the problem becomes one of finding the number of combinations of 12 elements taken 5 at a time. Hence, the required number is

$$C(12,5) = \frac{12(11)(10)(9)(8)}{5!} = 792$$

The symbol (n/r) is also used for $C(n,r)$.

EXERCISE SET 15.3

1. Prove that $C(n,r) = \dfrac{P(n,r)}{P(r,r)}$.   2. Prove that $C(n,r) = \dfrac{n!}{r!(n-r)!}$.

3. Prove that $C(n,r) = C(n,n-r)$.
4. Use Theorem (15.10) to evaluate: (a) $C(5,3)$; (b) $C(8,2)$; (c) $C(10,3)$; (d) $C(12,4)$.
5. Use the result of Prob. 3 to evaluate: (a) $C(20,17)$; (b) $C(28,25)$; (c) $C(42,39)$; (d) $C(100,98)$; (e) $C(120,118)$.
6. Find n if (a) $C(n+1,3) = 2C(n,2)$; (b) $C(n+1,n-1) = 15C(n,0)$.
7. In how many ways can a five-member committee be selected from a group of nine men?
8. How many different sums can be formed with a nickel, a dime, a quarter, a half-dollar, and a silver dollar if only three coins can be used?
9. How many subcommittees consisting of five Democrats and four Republicans can be formed from a committee of thirteen Democrats and eleven Republicans?
10. A standard deck contains fifty-two cards. Find the number of ways the cards can be dealt among four bridge players.
11. A group of fifteen men wish to play on a basketball team. Find the number of teams that can be formed.
12. A company has ten positions to be filled. Four of the positions must be filled by men and six by women. Nine men and eleven women apply. Find the number of ways the positions can be filled.
13. Prove that the number of subsets of a set S of n elements is given by

$$C(n,0) + C(n,1) + C(n,2) + \cdots + C(n,n)$$

14. Use the fact that $C(n,0)$ is defined to be 1 and show that the total number of combinations of n elements taken 1, 2, 3, ..., n at a time is $2^n - 1$. HINT: The first element of S can be treated in two ways. We can either take it or leave it. There are two ways of treating the second element. Hence, there are $2 \cdot 2$ ways of treating the first two elements, $2 \cdot 2 \cdot 2$ ways of treating the first three elements, and so on. Use the result of Prob. 13 to obtain

$$C(n,1) + C(n,2) + \cdots + C(n,n) = 2^n - 1$$

15. A test contains ten questions which must be answered True or False. Find the total number of ways the answers could be given.
16. Find the total number of different weights that can be formed with six objects weighing 1, 2, 4, 8, 16, and 32 oz respectively. Use the result of Prob. 14.

15.4 COMBINATIONS AND THE BINOMIAL THEOREM

In Sec. 14.5 we stated but did not prove the binomial theorem. We shall now state this important theorem and prove it.

THE BINOMIAL THEOREM. For any positive integer n, and any numbers a and b,

$$(a + b)^n = C(n,0)a^n + C(n,1)a^{n-1}b + C(n,2)a^{n-2}b^2 + \cdots$$
$$+ C(n,r)a^{n-r}b^r + \cdots + C(n,n)b^n$$

The product $(a + b)(a + b) = aa + ab + ba + bb$. Thus, the expansion of $(a + b)^2$ consists of the sum of all two-letter products that can be formed using only a's and b's. The product $(a + b)(a + b)(a + b)$ is equal to

$$(a + b)(aa + ab + ba + bb)$$

which is equal to

$$aaa + aab + aba + abb + baa + bab + bba + bbb$$

Hence, the expansion of $(a + b)^3$ consists of the sum of all three-letter products that can be obtained using only a's and b's. We can generalize this discussion in the form of a theorem.

THEOREM. The expansion of $(a + b)^n$, $n \in P$, consists of the sum of all n-letter products that can be obtained using only a's   **(15.11)** and b's.

If any n-letter product contains r of the b's, then it must contain $(n - r)$ a's and thus be of the form $a^{n-r}b^r$. If $r = 0$, this term becomes a^n, and if $r = n$, so that $n - r = 0$, this term becomes b^n. The expansion of $(a + b)^n$ is the sum of all terms of the form $a^{n-r}b^r$, where r is one of the numbers $0, 1, 2, 3, \ldots, n$. The expansion is therefore of the form

$$(a + b)^n = c_0 a^n + c_1 a^{n-1}b + c_2 a^{n-2}b^2 + \cdots + c_r a^{n-r}b^r + \cdots + c_n b^n$$

where the coefficients $c_0, c_1, c_2, \ldots, c_n$ are to be determined.

To find an expression for c_r, we consider the number of n-letter products that can be formed using $(n - r)$ of the a's and r of the b's. Each of these products is obtained by taking a from $n - r$ factors and b from the remaining r factors of $(a + b)^n$. Consequently, the number of terms of the form $a^{n-r}b^r$ is the number of ways in which r things can be selected from n things, that is, $C(n,r)$. Thus, $c_r = C(n,r)$. Hence,

$$(a + b)^n = C(n,0)a^n + C(n,1)a^{n-1}b + \cdots + C(n,r)a^{n-r}b^r + \cdots + C(n,n)b^n$$

Note that when these coefficients are evaluated, we have the binomial theorem as given in Theorem (14.18), that is

$$(a + b)^n = a^n + \frac{n}{1!} a^{n-1}b + \cdots$$

$$+ \frac{n(n + 1) \cdots (n - r + 1)}{r!} a^{n-r}b^r + \cdots + b^n$$

EXAMPLE 1.   Show that

$$C(n,1) + C(n,2) + \cdots + C(n,n) = 2^n - 1$$

Solution.   Let $a = b = 1$.   Then

$$(1 + 1)^n = 2^n = C(n,0)(1)^n + C(n,1)(1^{n-1})(1) + C(n,2)(1^{n-2})(1^2)$$
$$+ \cdots + C(n,n)(1^0)(1^n)$$
$$= 1 + C(n,1) + C(n,2) + \cdots + C(n,n)$$

Hence, $2^n - 1 = C(n,1) + C(n,2) + \cdots + C(n,n)$

EXAMPLE 2.   Find the fifth term in the expansion of $(2 - ix)^9$, where $i^2 = -1$.

Solution.   The fifth term is the $(r + 1)$th term.   Hence, $r = 4$.   Therefore, the fifth term is given by

$$C(9,4)(2^5)(-ix)^4 = 126(32)(x^4) = 4{,}032x^4$$

EXERCISE SET 15.4

1.   Prove that

$$C(n,0) - C(n,1) + C(n,2) - C(n,3) + \cdots + (-1)^n C(n,n) = 0$$

2.   Find the fifth term of $(1 + \tan x)^6$.
3.   By De Moivre's theorem, if $n \in P$,

$$(\cos t + i \sin t)^n = \cos nt + i \sin nt$$

Let $n = 3$ and expand the left member by the binomial theorem.   Equate the real parts and the imaginary parts of the left and right members to obtain the identities for $\sin 3t$ and $\cos 3t$.
4.   Use De Moivre's theorem and the binomial theorem to derive identities for $\sin 4t$ and $\cos 4t$.
5.   Derive identities for $\sin 5t$ and $\cos 5t$.
6.   A coin is flipped six times.   Find the total possible number of outcomes.   HINT: Expand $(h + t)^6$ and find the number of ways of obtaining six heads and no tails, five heads and one tail, four heads and two tails, and so on.
7.   How many different positive integers less than 10,000 can be formed from the digits 1, 2, 3, 4, 5, 6, 7, 8?
8.   A railway signal has three arms.   Each arm can be placed in four positions.   How many different signals can be made?

9. A man has eight friends. In how many ways can he invite one or more of them to dinner?

10. In how many ways can two dimes, three quarters, four half-dollars, and five silver dollars be distributed among fourteen persons so that each may receive a coin?

15.5 PROBABILITY

The theory of probability had its origin in games of chance (gambling). The foundations of the theory were laid in the seventeenth century by the mathematicians Pascal and Fermat. Since that time the theory of probability has increased in importance until today it is vital to our national survival. The insurance industry, one of the largest businesses in the world, is based on the ability to determine the likelihood of the occurrence of certain events. Probability methods are applied in such fields as quantum mechanics, the design of experiments, the interpretation of data, traffic control, the theory of strategy, and the allocation of equipment. A knowledge of the theory is vital in analyzing problems in communications and control, quantum mechanics, and kinetic theory.

Probabilities are associated with the outcomes of experiments. In each experiment there is a set of all conceivable outcomes, and we are interested in the chances that particular outcomes will result from the experiment. We shall begin with an intuitive approach.

Suppose a card is drawn from a deck of 52 bridge cards. What chance is there that the card is a king? To answer this question, we first make the following observations. There is no reason to think that any one of the 52 cards is "more likely" to be drawn than any other. Thus, the drawing of a card from the deck is a *trial* that can result in one of 52 "equally likely" outcomes. If a king is drawn, the trial is considered a success. If any other card is drawn, the trial is a failure. Hence, there are 4 possible successful outcomes out of 52 "equally likely" outcomes. It seems reasonable to expect that out of 52 such drawings, a king would be drawn approximately 4 times. We say that the chances of drawing a king are 4 in 52 and express this relationship as the ratio $4/52 = 1/13$. This ratio is the probability that a card drawn from a full deck is a king.

Such reasoning leads intuitively to the following statement, which is often taken as a definition of probability:

Intuitive definition
of probability

> If an experiment has n equally likely outcomes among which s are considered successes, then the probability of a success is s/n.

This statement, however, offers difficulties as a definition because it is circular. It uses the term "equally likely" in defining the word "probability." But what does "equally likely" mean? To say that an experiment has n equally likely outcomes is just another way of saying that the probability of any one outcome is the same as the probability of any other outcome. This means that we are using a notion of probability to give meaning to "equally likely." We need to break this circle.

The approach preferred by many is to define a probability function satisfying certain axioms. The theory can then be developed as an axiomatic system. The properties established by this axiomatic system are those readily suggested by experience, so that this mathematical structure is readily applicable.

Sample space

A *sample space* S for an experiment is the set of all possible results of the experiment. For example, if a coin is tossed, there are two possible outcomes, heads and tails, which we can represent by the letters H, T. Thus for the experiment of tossing a coin, the sample space is the set $\{H,T\}$. If a fair die is tossed on a table, there are six possible outcomes, which can be indicated by the list of numbers 1, 2, 3, 4, 5, 6. For the experiment of tossing a fair die, the sample space is the set $\{1,2,3,4,5,6\}$.

Outcome and event

Each element of a sample space for an experiment is called an *outcome*. A set of outcomes may all be successes, so we define an *event* to be a subset of the sample space.

Set function

We now define a real *set function* by assigning a unique real number to each subset A of a sample space S. The domain of this function is the set of all subsets of S and the range is a set of real numbers. For example, if $S = \{a,b\}$, the subsets of S are $\varnothing$, $\{a\}$, $\{b\}$, $\{a,b\}$. If to each of these subsets we assign a real number, as indicated by the following table, we have defined a real set function whose domain is the set of subsets of S and whose range is the set of real numbers $\{0, \frac{1}{2}, \frac{1}{4}\}$.

Subsets of S	$\varnothing$	$\{a\}$	$\{b\}$	$\{a,b\}$
Value of $f(S)$	0	½	¼	¼

We use the concept of a set function to define the probability of an event in a given sample space.

DEFINITION. If an event A is a subset of a sample space S, the probability of the event A, denoted by $p(A)$, is a value of a set function P satisfying the following axioms:

AXIOM 1. $0 \le P(A) \le 1$ for every subset $A \subseteq S$

(15.12)

AXIOM 2. $P(S) = 1$

AXIOM 3.   If A and B are disjoint subsets of S, then

$$P(A \cup B) = P(A) + P(B)$$

The function P is called a probability function. The range consists of real numbers between 0 and 1, inclusive, so that the probability of an event cannot be negative or greater than 1. We do not in general consider the probability of an event unless there is some doubt that it will occur. We find it advantageous, however, to consider events that are impossible and events that are certain as special cases of probability. If an event is logically certain, we assign to it the probability 1. If an event cannot logically occur, we assign the probability 0 to it.

Since the union of sets is associative, we can extend Axiom 3 to include any number of disjoint sets by use of mathematical induction. Thus,

$$P(A \cup B \cup C \cup \ldots) = P(A) + P(B) + P(C) + \cdots$$

When there are infinitely many outcomes in a sample space, the probability axioms may have to be reexamined. For example, what is the probability that a point of the interval from 0 to 1 on the real number line corresponds to the number ½ if the point is selected at random? In this course, we shall consider the theory of probability only as it applies to experiments having a finite number of outcomes.

Our intuition leads us to believe that in some cases each element of S should have the same probability. For example, in tossing a perfect coin, we assign the same probability to heads and tails. Thus, $P(H) = P(T)$. In this case, H and T are mutually exclusive events and $S = \{H\} \cup \{T\}$. Hence, by Axioms 2 and 3,

$$P(S) = P(H) + P(T) = 2P(T) = 1$$

or
$$P(T) = \frac{1}{2} \quad \text{and} \quad P(H) = \frac{1}{2}$$

In assigning probabilities to events, we either make assumptions that appear to fit the physical requirements, or we use estimates based on our experience. Suppose, for instance, that for some reason we consider good, we decide to assign the same probability P to each of the outcomes $s_1, s_2, \ldots, s_n$ of S. This is another way of saying that the outcomes are "equally likely." Hence,

$$P = P(s_1) = P(s_2) = \cdots = P(s_n)$$

Since
$$P(S) = P(s_1 \cup s_2 \cup \cdots \cup s_n)$$
$$= P(s_1) + P(s_2) + \cdots + P(s_n)$$
$$= 1$$

we have
$$nP = 1$$

and
$$P = \frac{1}{n}$$

Thus, $1/n$ is the probability of the occurrence of any one of a set of n equally likely, mutually exclusive events.

EXAMPLE 1. Find the probability of making a five in one throw of a fair die.

Solution. A fair die can fall on any one of its six faces. There is no reason to assume that any one of the six faces is more likely to come up than any other. Hence, the probability of throwing a five is ⅙.

Probability of an
event

Let A be a subset of S consisting of any k of the n elements of S. For convenience, let $A = \{s_1, s_2, \ldots, s_k\}$. Assume that the n elements of S are mutually exclusive and equally likely. Then the probability of the event A is defined to be the probability of the occurrence of some one of the k elements of A. Hence,

$$
\begin{aligned}
P(A) &= P(s_1 \cup s_2 \cup \cdots \cup s_k) \\
&= P(s_1) + P(s_2) + \cdots + P(s_k) \\
&= \frac{1}{n} + \frac{1}{n} + \cdots + \frac{1}{n} = \frac{k}{n}
\end{aligned}
$$

Since $k = n(A)$ and $n = n(S)$

$$
P(A) = \frac{n(A)}{n(S)} \tag{15.13}
$$

It is easy to show that Eq. (15.13) defines a probability function. It is left to the student to show that Axioms 1 to 3 are satisfied. Note that $n(A)$ is the number of distinct ways in which event A can occur. The number of distinct outcomes in the sample space is $n(S)$.

EXAMPLE 2. Find the probability of making a five in one throw of a pair of dice.

Solution. The number of ways one die can fall is 6. The other die can fall also in any one of 6 ways. By the Fundamental Principle, the two can fall in $6(6) = 36$ possible ways. Hence, $n(S) = 36$.

The ways in which a five can be made are (1,4), (2,3), (3,2), (4,1). This means that $n(A) = 4$. Hence, the probability of throwing a five with one roll of the dice is ⁴⁄₃₆ = ⅑.

Let A be a subset of a sample space S, and let A' be the complement of A with respect to S. This means that $S = A \cup A'$ and $A \cap A' = \emptyset$. Then

$$
P(S) = P(A \cup A') = 1
$$

Hence, $P(A) + P(A') = 1$

and $P(A') = 1 - P(A) \tag{15.14}$

Probability an
event will not
occur

If $P(A)$ is the probability that the event A will occur, then $P(A')$ is the probability that the event will *not* occur.

EXAMPLE 3. A pair of dice are rolled on a table. Find (1) the probability that the sum of the numbers is less than 5; (2) the probability that the sum is greater than or equal to 5.

Solution. (1) As in Example 2 above, $n(S) = 36$. Since

$$A = \{(1,1),(1,2),(1,3),(2,1),(2,2),(3,1)\}$$

it follows that $n(A) = 6$. Hence,

$$P(A) = \frac{6}{36} = \frac{1}{6}$$

(2) Since $P(A') = 1 - P(A) = 1 - \frac{1}{6} = \frac{5}{6}$, the probability that the sum is greater than or equal to 5 is $\frac{5}{6}$.

Relative frequency

Let n_1 be the number of trials (or observations) of an experiment in which the sample space is specified. Let s_1 be the number of times that a certain event A occurs. The number s_1 is then called the *frequency* of A in n_1 trials. The ratio s_1/n_1 is called the *relative frequency* of the occurrence of the event A for n_1 trials. The experiment is now repeated n_2 times, and the number of times s_2 that the event A occurs is observed. The relative frequency of A in n_2 trials is s_2/n_2. A series of such repetitions establishes a series of relative frequencies

$$\frac{n_1}{s_1}, \frac{n_2}{s_2}, \frac{n_3}{s_3}, \ldots$$

In many cases, experience shows that these relative frequencies differ very little from one set of repetitions to another. There seems to be some fixed number, denoted by $P(A)$, whose existence must be assumed, around which the relative frequencies seem to cluster. This number (whose existence we have assumed) is called the probability of the event A.

We say that the relative frequencies s_i/n_i are experimental approximations to the probability of the event A. For example, an imperfect die is rolled on a flat surface. The sample space is $S = \{1,2,3,4,5,6\}$. If we wish to estimate the probability that a three will turn up, we make many rolls of the die and find the relative frequency of the event of a three. We then take the relative frequency as an approximation of the probability that a three will turn up on a single roll of the die.

EXAMPLE 4. Experience has shown that a stamping machine turns out approximately 12 defective parts in every lot of 10,000. If a part is taken at random from a box containing 500 of the parts, what is the probability that it will be defective?

Solution. The relative frequency is 12/10,000. Hence, the probability that the part is defective is

$$P = \frac{12}{10,000} = 0.0012$$

Expectation

Let p be the probability of winning a prize which has a cash value of $\$v$. The *expectation* of winning the prize is defined to be the value of the prize multiplied by the probability of winning it, that is, pv.

EXAMPLE 5. A player who can throw an eleven in a single roll of a pair of dice receives a prize worth $\$72$. How much can he afford to pay for a chance at winning the prize?

Solution. We consider the expectation of winning a prize to be the price a player should be willing to pay for the privilege of playing the game. If he pays more, the "odds" are against him.

The probability that a player will throw an eleven is $\frac{1}{18}$. His expectation is then $\frac{1}{18}(72) = \$4$. This is the amount he can afford to pay for a chance at winning.

We find that combinations are quite useful in finding probabilities.

EXAMPLE 6. A bag contains five white, four red, and three black balls. (1) If three balls are drawn, find the probability that they are all white. (2) If six balls are drawn, find the probability that two are white, three are red, and one is black.

Solution. (1) There are $C(5,3)$ ways of drawing three white balls. Since the total number of ways of drawing any three balls is $C(12,3)$, the probability of drawing three white balls is

$$\frac{C(5,3)}{C(12,3)} = \frac{10}{220} = \frac{1}{22}$$

(2) The number of ways of drawing two white balls is $C(5,2)$; the number of ways of drawing three red balls is $C(4,3)$; the number of ways of drawing one black ball is $C(3,1)$. Hence, by the Fundamental Principle, there are in all $C(5,2) \cdot C(4,3) \cdot C(3,1)$ ways of drawing two white, three red, and one black. Since the total number of ways of drawing six balls from the bag is $C(12,6)$, the required probability is

$$\frac{C(5,2) \cdot C(4,3) \cdot C(3,1)}{C(12,6)} = \frac{120}{924} = \frac{10}{77}$$

The following exercises are of little practical value, but they offer some practice in applying the principles thus far considered.

EXERCISE SET 15.5

1. A bag contains six white, five red, and four black balls.

   (a) If two balls are drawn, find the probability that both are white.
   (b) If three balls are drawn, find the probability that all three are red.
   (c) If six balls are drawn, find the probability that there will be two balls of each color.
   (d) If seven balls are drawn, find the probability that four will be red and three will be white.
   (e) If nine balls are drawn, find the probability that two will be red, five white, and two black.

2. A deck of twenty-four tickets numbered 1, 2, 3, . . . , 24 is shuffled, then three tickets are drawn. Find the probability

   (a) That the three tickets are numbered 1, 2, and 3.
   (b) That either 1, 2, or 3 is among them.

3. What is the probability of throwing not more than five in a single throw of two dice?
4. Find the probability of throwing at least five in a single throw of two dice.
5. Three dice are rolled. What is the probability of throwing a ten?
6. Six people seat themselves at random at a round table. What is the probability that two particular persons will sit together?
7. If four cards are drawn from a deck of bridge cards, find the probability that they are of the same suit.
8. A committee of four people is to be selected at random from a group of eight men and twelve women. Find the probability that the committee will consist of two men and two women.
9. A committee of five is to be selected at random from ten men and eight women. What is the probability that there will be three men and two women on the committee?
10. One card is drawn from a fifty-two-card deck. Find the probability that it is either an ace or a king.
11. An automobile worth $3,000 is to be given to the holder of the stub of a ticket drawn at random from a container. If there are 60,000 tickets in the container, what is the expectation of a person holding 100 tickets?
12. One bill is drawn from a box in which five $1 bills, ten $5 bills, and twenty $10 bills have been shuffled. Find the expectation.
13. A stamping machine turns out 720 parts per hour. Experience has shown that there are approximately 20 defective parts per hour. Find the probability that a part picked at random will be defective.
14. In Prob. 13, find the number of defective parts expected in running the machine for 8 hr.

15.6 ADDITION LAWS

Compound event

   A *compound event* is any subset of a sample space S formed by the application of set operations to one or more events in S. Thus, if A and B are events in S, then $A \cup B$, $A \cap B$, and A' are compound events in the sample space. We have already proved (Theorem 15.14) that

$$P(A') = 1 - P(A)$$

Hence,
$$P(S') = 1 - P(S)$$

Since $\varnothing = S'$ and $P(S) = 1$, it follows that

$$P(\varnothing) = 0$$

If A and B are two events in S, and $n(A)$ and $n(B)$ denote the number of elements in sets A and B respectively, then

$$n(A \cup B) = n(A) + n(B) - n(A \cap B) \qquad (15.15)$$

This statement follows from the fact that in taking the number of elements in A plus the number of elements in B, we have counted the elements (if there are any) in the intersection twice, and hence we must subtract this number from the total in determining the number in $A \cup B$.   The principle is illustrated in Fig. 15.1.

If A and B are disjoint sets so that $A \cap B = \varnothing$, then A and B are said to be *mutually exclusive*.   In this case, $n(A \cup B) = n(A) + n(B)$.

Mutually exclusive events

> **THEOREM.**   If A and B are any events in a sample space S, then
>
> $$P(A \text{ or } B) = P(A \cup B) = P(A) + P(B) - P(A \cap B)$$
>
> **(15.16)**

Proof.   We shall prove this only for "equally likely" events, although the same method applies to probabilities defined by Axioms (15.12).

$$P(A \cup B) = \frac{n(A \cup B)}{n(S)} \qquad \text{why?}$$

$$= \frac{n(A) + n(B) - n(A \cap B)}{n(S)} \qquad \text{why?}$$

$$= \frac{n(A)}{n(S)} + \frac{n(B)}{n(S)} - \frac{n(A \cap B)}{n(S)} \qquad \text{why?}$$

$$= P(A) + P(B) - P(A \cap B) \qquad \text{why?}$$

FIGURE 15.1   (*a*) $n(A \cup B) = n(A) + n(B) - n(A \cap B)$;   (*b*) $n(A \cup B) = n(A) + n(B)$

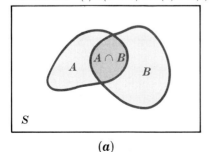

(*a*)

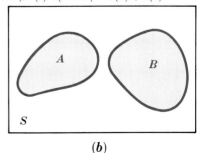

(*b*)

EXAMPLE 1. A card is drawn from a deck of fifty-two bridge cards. Find the probability that it will be a spade or a seven.

Solution. Let A be the set of spades and B the set of sevens. Then

$$P(A) = \frac{13}{52} = \frac{1}{4} \quad \text{and} \quad P(B) = \frac{4}{52} = \frac{1}{13}$$

The intersection $A \cap B$ is the single card seven of spades. Hence $n(A \cap B) = 1$, and $P(A \cap B) = \frac{1}{52}$. Then

$$P(A \cup B) = \frac{1}{4} + \frac{1}{13} - \frac{1}{52} = \frac{4}{13}$$

EXAMPLE 2. Find the probability of throwing a four in a single roll of a pair of dice.

Solution. A four can be made either by throwing a three and a one or by throwing a two and a two. These events are mutually exclusive. We know that $n(S) = 36$.

Let A be the event of throwing a three and a one, and let B be the event of throwing a two and a two. There are two ways of getting a three and a one, namely $(3,1)$, $(1,3)$. Hence, $n(A) = 2$ and $P(A) = \frac{2}{36}$. There is only one way of getting a two and a two, namely, $(2,2)$. Thus, $n(B) = 1$ and $P(B) = \frac{1}{36}$. Since the events are mutually exclusive, $n(A \cap B) = \varnothing$. Hence,

$$P(A \text{ or } B) = \frac{2}{36} + \frac{1}{36} = \frac{1}{12}$$

15.7 INDEPENDENT AND DEPENDENT EVENTS

In this section we shall consider the probability of occurrence of both of two events A and B, that is, the probability of A *and* B.

To denote the probability of event B, assuming that event A has occurred previously, we use the symbol $P(B|A)$. If the occurrence of event B is not affected by the prior occurrence or nonoccurrence of event A, then

$$P(B|A) = P(B)$$

We define *independent events* and *dependent events* as follows:

DEFINITION. If A and B are events in a sample space, and if $P(B|A) = P(B)$, then A and B are independent events. Two **(15.17)** events that are not independent are dependent events.

The following theorem is valid for both independent events and dependent events.

THEOREM. If A and B are events in a sample space and if $P(B|A)$ denotes the probability of the occurrence of B, given the prior occurrence of A, then (15.18)

$$P(A \text{ and } B) = P(A \cap B) = P(A) \cdot P(B|A)$$

Proof. Again we shall give the proof only for equally likely events.

$$P(A \cap B) = \frac{n(A \cap B)}{n(S)} \qquad \text{why?}$$

$$= \frac{n(A \cap B)}{n(A)} \cdot \frac{n(A)}{n(S)} \qquad \text{why?}$$

$$= P(B|A) \cdot P(A) \qquad \text{why?}$$

COROLLARY. If A and B are independent events in a sample space, then (15.19)

$$P(A \cap B) = P(A) \cdot P(B)$$

EXAMPLE 1. Two cards are drawn from a bridge deck of fifty-two cards. Find the probability that two kings will be drawn (1) if the first card is returned to the deck and the deck is reshuffled before the second card is drawn, and (2) if the first card is not returned to the deck before the second drawing.

Solution. (1) If the first card is returned to the deck and the deck then shuffled, the events are independent.

If the first card is a king, the probability of drawing two kings is

$$P(A \text{ and } B) = P(A \cap B)$$
$$= P(A) \cdot P(B)$$
$$= \frac{4}{52} \cdot \frac{4}{52} = \frac{1}{169}$$

(2) If the first card is not replaced in the deck before the second card is drawn, the second draw is dependent on the first. Assuming that the first draw is a king, the probability of drawing two kings is

$$P(A \text{ and } B) = P(A \cap B) = P(A) \cdot P(B|A)$$
$$= \frac{4}{52} \cdot \frac{3}{51} = \frac{1}{221}$$

Corollary (15.19) can be extended to apply to any finite number of independent events A, B, C, ... , N so that

$$P(A \cap B \cap C \cdots \cap N) = P(A) \cdot P(B) \cdot P(C) \cdot \cdots \cdot P(N)$$

15.8 REPEATED TRIALS. BINOMIAL DISTRIBUTION

If the probability $P(A)$ of an event in any given trial is known, the probability of the occurrence of the event exactly k times in n trials can be found, provided the n trials are independent.

Let $P(A')$ be the probability of the nonoccurrence of A in one trial. Then, by the extension of Corollary (15.19), the probability that A occurs k times and fails to occur $n - k$ times *in a given order* is the product of k factors $P(A)$ and $n - k$ factors $P(A')$, or

$$[P(A)]^k[P(A')]^{n-k}$$

Now, k occurrences may come out of n trials in $C(n,k)$ ways. Since the different ways have the same probability and are mutually exclusive, the probability that A occurs exactly k times in n trials is

$$C(n,k)[P(A)]^k[P(A')]^{n-k}$$

Denoting $P(A)$ by p and $P(A')$ by q, the probability can be expressed as

$$C(n,k)p^kq^{n-k}$$

We observe that $C(n,k)p^kq^{n-k}$ is the $(k + 1)$th term of the binomial expansion

$$(p + q)^n = C(n,0)p^nq^0 + C(n,1)p^{n-1}q + \cdots$$
$$+ C(n,k)p^{n-k}q^k + \cdots + C(n,n)p^0q^n$$

The successive terms of this expansion give the probabilities that the event will occur in exactly n, $n - 1$, ... , k, ... , 3, 2, 1, 0 times in n trials.

Probability of k successes in n trials

The probability that an event will occur *at least* k times in n trials is determined as follows. The event will occur at least k times in n trials if it occurs n, $n - 1$, $n - 2$, ... , k times. Since these events are mutually exclusive, the probability is the sum of the probabilities

$$C(n,0)p^nq^0 + C(n,1)p^{n-1}q + \cdots + C(n,n - k)p^kq^{n-k}$$

Since $C(n,k) = C(n,n - k)$, we can write the probability that an event will occur at least k times in n trials as

$$C(n,n)p^n + C(n,n - 1)p^{n-1}q + \cdots + C(n,k)p^kq^{n-k}$$

EXAMPLE 1. Find the probability of throwing exactly two aces in five throws of a perfect die.

Solution. Since there is only one way to throw an ace, the probability that an ace will come up is $p = \frac{1}{6}$ and $q = 1 - \frac{1}{6} = \frac{5}{6}$. Hence, the probability is

$$C(5,2)p^2q^3 = 10\left(\frac{1}{6}\right)^2\left(\frac{5}{6}\right)^3 = \frac{625}{3,888}$$

EXAMPLE 2. Find the probability of throwing at least two "tails" in four flips of a perfect coin.

Solution. The probability is

$$p^4 + C(4,1)p^3q + C(4,2)p^2q^2 = \left(\frac{1}{2}\right)^4 + 4\left(\frac{1}{2}\right)^3\left(\frac{1}{2}\right) + 6\left(\frac{1}{2}\right)^2\left(\frac{1}{2}\right)^2$$

$$= \frac{11}{16}$$

Let us denote the probability that an event A will occur k times in n trials by $P_{k,n}$. If the trials are independent, then

$$P_{k,n} = C(n,k)p^kq^{n-k}$$

Binomial distribution   The function defined by this equation is called the *binomial distribution.* It is an example of a class of functions called *probability distribution functions* that are used extensively in statistics.

EXERCISE SET 15.6

1. A box contains eight red, five white, and two black marbles. Find the probability of drawing a white or black marble in a single draw.
2. A box contains six red, five white, and four blue balls. Three balls are drawn at random, and after each drawing the ball is replaced before the next one is drawn. Find the probability that the first ball drawn is red, the second white, and the third blue.
3. Find the probability of throwing a five or a six in a single throw of a die.
4. A box contains five red, six white, and nine blue marbles. Two marbles are taken at random without being replaced in the box. Find the probability that both balls will be red.
5. Eleven persons are to be seated at a round table. What is the probability that two particular persons will sit together?
6. A committee of five is to be selected at random from ten boys and eight girls. Find the probability that there will be three boys and two girls on the committee.
7. Find the probability of throwing at least a four in two throws with a die.
8. The probability that a certain man will live ten years is $\frac{1}{4}$. The probability that his wife will live ten years is $\frac{1}{5}$. What is the probability that at least one will live ten years?
9. Ten coins are tossed. What is the probability that at least six of them will be tails?
10. Find the probability of throwing exactly four sixes in six throws with a single die.
11. One box contains five silver dollars and seven nickels. Another box contains three silver dollars and twelve nickels. Find the probability of getting a dollar by drawing a single coin from one of the boxes taken at random.

12. A coin is tossed eight times. Find the probability that heads will come up at least five times.

13. Find the probability of throwing ten with a pair of dice exactly three times in four trials.

14. If two coins are tossed five times, find the probability that there will be five heads and five tails.

15. Each of four persons draws a card from a standard bridge deck. What is the probability that there will be one of each suit?

16. A box contains five white and three black balls. If four balls are drawn and not replaced, what is the probability the balls drawn are alternately of different colors?

17. The probability of a certain event is $\frac{3}{7}$, and the probability of another, independent of the first, is $\frac{6}{11}$. Find the probability that at least one of the events will happen.

18. A bag contains six balls. A person takes one out and then replaces it. After he has done this six times, find the probability that he has had in his hand every ball in the bag.

19. Show that $P(B|A) = \dfrac{n(A \cap B)}{n(A)}$. [This was used in the proof of Theorem (15.18).]

20. If A and B are independent events, show that $P(B|A') = P(B\,|\,A) = P(B)$.

21. Show that if $P(B|A') = P(B)$, then A and B are independent events.

TABLES

TABLE I   Four-place Values of Functions of Numbers

t	$\sin t$	$\cos t$	$\tan t$	$\cot t$	$\sec t$	$\csc t$
.00	.0000	1.0000	.0000	. . . . .	1.000	. . . . .
.01	.0100	1.0000	.0100	99.997	1.000	100.00
.02	.0200	.9998	.0200	49.993	1.000	50.00
.03	.0300	.9996	.0300	33.323	1.000	33.34
.04	.0400	.9992	.0400	24.987	1.001	25.01
.05	.0500	.9988	.0500	19.983	1.001	20.01
.06	.0600	.9982	.0601	16.647	1.002	16.68
.07	.0699	.9976	.0701	14.262	1.002	14.30
.08	.0799	.9968	.0802	12.473	1.003	12.51
.09	.0899	.9960	.0902	11.081	1.004	11.13
.10	.0998	.9950	.1003	9.967	1.005	10.02
.11	.1098	.9940	.1104	9.054	1.006	9.109
.12	.1197	.9928	.1206	8.293	1.007	8.353
.13	.1296	.9916	.1307	7.649	1.009	7.714
.14	.1395	.9902	.1409	7.096	1.010	7.166
.15	.1494	.9888	.1511	6.617	1.011	6.692
.16	.1593	.9872	.1614	6.197	1.013	6.277
.17	.1692	.9856	.1717	5.826	1.015	5.911
.18	.1790	.9838	.1820	5.495	1.016	5.586
.19	.1889	.9820	.1923	5.200	1.018	5.295
.20	.1987	.9801	.2027	4.933	1.020	5.033
.21	.2085	.9780	.2131	4.692	1.022	4.797
.22	.2182	.9759	.2236	4.472	1.025	4.582
.23	.2280	.9737	.2341	4.271	1.027	4.386
.24	.2377	.9713	.2447	4.086	1.030	4.207
.25	.2474	.9689	.2553	3.916	1.032	4.042
.26	.2571	.9664	.2660	3.759	1.035	3.890
.27	.2667	.9638	.2768	3.613	1.038	3.749
.28	.2764	.9611	.2876	3.478	1.041	3.619
.29	.2860	.9582	.2984	3.351	1.044	3.497
.30	.2955	.9553	.3093	3.233	1.047	3.384
.31	.3051	.9523	.3203	3.122	1.050	3.278
.32	.3146	.9492	.3314	3.018	1.053	3.179
.33	.3240	.9460	.3425	2.920	1.057	3.086
.34	.3335	.9428	.3537	2.827	1.061	2.999
.35	.3429	.9394	.3650	2.740	1.065	2.916
.36	.3523	.9359	.3764	2.657	1.068	2.839
.37	.3616	.9323	.3879	2.578	1.073	2.765
.38	.3709	.9287	.3994	2.504	1.077	2.696
.39	.3802	.9249	.4111	2.433	1.081	2.630
t	$\sin t$	$\cos t$	$\tan t$	$\cot t$	$\sec t$	$\csc t$

By permission from Alvin K. Bettinger and John A. Englund, *Algebra and Trigonometry*, International Textbook Company, Scranton, Pa., 1960.

TABLE I   (*continued*)

t	$\sin t$	$\cos t$	$\tan t$	$\cot t$	$\sec t$	$\csc t$
.40	.3894	.9211	.4228	2.365	1.086	2.568
.41	.3986	.9171	.4346	2.301	1.090	2.509
.42	.4078	.9131	.4466	2.239	1.095	2.452
.43	.4169	.9090	.4586	2.180	1.100	2.399
.44	.4259	.9048	.4708	2.124	1.105	2.348
.45	.4350	.9004	.4831	2.070	1.111	2.299
.46	.4439	.8961	.4954	2.018	1.116	2.253
.47	.4529	.8916	.5080	1.969	1.122	2.208
.48	.4618	.8870	.5206	1.921	1.127	2.166
.49	.4706	.8823	.5334	1.875	1.133	2.125
.50	.4794	.8776	.5463	1.830	1.139	2.086
.51	.4882	.8727	.5594	1.788	1.146	2.048
.52	.4969	.8678	.5726	1.747	1.152	2.013
.53	.5055	.8628	.5859	1.707	1.159	1.978
.54	.5141	.8577	.5994	1.668	1.166	1.945
.55	.5227	.8525	.6131	1.631	1.173	1.913
.56	.5312	.8473	.6269	1.595	1.180	1.883
.57	.5396	.8419	.6410	1.560	1.188	1.853
.58	.5480	.8365	.6552	1.526	1.196	1.825
.59	.5564	.8309	.6696	1.494	1.203	1.797
.60	.5646	.8253	.6841	1.462	1.212	1.771
.61	.5729	.8196	.6989	1.431	1.220	1.746
.62	.5810	.8139	.7139	1.401	1.229	1.721
.63	.5891	.8080	.7291	1.372	1.238	1.697
.64	.5972	.8021	.7445	1.343	1.247	1.674
.65	.6052	.7961	.7602	1.315	1.256	1.652
.66	.6131	.7900	.7761	1.288	1.266	1.631
.67	.6210	.7838	.7923	1.262	1.276	1.610
.68	.6288	.7776	.8087	1.237	1.286	1.590
.69	.6365	.7712	.8253	1.212	1.297	1.571
.70	.6442	.7648	.8423	1.187	1.307	1.552
.71	.6518	.7584	.8595	1.163	1.319	1.534
.72	.6594	.7518	.8771	1.140	1.330	1.517
.73	.6669	.7452	.8949	1.117	1.342	1.500
.74	.6743	.7385	.9131	1.095	1.354	1.483
.75	.6816	.7317	.9316	1.073	1.367	1.467
.76	.6889	.7248	.9505	1.052	1.380	1.452
.77	.6961	.7179	.9697	1.031	1.393	1.437
.78	.7033	.7109	.9893	1.011	1.407	1.422
.79	.7104	.7038	1.009	.9908	1.421	1.408
t	$\sin t$	$\cos t$	$\tan t$	$\cot t$	$\sec t$	$\csc t$

TABLE I (*continued*)

t	$\sin t$	$\cos t$	$\tan t$	$\cot t$	$\sec t$	$\csc t$
.80	.7174	.6967	1.030	.9712	1.435	1.394
.81	.7243	.6895	1.050	.9520	1.450	1.381
.82	.7311	.6822	1.072	.9331	1.466	1.368
.83	.7379	.6749	1.093	.9146	1.482	1.355
.84	.7446	.6675	1.116	.8964	1.498	1.343
.85	.7513	.6600	1.138	.8785	1.515	1.331
.86	.7578	.6524	1.162	.8609	1.533	1.320
.87	.7643	.6448	1.185	.8437	1.551	1.308
.88	.7707	.6372	1.210	.8267	1.569	1.297
.89	.7771	.6294	1.235	.8100	1.589	1.287
.90	.7833	.6216	1.260	.7936	1.609	1.277
.91	.7895	.6137	1.286	.7774	1.629	1.267
.92	.7956	.6058	1.313	.7615	1.651	1.257
.93	.8016	.5978	1.341	.7458	1.673	1.247
.94	.8076	.5898	1.369	.7303	1.696	1.238
.95	.8134	.5817	1.398	.7151	1.719	1.229
.96	.8192	.5735	1.428	.7001	1.744	1.221
.97	.8249	.5653	1.459	.6853	1.769	1.212
.98	.8305	.5570	1.491	.6707	1.795	1.204
.99	.8360	.5487	1.524	.6563	1.823	1.196
1.00	.8415	.5403	1.557	.6421	1.851	1.188
1.01	.8468	.5319	1.592	.6281	1.880	1.181
1.02	.8521	.5234	1.628	.6142	1.911	1.174
1.03	.8573	.5148	1.665	.6005	1.942	1.166
1.04	.8624	.5062	1.704	.5870	1.975	1.160
1.05	.8674	.4976	1.743	.5736	2.010	1.153
1.06	.8724	.4889	1.784	.5604	2.046	1.146
1.07	.8772	.4801	1.827	.5473	2.083	1.140
1.08	.8820	.4713	1.871	.5344	2.122	1.134
1.09	.8866	.4625	1.917	.5216	2.162	1.128
1.10	.8912	.4536	1.965	.5090	2.205	1.122
1.11	.8957	.4447	2.014	.4964	2.249	1.116
1.12	.9001	.4357	2.066	.4840	2.295	1.111
1.13	.9044	.4267	2.120	.4718	2.344	1.106
1.14	.9086	.4176	2.176	.4596	2.395	1.101
1.15	.9128	.4085	2.234	.4475	2.448	1.096
1.16	.9168	.3993	2.296	.4356	2.504	1.091
1.17	.9208	.3902	2.360	.4237	2.563	1.086
1.18	.9246	.3809	2.427	.4120	2.625	1.082
1.19	.9284	.3717	2.498	.4003	2.691	1.077
t	$\sin t$	$\cos t$	$\tan t$	$\cot t$	$\sec t$	$\csc t$

TABLE I   (*continued*)

t	sin t	cos t	tan t	cot t	sec t	csc t
1.20	.9320	.3624	2.572	.3888	2.760	1.073
1.21	.9356	.3530	2.650	.3773	2.833	1.069
1.22	.9391	.3436	2.733	.3659	2.910	1.065
1.23	.9425	.3342	2.820	.3546	2.992	1.061
1.24	.9458	.3248	2.912	.3434	3.079	1.057
1.25	.9490	.3153	3.010	.3323	3.171	1.054
1.26	.9521	.3058	3.113	.3212	3.270	1.050
1.27	.9551	.2963	3.224	.3102	3.375	1.047
1.28	.9580	.2867	3.341	.2993	3.488	1.044
1.29	.9608	.2771	3.467	.2884	3.609	1.041
1.30	.9636	.2675	3.602	.2776	3.738	1.038
1.31	.9662	.2579	3.747	.2669	3.878	1.035
1.32	.9687	.2482	3.903	.2562	4.029	1.032
1.33	.9711	.2385	4.072	.2456	4.193	1.030
1.34	.9735	.2288	4.256	.2350	4.372	1.027
1.35	.9757	.2190	4.455	.2245	4.566	1.025
1.36	.9779	.2092	4.673	.2140	4.779	1.023
1.37	.9799	.1994	4.913	.2035	5.014	1.021
1.38	.9819	.1896	5.177	.1931	5.273	1.018
1.39	.9837	.1798	5.471	.1828	5.561	1.017
1.40	.9854	.1700	5.798	.1725	5.883	1.015
1.41	.9871	.1601	6.165	.1622	6.246	1.013
1.42	.9887	.1502	6.581	.1519	6.657	1.011
1.43	.9901	.1403	7.055	.1417	7.126	1.010
1.44	.9915	.1304	7.602	.1315	7.667	1.009
1.45	.9927	.1205	8.238	.1214	8.299	1.007
1.46	.9939	.1106	8.989	.1113	9.044	1.006
1.47	.9949	.1006	9.887	.1011	9.938	1.005
1.48	.9959	.0907	10.983	.0910	11.029	1.004
1.49	.9967	.0807	12.350	.0810	12.390	1.003
1.50	.9975	.0707	14.101	.0709	14.137	1.003
1.51	.9982	.0608	16.428	.0609	16.458	1.002
1.52	.9987	.0508	19.670	.0508	19.695	1.001
1.53	.9992	.0408	24.498	.0408	24.519	1.001
1.54	.9995	.0308	32.461	.0308	32.476	1.000
1.55	.9998	.0208	48.078	.0208	48.089	1.000
1.56	.9999	.0108	92.620	.0108	92.626	1.000
1.57	1.0000	.0008	1255.8	.0008	1255.8	1.000
1.58	1.0000	−.0092	−108.65	−.0092	−108.65	1.000
1.59	.9998	−.0192	−52.067	−.0192	−52.08	1.000
1.60	.9996	−.0292	−34.233	−.0292	−34.25	1.000
t	sin t	cos t	tan t	cot t	sec t	csc t

TABLE II Four-place Values of Functions

→	Sin	Cos	Tan	Cot	Sec	Csc	
0°00′	.0000	1.000	.0000	——	1.000	——	**90°00′**
10′	029	000	029	343.8	000	343.8	89°50′
20′	058	000	058	171.9	000	171.9	40′
30′	.0087	1.000	.0087	114.6	1.000	114.6	30′
40′	116	.9999	116	85.94	000	85.95	20′
0°50′	145	999	145	68.75	000	68.76	10′
1°00′	.0175	.9998	.0175	57.29	1.000	57.30	**89°00′**
10′	204	998	204	49.10	000	49.11	88°50′
20′	233	997	233	42.96	000	42.98	40′
30′	.0262	.9997	.0262	38.19	1.000	38.20	30′
40′	291	996	291	34.37	000	34.38	20′
1°50′	320	995	320	31.24	001	31.26	10′
2°00′	.0349	.9994	.0349	28.64	1.001	28.65	**88°00′**
10′	378	993	378	26.43	001	26.45	87°50′
20′	407	992	407	24.54	001	24.56	40′
30′	.0436	.9990	.0437	22.90	1.001	22.93	30′
40′	465	989	466	21.47	001	21.49	20′
2°50′	494	988	495	20.21	001	20.23	10′
3°00′	.0523	.9986	.0524	19.08	1.001	19.11	**87°00′**
10′	552	985	553	18.07	002	18.10	86°50′
20′	581	983	582	17.17	002	17.20	40′
30′	.0610	.9981	.0612	16.35	1.002	16.38	30′
40′	640	980	641	15.60	002	15.64	20′
3°50′	669	978	670	14.92	002	14.96	10′
4°00′	.0698	.9976	.0699	14.30	1.002	14.34	**86°00′**
10′	727	974	729	13.73	003	13.76	85°50′
20′	756	971	758	13.20	003	13.23	40′
30′	.0785	.9969	.0787	12.71	1.003	12.75	30′
40′	814	967	816	12.25	003	12.29	20′
4°50′	843	964	846	11.83	004	11.87	10′
5°00′	.0872	.9962	.0875	11.43	1.004	11.47	**85°00′**
10′	901	959	904	11.06	004	11.10	84°50′
20′	929	957	934	10.71	004	10.76	40′
30′	.0958	.9954	.0963	10.39	1.005	10.43	30′
40′	.0987	951	.0992	10.08	005	10.13	20′
5°50′	.1016	948	.1022	9.788	005	9.839	10′
6°00′	.1045	.9945	.1051	9.514	1.006	9.567	**84°00′**
	Cos	Sin	Cot	Tan	Csc	Sec	←

By permission from Alvin K. Bettinger and John A. Englund, *Algebra and Trigonometry*, International Textbook Company, Scranton, Pa., 1960.

TABLE II (*continued*)

⟶	Sin	Cos	Tan	Cot	Sec	Csc	
6°00′	.1045	.9945	.1051	9.514	1.006	9.567	**84°00′**
10′	074	942	080	255	006	309	83°50′
20′	103	939	110	9.010	006	9.065	40′
30′	.1132	.9936	.1139	8.777	1.006	8.834	30′
40′	161	932	169	556	007	614	20′
6°50′	190	929	198	345	007	405	10′
7°00′	.1219	.9925	.1228	8.144	1.008	8.206	**83°00′**
10′	248	922	257	7.953	008	8.016	82°50′
20′	276	918	287	770	008	7.834	40′
30′	.1305	.9914	.1317	7.596	1.009	7.661	30′
40′	334	911	346	429	009	496	20′
7°50′	363	907	376	269	009	337	10′
8°00′	.1392	.9903	.1405	7.115	1.010	7.185	**82°00′**
10′	421	899	435	6.968	010	7.040	81°50′
20′	449	894	465	827	011	6.900	40′
30′	.1478	.9890	.1495	6.691	1.011	6.765	30′
40′	507	886	524	561	012	636	20′
8°50′	536	881	554	435	012	512	10′
9°00′	.1564	.9877	.1584	6.314	1.012	6.392	**81°00′**
10′	593	872	614	197	013	277	80°50′
20′	622	868	644	6.084	013	166	40′
30′	.1650	.9863	.1673	5.976	1.014	6.059	30′
40′	679	858	703	871	014	5.955	20′
9°50′	708	853	733	769	015	855	10′
10°00′	.1736	.9848	.1763	5.671	1.015	5.759	**80°00′**
10′	765	843	793	576	016	665	79°50′
20′	794	838	823	485	016	575	40′
30′	.1822	.9833	.1853	5.396	1.017	5.487	30′
40′	851	827	883	309	018	403	20′
10°50′	880	822	914	226	018	320	10′
11°00′	.1908	.9816	.1944	5.145	1.019	5.241	**79°00′**
10′	937	811	.1974	5.066	019	164	78°50′
20′	965	805	.2004	4.989	020	089	40′
30′	.1994	.9799	.2035	4.915	1.020	5.016	30′
40′	.2022	793	065	843	021	4.945	20′
11°50′	051	787	095	773	022	876	10′
12°00′	.2079	.9781	.2126	4.705	1.022	4.810	**78°00′**
	Cos	Sin	Cot	Tan	Csc	Sec	⟵

TABLE II (*continued*)

$\longrightarrow$	Sin	Cos	Tan	Cot	Sec	Csc	
12°00′	.2079	.9781	.2126	4.705	1.022	4.810	**78°00′**
10′	108	775	156	638	023	745	77°50′
20′	136	769	186	574	024	682	40′
30′	.2164	.9763	.2217	4.511	1.024	4.620	30′
40′	193	757	247	449	025	560	20′
12°50′	221	750	278	390	026	502	10′
13°00′	.2250	.9744	.2309	4.331	1.026	4.445	**77°00′**
10′	278	737	339	275	027	390	76°50′
20′	306	730	370	219	028	336	40′
30′	.2334	.9724	.2401	4.165	1.028	4.284	30′
40′	363	717	432	113	029	232	20′
13°50′	391	710	462	061	030	182	10′
14°00′	.2419	.9703	.2493	4.011	1.031	4.134	**76°00′**
10′	447	696	524	3.962	031	086	75°50′
20′	476	689	555	914	032	4.039	40′
30′	.2504	.9681	.2586	3.867	1.033	3.994	30′
40′	532	674	617	821	034	950	20′
14°50′	560	667	648	776	034	906	10′
15°00′	.2588	.9659	.2679	3.732	1.035	3.864	**75°00′**
10′	616	652	711	689	036	822	74°50′
20′	644	644	742	647	037	782	40′
30′	.2672	.9636	.2773	3.606	1.038	3.742	30′
40′	700	628	805	566	039	703	20′
15°50′	728	621	836	526	039	665	10′
16°00′	.2756	.9613	.2867	3.487	1.040	3.628	**74°00′**
10′	784	605	899	450	041	592	73°50′
20′	812	596	931	412	042	556	40′
30′	.2840	.9588	.2962	3.376	1.043	3.521	30′
40′	868	580	.2994	340	044	487	20′
16°50′	896	572	.3026	305	045	453	10′
17°00′	.2924	.9563	.3057	3.271	1.046	3.420	**73°00′**
10′	952	555	089	237	047	388	72°50′
20′	.2979	546	121	204	048	356	40′
30′	.3007	.9537	.3153	3.172	1.049	3.326	30′
40′	035	528	185	140	049	295	20′
17°50′	062	520	217	108	050	265	10′
18°00′	.3090	.9511	.3249	3.078	1.051	3.236	**72°00′**
	Cos	Sin	Cot	Tan	Csc	Sec	$\longleftarrow$

TABLE II   (*continued*)

⟶	Sin	Cos	Tan	Cot	Sec	Csc	
18°00′	.3090	.9511	.3249	3.078	1.051	3.236	**72°00′**
10′	118	502	281	047	052	207	71°50′
20′	145	492	314	3.018	053	179	40′
30′	.3173	.9483	.3346	2.989	1.054	3.152	30′
40′	201	474	378	960	056	124	20′
18°50′	228	465	411	932	057	098	10′
19°00′	.3256	.9455	.3443	2.904	1.058	3.072	**71°00′**
10′	283	446	476	877	059	046	70°50′
20′	311	436	508	850	060	3.021	40′
30′	.3338	.9426	.3541	2.824	1.061	2.996	30′
40′	365	417	574	798	062	971	20′
19°50′	393	407	607	773	063	947	10′
20°00′	.3420	.9397	.3640	2.747	1.064	2.924	**70°00′**
10′	448	387	673	723	065	901	69°50′
20′	475	377	706	699	066	878	40′
30′	.3502	.9367	.3739	2.675	1.068	2.855	30′
40′	529	356	772	651	069	833	20′
20°50′	557	346	805	628	070	812	10′
21°00′	.3584	.9336	.3839	2.605	1.071	2.790	**69°00′**
10′	611	325	872	583	072	769	68°50′
20′	638	315	906	560	074	749	40′
30′	.3665	.9304	.3939	2.539	1.075	2.729	30′
40′	692	293	.3973	517	076	709	20′
21°50′	719	283	.4006	496	077	689	10′
22°00′	.3746	.9272	.4040	2.475	1.079	2.669	**68°00′**
10′	773	261	074	455	080	650	67°50′
20′	800	250	108	434	081	632	40′
30′	.3827	.9239	.4142	2.414	1.082	2.613	30′
40′	854	228	176	394	084	595	20′
22°50′	881	216	210	375	085	577	10′
23°00′	.3907	.9205	.4245	2.356	1.086	2.559	**67°00′**
10′	934	194	279	337	088	542	66°50′
20′	961	182	314	318	089	525	40′
30′	.3987	.9171	.4348	2.300	1.090	2.508	30′
40′	.4014	159	383	282	092	491	20′
23°50′	041	147	417	264	093	475	10′
24°00′	.4067	.9135	.4452	2.246	1.095	2.459	**66°00′**
	Cos	Sin	Cot	Tan	Csc	Sec	⟵

TABLE II   (*continued*)

$\longrightarrow$	Sin	Cos	Tan	Cot	Sec	Csc	
24°00′	.4067	.9135	.4452	2.246	1.095	2.459	**66°00′**
10′	094	124	487	229	096	443	65°50′
20′	120	112	522	211	097	427	40′
30′	.4147	.9100	.4557	2.194	1.099	2.411	30′
40′	173	088	592	177	100	396	20′
24°50′	200	075	628	161	102	381	10′
25°00′	.4226	.9063	.4663	2.145	1.103	2.366	**65°00′**
10′	253	051	699	128	105	352	64°50′
20′	279	038	734	112	106	337	40′
30′	.4305	.9026	.4770	2.097	1.108	2.323	30′
40′	331	013	806	081	109	309	20′
25°50′	358	.9001	841	066	111	295	10′
26°00′	.4384	.8988	.4877	2.050	1.113	2.281	**64°00′**
10′	410	975	913	035	114	268	63°50′
20′	436	962	950	020	116	254	40′
30′	.4462	.8949	.4986	2.006	1.117	2.241	30′
40′	488	936	.5022	1.991	119	228	20′
26°50′	514	923	059	977	121	215	10′
27°00′	.4540	.8910	.5095	1.963	1.122	2.203	**63°00′**
10′	566	897	132	949	124	190	62°50′
20′	592	884	169	935	126	178	40′
30′	.4617	.8870	.5206	1.921	1.127	2.166	30′
40′	643	857	243	907	129	154	20′
27°50′	669	843	280	894	131	142	10′
28°00′	.4695	.8829	.5317	1.881	1.133	2.130	**62°00′**
10′	720	816	354	868	134	118	61°50′
20′	746	802	392	855	136	107	40′
30′	.4772	.8788	.5430	1.842	1.138	2.096	30′
40′	797	774	467	829	140	085	20′
28°50′	823	760	505	816	142	074	10′
29°00′	.4848	.8746	.5543	1.804	1.143	2.063	**61°00′**
10′	874	732	581	792	145	052	60°50′
20′	899	718	619	780	147	041	40′
30′	.4924	.8704	.5658	1.767	1.149	2.031	30′
40′	950	689	696	756	151	020	20′
29°50′	.4975	675	735	744	153	010	10′
30°00′	.5000	.8660	.5774	1.732	1.155	2.000	**60°00′**
	Cos	Sin	Cot	Tan	Csc	Sec	$\longleftarrow$

TABLE II *(continued)*

⟶	Sin	Cos	Tan	Cot	Sec	Csc	
30°00′	.5000	.8660	.5774	1.732	1.155	2.000	**60°00′**
10′	025	646	812	720	157	1.990	59°50′
20′	050	631	851	709	159	980	40′
30′	.5075	.8616	.5890	1.698	1.161	1.970	30′
40′	100	601	930	686	163	961	20′
30°50′	125	587	.5969	675	165	951	10′
31°00′	.5150	.8572	.6009	1.664	1.167	1.942	**59°00′**
10′	175	557	048	653	169	932	58°50′
20′	200	542	088	643	171	923	40′
30′	.5225	.8526	.6128	1.632	1.173	1.914	30′
40′	250	511	168	621	175	905	20′
31°50′	275	496	208	611	177	896	10′
32°00′	.5299	.8480	.6249	1.600	1.179	1.887	**58°00′**
10′	324	465	289	590	181	878	57°50′
20′	348	450	330	580	184	870	40′
30′	.5373	.8434	.6371	1.570	1.186	1.861	30′
40′	398	418	412	560	188	853	20′
32°50′	422	403	453	550	190	844	10′
33°00′	.5446	.8387	.6494	1.540	1.192	1.836	**57°00′**
10′	471	371	536	530	195	828	56°50′
20′	495	355	577	520	197	820	40′
30′	.5519	.8339	.6619	1.511	1.199	1.812	30′
40′	544	323	661	501	202	804	20′
33°50′	568	307	703	492	204	796	10′
34°00′	.5592	.8290	.6745	1.483	1.206	1.788	**56°00′**
10′	616	274	787	473	209	781	55°50′
20′	640	258	830	464	211	773	40′
30′	.5664	.8241	.6873	1.455	1.213	1.766	30′
40′	688	225	916	446	216	758	20′
34°50′	712	208	.6959	437	218	751	10′
35°00′	.5736	.8192	.7002	1.428	1.221	1.743	**55°00′**
10′	760	175	046	419	223	736	54°50′
20′	783	158	089	411	226	729	40′
30′	.5807	.8141	.7133	1.402	1.228	1.722	30′
40′	831	124	177	393	231	715	20′
35°50′	854	107	221	385	233	708	10′
36°00′	.5878	.8090	.7265	1.376	1.236	1.701	**54°00′**
	Cos	Sin	Cot	Tan	Csc	Sec	⟵

TABLE II   (*continued*)

⟶	Sin	Cos	Tan	Cot	Sec	Csc	
36°00′	.5878	.8090	.7265	1.376	1.236	1.701	**54°00′**
10′	901	073	310	368	239	695	53°50′
20′	925	056	355	360	241	688	40′
30′	.5948	.8039	.7400	1.351	1.244	1.681	30′
40′	972	021	445	343	247	675	20′
36°50′	.5995	.8004	490	335	249	668	10′
37°00′	.6018	.7986	.7536	1.327	1.252	1.662	**53°00′**
10′	041	969	581	319	255	655	52°50′
20′	065	951	627	311	258	649	40′
30′	.6088	.7934	.7673	1.303	1.260	1.643	30′
40′	111	916	720	295	263	636	20′
37°50′	134	898	766	288	266	630	10′
38°00′	.6157	.7880	.7813	1.280	1.269	1.624	**52°00′**
10′	180	862	860	272	272	618	51°50′
20′	202	844	907	265	275	612	40′
30′	.6225	.7826	.7954	1.257	1.278	1.606	30′
40′	248	808	.8002	250	281	601	20′
38°50′	271	790	050	242	284	595	10′
39°00′	.6293	.7771	.8098	1.235	1.287	1.589	**51°00′**
10′	316	753	146	228	290	583	50°50′
20′	338	735	195	220	293	578	40′
30′	.6361	.7716	.8243	1.213	1.296	1.572	30′
40′	383	698	292	206	299	567	20′
39°50′	406	679	342	199	302	561	10′
40°00′	.6428	.7660	.8391	1.192	1.305	1.556	**50°00′**
10′	450	642	441	185	309	550	49°50′
20′	472	623	491	178	312	545	40′
30′	.6494	.7604	.8541	1.171	1.315	1.540	30′
40′	517	585	591	164	318	535	20′
40°50′	539	566	642	157	322	529	10′
41°00′	.6561	.7547	.8693	1.150	1.325	1.524	**49°00′**
10′	583	528	744	144	328	519	48°50′
20′	604	509	796	137	332	514	40′
30′	.6626	.7490	.8847	1.130	1.335	1.509	30′
40′	648	470	899	124	339	504	20′
41°50′	670	451	.8952	117	342	499	10′
42°00′	.6691	.7431	.9004	1.111	1.346	1.494	**48°00′**
	Cos	Sin	Cot	Tan	Csc	Sec	⟵

TABLE II   (*continued*)

⟶	Sin	Cos	Tan	Cot	Sec	Csc	
42°00′	.6691	.7431	.9004	1.111	1.346	1.494	**48°00′**
10′	713	412	057	104	349	490	47°50′
20′	734	392	110	098	353	485	40′
30′	.6756	.7373	.9163	1.091	1.356	1.480	30′
40′	777	353	217	085	360	476	20′
42°50′	799	333	271	079	364	471	10′
43°00′	.6820	.7314	.9325	1.072	1.367	1.466	**47°00′**
10′	841	294	380	066	371	462	46°50′
20′	862	274	435	060	375	457	40′
30′	.6884	.7254	.9490	1.054	1.379	1.453	30′
40′	905	234	545	048	382	448	20′
43°50′	926	214	601	042	386	444	10′
44°00′	.6947	.7193	.9657	1.036	1.390	1.440	**46°00′**
10′	967	173	713	030	394	435	45°50′
20′	.6988	153	770	024	398	431	40′
30′	.7009	.7133	.9827	1.018	1.402	1.427	30′
40′	030	112	884	012	406	423	20′
44°50′	050	092	.9942	006	410	418	10′
45°00′	.7071	.7071	1.000	1.000	1.414	1.414	**45°00′**
	Cos	Sin	Cot	Tan	Csc	Sec	⟵

TABLE III   Logarithms of Numbers from 1.00 to 9.99

N	0	1	2	3	4	5	6	7	8	9
1.0	0.0000	0.004321	0.008600	0.01284	0.01703	0.02119	0.02531	0.02938	0.03342	0.03743
1.1	0.04139	0.04532	0.04922	0.05308	0.05690	0.06070	0.06446	0.06819	0.07188	0.07555
1.2	0.07918	0.08279	0.08636	0.08991	0.09342	0.09691	0.1004	0.1038	0.1072	0.1106
1.3	0.1139	0.1173	0.1206	0.1239	0.1271	0.1303	0.1335	0.1367	0.1399	0.1430
1.4	0.1461	0.1492	0.1523	0.1553	0.1584	0.1614	0.1644	0.1673	0.1703	0.1732
1.5	0.1761	0.1790	0.1818	0.1847	0.1875	0.1903	0.1931	0.1959	0.1987	0.2014
1.6	0.2041	0.2068	0.2095	0.2122	0.2148	0.2175	0.2201	0.2227	0.2253	0.2279
1.7	0.2304	0.2330	0.2355	0.2380	0.2405	0.2430	0.2455	0.2480	0.2504	0.2529
1.8	0.2553	0.2577	0.2601	0.2625	0.2648	0.2673	0.2695	0.2718	0.2742	0.2765
1.9	0.2788	0.2810	0.2833	0.2856	0.2878	0.2900	0.2923	0.2945	0.2967	0.2989
2.0	0.3010	0.3032	0.3054	0.3075	0.3096	0.3118	0.3139	0.3160	0.3181	0.3201
2.1	0.3222	0.3243	0.3263	0.3284	0.3304	0.3324	0.3345	0.3365	0.3385	0.3404
2.2	0.3424	0.3444	0.3464	0.3483	0.3502	0.3522	0.3541	0.3560	0.3579	0.3598
2.3	0.3617	0.3636	0.3655	0.3674	0.3692	0.3711	0.3729	0.3747	0.3766	0.3784
2.4	0.3802	0.3820	0.3838	0.3856	0.3874	0.3892	0.3909	0.3927	0.3945	0.3962
2.5	0.3979	0.3997	0.4014	0.4031	0.4048	0.4065	0.4082	0.4099	0.4116	0.4133
2.6	0.4150	0.4166	0.4183	0.4200	0.4216	0.4232	0.4249	0.4265	0.4281	0.4298
2.7	0.4314	0.4330	0.4346	0.4362	0.4378	0.4393	0.4409	0.4425	0.4440	0.4456
2.8	0.4472	0.4487	0.4502	0.4518	0.4533	0.4548	0.4564	0.4579	0.4594	0.4609
2.9	0.4624	0.4639	0.4654	0.4669	0.4683	0.4698	0.4713	0.4728	0.4742	0.4757
3.0	0.4771	0.4786	0.4800	0.4814	0.4829	0.4843	0.4857	0.4871	0.4886	0.4900
3.1	0.4914	0.4928	0.4942	0.4955	0.4969	0.4983	0.4997	0.5011	0.5024	0.5038
3.2	0.5051	0.5065	0.5079	0.5092	0.5105	0.5119	0.5132	0.5145	0.5159	0.5172
3.3	0.5185	0.5198	0.5211	0.5224	0.5237	0.5250	0.5263	0.5276	0.5289	0.5302
3.4	0.5315	0.5328	0.5340	0.5353	0.5366	0.5378	0.5391	0.5403	0.5416	0.5428
3.5	0.5441	0.5453	0.5465	0.5478	0.5490	0.5502	0.5514	0.5527	0.5539	0.5551
3.6	0.5563	0.5575	0.5587	0.5599	0.5611	0.5623	0.5635	0.5647	0.5658	0.5670
3.7	0.5682	0.5694	0.5705	0.5717	0.5729	0.5740	0.5752	0.5763	0.5775	0.5786
3.8	0.5798	0.5809	0.5821	0.5832	0.5843	0.5855	0.5866	0.5877	0.5888	0.5899
3.9	0.5911	0.5922	0.5933	0.5944	0.5955	0.5966	0.5977	0.5988	0.5999	0.6010
4.0	0.6021	0.6031	0.6042	0.6053	0.6064	0.6075	0.6085	0.6096	0.6107	0.6117
4.1	0.6128	0.6138	0.6149	0.6160	0.6170	0.6180	0.6191	0.6201	0.6212	0.6222
4.2	0.6232	0.6243	0.6253	0.6263	0.6274	0.6284	0.6294	0.6304	0.6314	0.6325
4.3	0.6335	0.6345	0.6355	0.6365	0.6375	0.6385	0.6395	0.6405	0.6415	0.6425
4.4	0.6435	0.6444	0.6454	0.6464	0.6474	0.6484	0.6493	0.6503	0.6513	0.6522
4.5	0.6532	0.6542	0.6551	0.6561	0.6571	0.6580	0.6590	0.6599	0.6609	0.6618
4.6	0.6628	0.6637	0.6646	0.6656	0.6665	0.6675	0.6684	0.6693	0.6702	0.6712
4.7	0.6721	0.6730	0.6739	0.6749	0.6758	0.6767	0.6776	0.6785	0.6794	0.6803
4.8	0.6812	0.6821	0.6830	0.6839	0.6848	0.6857	0.6866	0.6875	0.6884	0.6893
4.9	0.6902	0.6911	0.6920	0.6928	0.6937	0.6946	0.6955	0.6964	0.6972	0.6981
5.0	0.6990	0.6998	0.7007	0.7016	0.7024	0.7033	0.7042	0.7050	0.7059	0.7067
5.1	0.7076	0.7084	0.7093	0.7101	0.7110	0.7118	0.7126	0.7135	0.7143	0.7152
5.2	0.7160	0.7168	0.7177	0.7185	0.7193	0.7202	0.7210	0.7218	0.7226	0.7235
5.3	0.7243	0.7251	0.7259	0.7267	0.7275	0.7284	0.7292	0.7300	0.7308	0.7316
5.4	0.7324	0.7332	0.7340	0.7348	0.7356	0.7364	0.7372	0.7380	0.7388	0.7396

By permission from Thomas L. Wade and Howard E. Taylor, *Fundamental Mathematics*, McGraw-Hill Book Company, New York, 1960.

TABLE III   (*continued*)

N	0	1	2	3	4	5	6	7	8	9
5.5	0.7404	0.7412	0.7419	0.7427	0.7435	0.7443	0.7451	0.7459	0.7466	0.7474
5.6	0.7482	0.7490	0.7497	0.7505	0.7513	0.7520	0.7528	0.7536	0.7543	0.7551
5.7	0.7559	0.7566	0.7574	0.7582	0.7589	0.7597	0.7604	0.7612	0.7619	0.7627
5.8	0.7634	0.7642	0.7649	0.7657	0.7664	0.7672	0.7679	0.7686	0.7694	0.7701
5.9	0.7709	0.7716	0.7723	0.7731	0.7738	0.7745	0.7752	0.7760	0.7767	0.7774
6.0	0.7782	0.7789	0.7796	0.7803	0.7810	0.7818	0.7825	0.7832	0.7839	0.7846
6.1	0.7853	0.7860	0.7868	0.7875	0.7882	0.7889	0.7896	0.7903	0.7910	0.7917
6.2	0.7924	0.7931	0.7938	0.7945	0.7952	0.7959	0.7966	0.7973	0.7980	0.7987
6.3	0.7993	0.8000	0.8007	0.8014	0.8021	0.8028	0.8035	0.8041	0.8048	0.8055
6.4	0.8062	0.8069	0.8075	0.8082	0.8089	0.8096	0.8102	0.8109	0.8116	0.8122
6.5	0.8129	0.8136	0.8142	0.8149	0.8156	0.8162	0.8169	0.8176	0.8182	0.8189
6.6	0.8195	0.8202	0.8209	0.8215	0.8222	0.8228	0.8235	0.8241	0.8248	0.8254
6.7	0.8261	0.8267	0.8274	0.8280	0.8287	0.8293	0.8299	0.8306	0.8312	0.8319
6.8	0.8325	0.8331	0.8338	0.8344	0.8351	0.8357	0.8363	0.8370	0.8376	0.8382
6.9	0.8388	0.8395	0.8401	0.8407	0.8414	0.8420	0.8426	0.8432	0.8439	0.8445
7.0	0.8451	0.8457	0.8463	0.8470	0.8476	0.8482	0.8488	0.8494	0.8500	0.8506
7.1	0.8513	0.8519	0.8525	0.8531	0.8537	0.8543	0.8549	0.8555	0.8561	0.8567
7.2	0.8573	0.8579	0.8585	0.8591	0.8597	0.8603	0.8609	0.8615	0.8621	0.8627
7.3	0.8633	0.8639	0.8645	0.8651	0.8657	0.8663	0.8669	0.8675	0.8681	0.8686
7.4	0.8692	0.8698	0.8704	0.8710	0.8716	0.8722	0.8727	0.8733	0.8739	0.8745
7.5	0.8751	0.8756	0.8762	0.8768	0.8774	0.8779	0.8785	0.8791	0.8797	0.8802
7.6	0.8808	0.8814	0.8820	0.8825	0.8831	0.8837	0.8842	0.8848	0.8854	0.8859
7.7	0.8865	0.8871	0.8876	0.8882	0.8887	0.8893	0.8899	0.8904	0.8910	0.8915
7.8	0.8921	0.8927	0.8932	0.8938	0.8943	0.8949	0.8954	0.8960	0.8965	0.8971
7.9	0.8976	0.8982	0.8987	0.8993	0.8998	0.9004	0.9009	0.9015	0.9020	0.9025
8.0	0.9031	0.9036	0.9042	0.9047	0.9053	0.9058	0.9063	0.9069	0.9074	0.9079
8.1	0.9085	0.9090	0.9096	0.9101	0.9106	0.9112	0.9117	0.9122	0.9128	0.9133
8.2	0.9138	0.9143	0.9149	0.9154	0.9159	0.9165	0.9170	0.9175	0.9180	0.9186
8.3	0.9191	0.9196	0.9201	0.9206	0.9212	0.9217	0.9222	0.9227	0.9232	0.9238
8.4	0.9243	0.9248	0.9253	0.9258	0.9263	0.9269	0.9274	0.9279	0.9284	0.9289
8.5	0.9294	0.9299	0.9304	0.9309	0.9315	0.9320	0.9325	0.9330	0.9335	0.9340
8.6	0.9345	0.9350	0.9355	0.9360	0.9365	0.9370	0.9375	0.9380	0.9385	0.9390
8.7	0.9395	0.9400	0.9405	0.9410	0.9415	0.9420	0.9425	0.9430	0.9435	0.9440
8.8	0.9445	0.9450	0.9455	0.9460	0.9465	0.9469	0.9474	0.9479	0.9484	0.9489
8.9	0.9494	0.9499	0.9504	0.9509	0.9513	0.9518	0.9523	0.9528	0.9533	0.9538
9.0	0.9542	0.9547	0.9552	0.9557	0.9562	0.9566	0.9571	0.9576	0.9581	0.9586
9.1	0.9590	0.9595	0.9600	0.9605	0.9609	0.9614	0.9619	0.9624	0.9628	0.9633
9.2	0.9638	0.9643	0.9647	0.9652	0.9657	0.9661	0.9666	0.9671	0.9675	0.9680
9.3	0.9685	0.9689	0.9694	0.9699	0.9703	0.9708	0.9713	0.9717	0.9722	0.9727
9.4	0.9731	0.9736	0.9741	0.9745	0.9750	0.9754	0.9759	0.9763	0.9768	0.9773
9.5	0.9777	0.9782	0.9786	0.9791	0.9795	0.9800	0.9805	0.9809	0.9814	0.9818
9.6	0.9823	0.9827	0.9832	0.9836	0.9841	0.9845	0.9850	0.9854	0.9859	0.9863
9.7	0.9868	0.9872	0.9877	0.9881	0.9886	0.9890	0.9894	0.9899	0.9903	0.9908
9.8	0.9912	0.9917	0.9921	0.9926	0.9930	0.9934	0.9939	0.9943	0.9948	0.9952
9.9	0.9956	0.9961	0.9965	0.9969	0.9974	0.9978	0.9983	0.9987	0.9991	0.9996

TABLE IV FOUR-PLACE LOGARITHMS OF FUNCTIONS°

→	L Sin	L Tan	L Cot	L Cos	
0°00′				10.0000	**90°00′**
10′	7.4637	7.4637	12.5363	.0000	89°50′
20′	.7648	.7648	.2352	.0000	40′
30′	7.9408	7.9409	12.0591	.0000	30′
40′	8.0658	8.0658	11.9342	.0000	20′
0°50′	.1627	.1627	.8373	10.0000	10′
1°00′	8.2419	8.2419	11.7581	9.9999	**89°00′**
10′	.3088	.3089	.6911	.9999	88°50′
20′	.3668	.3669	.6331	.9999	40′
30′	.4179	.4181	.5819	.9999	30′
40′	.4637	.4638	.5362	.9998	20′
1°50′	.5050	.5053	.4947	.9998	10′
2°00′	8.5428	8.5431	11.4569	9.9997	**88°00′**
10′	.5776	.5779	.4221	.9997	87°50′
20′	.6097	.6101	.3899	.9996	40′
30′	.6397	.6401	.3599	.9996	30′
40′	.6677	.6682	.3318	.9995	20′
2°50′	.6940	.6945	.3055	.9995	10′
3°00′	8.7188	8.7194	11.2806	9.9994	**87°00′**
10′	.7423	.7429	.2571	.9993	86°50′
20′	.7645	.7652	.2348	.9993	40′
30′	.7857	.7865	.2135	.9992	30′
40′	.8059	.8067	.1933	.9991	20′
3°50′	.8251	.8261	.1739	.9990	10′
4°00′	8.8436	8.8446	11.1554	9.9989	**86°00′**
10′	.8613	.8624	.1376	.9989	85°50′
20′	.8783	.8795	.1205	.9988	40′
30′	.8946	.8960	.1040	.9987	30′
40′	.9104	.9118	.0882	.9986	20′
4°50′	.9256	.9272	.0728	.9985	10′
5°00′	8.9403	8.9420	11.0580	9.9983	**85°00′**
10′	.9545	.9563	.0437	.9982	84°50′
20′	.9682	.9701	.0299	.9981	40′
30′	.9816	.9836	.0164	.9980	30′
40′	8.9945	8.9966	11.0034	.9979	20′
5°50′	9.0070	9.0093	10.9907	.9977	10′
6°00′	9.0192	9.0216	10.9784	9.9976	**84°00′**
	L Cos	L Cot	L Tan	L Sin	←

By permission from Alvin K. Bettinger and John A. Englund, *Algebra and Trigonometry*, International Textbook Company, Scranton, Pa., 1960.

° Subtract 10 from each entry; example: Log sin 0°20′ = 7.7648 − 10.

TABLE IV° (*continued*)

→	L Sin	L Tan	L Cot	L Cos	
6°00′	9.0192	9.0216	10.9784	9.9976	**84°00′**
10′	.0311	.0336	.9664	.9975	83°50′
20′	.0426	.0453	.9547	.9973	40′
30′	.0539	.0567	.9433	.9972	30′
40′	.0648	.0678	.9322	.9971	20′
6°50′	.0755	.0786	.9214	.9969	10′
7°00′	9.0859	9.0891	10.9109	9.9968	**83°00′**
10′	.0961	.0995	.9005	.9966	82°50′
20′	.1060	.1096	.8904	.9964	40′
30′	.1157	.1194	.8806	.9963	30′
40′	.1252	.1291	.8709	.9961	20′
7°50′	.1345	.1385	.8615	.9959	10′
8°00′	9.1436	9.1478	10.8522	9.9958	**82°00′**
10′	.1525	.1569	.8431	.9956	81°50′
20′	.1612	.1658	.8342	.9954	40′
30′	.1697	.1745	.8255	.9952	30′
40′	.1781	.1831	.8169	.9950	20′
8°50′	.1863	.1915	.8085	.9948	10′
9°00′	9.1943	9.1997	10.8003	9.9946	**81°00′**
10′	.2022	.2078	.7922	.9944	80°50′
20′	.2100	.2158	.7842	.9942	40′
30′	.2176	.2236	.7764	.9940	30′
40′	.2251	.2313	.7687	.9938	20′
9°50′	.2324	.2389	.7611	.9936	10′
10°00′	9.2397	9.2463	10.7537	9.9934	**80°00′**
10′	.2468	.2536	.7464	.9931	79°50′
20′	.2538	.2609	.7391	.9929	40′
30′	.2606	.2680	.7320	.9927	30′
40′	.2674	.2750	.7250	.9924	20′
10°50′	.2740	.2819	.7181	.9922	10′
11°00′	9.2806	9.2887	10.7113	9.9919	**79°00′**
10′	.2870	.2953	.7047	.9917	78°50′
20′	.2934	.3020	.6980	.9914	40′
30′	.2997	.3085	.6915	.9912	30′
40′	.3058	.3149	.6851	.9909	20′
11°50′	.3119	.3212	.6788	.9907	10′
12°00′	9.3179	9.3275	10.6725	9.9904	**78°00′**
	L Cos	L Cot	L Tan	L Sin	←

° Subtract 10 from each entry; example: Log tan 6°40′ = 9.0678 − 10.

TABLE IV° *(continued)*

⟶	L Sin	L Tan	L Cot	L Cos	
12°00′	9.3179	9.3275	10.6725	9.9904	**78°00′**
10′	.3238	.3336	.6664	.9901	77°50′
20′	.3296	.3397	.6603	.9899	40′
30′	.3353	.3458	.6542	.9896	30′
40′	.3410	.3517	.6483	.9893	20′
12°50′	.3466	.3576	.6424	.9890	10′
13°00′	9.3521	9.3634	10.6366	9.9887	**77°00′**
10′	.3575	.3691	.6309	.9884	76°50′
20′	.3629	.3748	.6252	.9881	40′
30′	.3682	.3804	.6196	.9878	30′
40′	.3734	.3859	.6141	.9875	20′
13°50′	.3786	.3914	.6086	.9872	10′
14°00′	9.3837	9.3968	10.6032	9.9869	**76°00′**
10′	.3887	.4021	.5979	.9866	75°50′
20′	.3937	.4074	.5926	.9863	40′
30′	.3986	.4127	.5873	.9859	30′
40′	.4035	.4178	.5822	.9856	20′
14°50′	.4083	.4230	.5770	.9853	10′
15°00′	9.4130	9.4281	10.5719	9.9849	**75°00′**
10′	.4177	.4331	.5669	.9846	74°50′
20′	.4223	.4381	.5619	.9843	40′
30′	.4269	.4430	.5570	.9839	30′
40′	.4314	.4479	.5521	.9836	20′
15°50′	.4359	.4527	.5473	.9832	10′
16°00′	9.4403	9.4575	10.5425	9.9828	**74°00′**
10′	.4447	.4622	.5378	.9825	73°50′
20′	.4491	.4669	.5331	.9821	40′
30′	.4533	.4716	.5284	.9817	30′
40′	.4576	.4762	.5238	.9814	20′
16°50′	.4618	.4808	.5192	.9810	10′
17°00′	9.4659	9.4853	10.5147	9.9806	**73°00′**
10′	.4700	.4898	.5102	.9802	72°50′
20′	.4741	.4943	.5057	.9798	40′
30′	.4781	.4987	.5013	.9794	30′
40′	.4821	.5031	.4969	.9790	20′
17°50′	.4861	.5075	.4925	.9786	10′
18°00′	9.4900	9.5118	10.4882	9.9782	**72°00′**
	L Cos	L Cot	L Tan	L Sin	⟵

° Subtract 10 from each entry; example: Log cot 12°50′ = 10.6424 − 10.

TABLE IV°   (*continued*)

⟶	L Sin	L Tan	L Cot	L Cos	
18°00′	9.4900	9.5118	10.4882	9.9782	**72°00′**
10′	.4939	.5161	.4839	.9778	71°50′
20′	.4977	.5203	.4797	.9774	40′
30′	.5015	.5245	.4755	.9770	30′
40′	.5052	.5287	.4713	.9765	20′
18°50′	.5090	.5329	.4671	.9761	10′
19°00′	9.5126	9.5370	10.4630	9.9757	**71°00′**
10′	.5163	.5411	.4589	.9752	70°50′
20′	.5199	.5451	.4549	.9748	40′
30′	.5235	.5491	.4509	.9743	30′
40′	.5270	.5531	.4469	.9739	20′
19°50′	.5306	.5571	.4429	.9734	10′
20°00′	9.5341	9.5611	10.4389	9.9730	**70°00′**
10′	.5375	.5650	.4350	.9725	69°50′
20′	.5409	.5689	.4311	.9721	40′
30′	.5443	.5727	.4273	.9716	30′
40′	.5477	.5766	.4234	.9711	20′
20°50′	.5510	.5804	.4196	.9706	10′
21°00′	9.5543	9.5842	10.4158	9.9702	**69°00′**
10′	.5576	.5879	.4121	.9697	68°50′
20′	.5609	.5917	.4083	.9692	40′
30′	.5641	.5954	.4046	.9687	30′
40′	.5673	.5991	.4009	.9682	20′
21°50′	.5704	.6028	.3972	.9677	10′
22°00′	9.5736	9.6064	10.3936	9.9672	**68°00′**
10′	.5767	.6100	.3900	.9667	67°50′
20′	.5798	.6136	.3864	.9661	40′
30′	.5828	.6172	.3828	.9656	30′
40′	.5859	.6208	.3792	.9651	20′
22°50′	.5889	.6243	.3757	.9646	10′
23°00′	9.5919	9.6279	10.3721	9.9640	**67°00′**
10′	.5948	.6314	.3686	.9635	66°50′
20′	.5978	.6348	.3652	.9629	40′
30′	.6007	.6383	.3617	.9624	30′
40′	.6036	.6417	.3583	.9618	20′
23°50′	.6065	.6452	.3548	.9613	10′
24°00′	9.6093	9.6486	10.3514	9.9607	**66°00′**
	L Cos	L Cot	L Tan	L Sin	⟵

° Substract 10 from each entry; example: Log cos 18°20′ = 9.9774 − 10.

TABLE IV° (*continued*)

$\longrightarrow$	L Sin	L Tan	L Cot	L Cos	
24°00′	9.6093	9.6486	10.3514	9.9607	**66°00′**
10′	.6121	.6520	.3480	.9602	65°50′
20′	.6149	.6553	.3447	.9596	40′
30′	.6177	.6587	.3413	.9590	30′
40′	.6205	.6620	.3380	.9584	20′
24°50′	.6232	.6654	.3346	.9579	10′
25°00′	9.6259	9.6687	10.3313	9.9573	**65°00′**
10′	.6286	.6720	.3280	.9567	64°50′
20′	.6313	.6752	.3248	.9561	40′
30′	.6340	.6785	.3215	.9555	30′
40′	.6366	.6817	.3183	.9549	20′
25°50′	.6392	.6850	.3150	.9543	10′
26°00′	9.6418	9.6882	10.3118	9.9537	**64°00′**
10′	.6444	.6914	.3086	.9530	63°50′
20′	.6470	.6946	.3054	.9524	40′
30′	.6495	.6977	.3023	.9518	30′
40′	.6521	.7009	.2991	.9512	20′
26°50′	.6546	.7040	.2960	.9505	10′
27°00′	9.6570	9.7072	10.2928	9.9499	**63°00′**
10′	.6595	.7103	.2897	.9492	62°50′
20′	.6620	.7134	.2866	.9486	40′
30′	.6644	.7165	.2835	.9479	30′
40′	.6668	.7196	.2804	.9473	20′
27°50′	.6692	.7226	.2774	.9466	10′
28°00′	9.6716	9.7257	10.2743	9.9459	**62°00′**
10′	.6740	.7287	.2713	.9453	61°50′
20′	.6763	.7317	.2683	.9446	40′
30′	.6787	.7348	.2652	.9439	30′
40′	.6810	.7378	.2622	.9432	20′
28°50′	.6833	.7408	.2592	.9425	10′
29°00′	9.6856	9.7438	10.2562	9.9418	**61°00′**
10′	.6878	.7467	.2533	.9411	60°50′
20′	.6901	.7497	.2503	.9404	40′
30′	.6923	.7526	.2474	.9397	30′
40′	.6946	.7556	.2444	.9390	20′
29°50′	.6968	.7585	.2415	.9383	10′
30°00′	9.6990	9.7614	10.2386	9.9375	**60°00′**
	L Cos	L Cot	L Tan	L Sin	$\longleftarrow$

° Subtract 10 from each entry; example: Log tan 65°10′ = 10.3346 − 10.

TABLE IV° *(continued)*

⟶	L Sin	L Tan	L Cot	L Cos	
30°00′	9.6990	9.7614	10.2386	9.9375	**60°00′**
10′	.7012	.7644	.2356	.9368	59°50′
20′	.7033	.7673	.2327	9361	40′
30′	.7055	.7701	.2299	.9353	30′
40′	.7076	.7730	.2270	.9346	20′
30°50′	.7097	.7759	.2241	.9338	10′
31°00′	9.7118	9.7788	10.2212	9.9331	**59°00′**
10′	.7139	.7816	.2184	.9323	58°50′
20′	.7160	.7845	.2155	.9315	40′
30′	.7181	.7873	.2127	.9308	30′
40′	.7201	.7902	.2098	.9300	20′
31°50′	.7222	.7930	.2070	.9292	10′
32°00′	9.7242	9.7958	10.2042	9.9284	**58°00′**
10′	.7262	.7986	.2014	.9276	57°50′
20′	.7282	.8014	.1986	.9268	40′
30′	.7302	.8042	.1958	.9260	30′
40′	.7322	.8070	.1930	.9252	20′
32°50′	.7342	.8097	.1903	.9244	10′
33°00′	9.7361	9.8125	10.1875	9.9236	**57°00′**
10′	.7380	.8153	.1847	.9228	56°50′
20′	.7400	.8180	.1820	.9219	40′
30′	.7419	.8208	.1792	.9211	30′
40′	.7438	.8235	.1765	.9203	20′
33°50′	.7457	.8263	.1737	.9194	10′
34°00′	9.7476	9.8290	10.1710	9.9186	**56°00′**
10′	.7494	.8317	.1683	.9177	55°50′
20′	.7513	.8344	.1656	.9169	40′
30′	.7531	.8371	.1629	.9160	30′
40′	.7550	.8398	.1602	.9151	20′
34°50′	.7568	.8425	.1575	.9142	10′
35°00′	9.7586	9.8452	10.1548	9.9134	**55°00′**
10′	.7604	.8479	.1521	.9125	54°50′
20′	.7622	.8506	.1494	.9116	40′
30′	.7640	.8533	.1467	.9107	30′
40′	.7657	.8559	.1441	.9098	20′
35°50′	.7675	.8586	.1414	.9089	10′
36°00′	9.7692	9.8613	10.1387	9.9080	**54°00′**
	L Cos	L Cot	L Tan	L Sin	⟵

° Subtract 10 from each entry; example: Log sin 31°10′ = 9.7139 − 10.

TABLE IV° *(continued)*

⟶	L Sin	L Tan	L Cot	L Cos	
36°00′	9.7692	9.8613	10.1387	9.9080	**54°00′**
10′	.7710	.8639	.1361	.9070	53°50′
20′	.7727	.8666	.1334	.9061	40′
30′	.7744	.8692	.1308	.9052	30′
40′	.7761	.8718	.1282	.9042	20′
36°50′	.7778	.8745	.1255	.9033	10′
37°00′	9.7795	9.8771	10.1229	9.9023	**53°00′**
10′	.7811	.8797	.1203	.9014	52°50′
20′	.7828	.8824	.1176	.9004	40′
30′	.7844	.8850	.1150	.8995	30′
40′	.7861	.8876	.1124	.8985	20′
37°50′	.7877	.8902	.1098	.8975	10′
38°00′	9.7893	9.8928	10.1072	9.8965	**52°00′**
10′	.7910	.8954	.1046	.8955	51°50′
20′	.7926	.8980	.1020	.8945	40′
30′	.7941	.9006	.0994	.8935	30′
40′	.7957	.9032	.0968	.8925	20′
38°50′	.7973	.9058	.0942	.8915	10′
39°00′	9.7989	9.9084	10.0916	9.8905	**51°00′**
10′	.8004	.9110	.0890	.8895	50°50′
20′	.8020	.9135	.0865	.8884	40′
30′	.8035	.9161	.0839	.8874	30′
40′	.8050	.9187	.0813	.8864	20′
39°50′	.8066	.9212	.0788	.8853	10′
40°00′	9.8081	9.9238	10.0762	9.8843	**50°00′**
10′	.8096	.9264	.0736	.8832	49°50′
20′	.8111	.9289	.0711	.8821	40′
30′	.8125	.9315	.0685	.8810	30′
40′	.8140	.9341	.0659	.8800	20′
40°50′	.8155	.9366	.0634	.8789	10′
41°00′	9.8169	9.9392	10.0608	9.8778	**49°00′**
10′	.8184	.9417	.0583	.8767	48°50′
20′	.8198	.9443	.0557	.8756	40′
30′	.8213	.9468	.0532	.8745	30′
40′	.8227	.9494	.0506	.8733	20′
41°50′	.8241	.9519	.0481	.8722	10′
42°00′	9.8255	9.9544	10.0456	9.8711	**48°00′**
	L Cos	L Cot	L Tan	L Sin	⟵

° Subtract 10 from each entry; example: Log cos 40°20′ = 9.8821 − 10.

TABLE IV° *(continued)*

⟶	L Sin	L Tan	L Cot	L Cos	
42°00′	9.8255	9.9544	10.0456	9.8711	**48°00′**
10′	.8269	.9570	.0430	.8699	47°50′
20′	.8283	.9595	.0405	.8688	40′
30′	.8297	.9621	.0379	.8676	30′
40′	.8311	.9646	.0354	.8665	20′
42°50′	.8324	.9671	.0329	.8653	10′
43°00′	9.8338	9.9697	10.0303	9.8641	**47°00′**
10′	.8351	.9722	.0278	.8629	46°50′
20′	.8365	.9747	.0253	.8618	40′
30′	.8378	.9772	.0228	.8606	30′
40′	.8391	.9798	.0202	.8594	20′
43°50′	.8405	.9823	.0177	.8582	10′
44°00′	9.8418	9.9848	10.0152	9.8569	**46°00′**
10′	.8431	.9874	.0126	.8557	45°50′
20′	.8444	.9899	.0101	.8545	40′
30′	.8457	.9924	.0076	.8532	30′
40′	.8469	.9949	.0051	.8520	20′
44°50′	.8482	.9975	.0025	.8507	10′
45°00′	9.8495	10.0000	10.0000	9.8495	**45°00′**
	L Cos	L Cot	L Tan	L Sin	⟵

° Subtract 10 from each entry; example: Log tan 47°50′ = 10.0430 − 10.

ANSWERS

CHAPTER 1

EXERCISE SET 1.1

1. $\{1,2,3,\ldots,11\}$ **3.** $\{4,8,12,\ldots,40\}$ **5.** $\{\frac{3}{1},\frac{3}{2},\frac{3}{3},\ldots,\frac{3}{8}\}$ **7.** $B = \{$positive integers that are multiples of 3 and are less than 15$\}$ **9.** $D = \{$articles of the English language$\}$ **11.** False   **13.** True   **15.** True   **17.** Any three of $\{\ \}$, $\{a\}$, $\{b\}$, $\{a,b\}$
19. $\{\ \}$, $\{1\}$, $\{2\}$, $\{3\}$, $\{4\}$, $\{1,2\}$, $\{1,3\}$, $\{1,4\}$, $\{2,3\}$, $\{2,4\}$, $\{3,4\}$, $\{1,2,3\}$, $\{1,2,4\}$, $\{1,3,4\}$, $\{2,3,4\}$, $\{1,2,3,4\}$ **21.** No   **23.** Yes   **25.** (a) Infinite; (c) finite; (e) finite; (g) finite

EXERCISE SET 1.2

1. (a) $\{c\}$; (c) $\{a,d\}$; (e) $\{a,c,d,e,f,g,h,i,j\}$; (g) $\{c,e,j\}$; (i) $\{a,b,c,d\}$; (k) $\{c,e,j\}$; (m) $\{e,f,g,h,i,j\}$; (o) $\{b\}$; (q) $\{b\}$ **2.** (a) $\{2,4,5,6,7,8\}$; (c) $\{0,1,2,3\}$; (e) $\{0,1,2,3,4,5,6,7,8,9\}$; (g) $\{4,5,6,7,8,9\}$; (i) $\{0,1,2,3,4,9\}$; (k) $\{0,1,2,3,4,5,6,7,8,9\}$; (m) $\{6,8\}$; (o) $\{0,1,2,3,6,8\}$
3. (a) $\{q,t,u\}$, yes; (c) $\{v,w\}$, yes

EXERCISE SET 1.3

1. (a) $\{(w,y),(w,z),(x,y),(x,z)\}$; (b) $\{(y,w),(y,x),(z,w),(z,x)\}$; no
3. $\{(1,1),(1,2),(1,3),(2,1),(2,2),(2,3),(3,1),(3,2),(3,3)\}$
5. $\{(3,3),(3,5),(3,7),(5,3),(5,5),(5,7),(7,3),(7,5),(7,7)\}$ **7.** 6   **9.** 7
10. (a) 5; (c) 7; (e) 7   **13.** 9   **15.** 43   **17.** 34

CHAPTER 2

EXERCISE SET 2.1

1. Commutative law   **3.** Identity axiom   **5.** Identity axiom   **7.** Associative law
9. Identity   **11.** Identity   **13.** Inverse axiom   **15.** Inverse axiom

17. Commutative law and inverse axiom 19. Identity and distributive
21. Distributive law 23. 3 25. −11 27. 5 29. −3 31. −8 33. −8
35. 21 37. 12 39. −7 41. 3 43. −8 45. −$\frac{3}{2}$ 47. −$3x − 2y$
49. −$11a + 9b$ 51. $(3x − 5y)$ 53. $\frac{1}{3}$ 55. $\frac{3}{2}$ 57. $\frac{2}{3}$ 59. $\frac{5}{24}$
61. $1/(2a − 3b)$ 63. $7b/3a$ 65. $(x + y)/2$ 67. 1.7 69. 2 71. $\frac{59}{21}$
73. $\frac{9}{35}$ 75. −$\frac{20}{18}$ 77. −$\frac{56}{3}$

EXERCISE SET 2.2

5. $\{x \mid x > \frac{1}{2}\}$ 7. $\{x \mid x > 3\}$

EXERCISE SET 2.3

1. (a) 9; (c) 4; (e) 13; (g) 24; (i) 56; (k) $\frac{3}{4}$ 3. $\{-2,6\}$
5. (a) $\{-10,4\}$; (c) $\{x \mid -2 < x\} \cap \{x \mid x < 4\}$; (e) $\{x \mid -\frac{1}{2} < x < \frac{9}{2}\}$;
(g) $\{x \mid x < -1\} \cup \{x \mid x > 5\}$; (i) $\{x \mid -\frac{3}{5} \leq x \leq 1\}$; (k) $\{x \mid -2 \leq x \leq \frac{10}{7}\}$;
(m) $\{x \mid -\frac{7}{2} \leq x \leq 4\}$; (o) $\{x \mid -7 \leq x \leq 2\}$; (q) $\{x \mid x \leq \frac{1}{3}\} \cup \{x \mid x \geq 1\}$

EXERCISE SET 2.4

1. $6x$ 3. $3xy^2 + 2x^4 + 6x^2y^2 + x^2y$ 5. No similar terms 7. −$2x$ 9. $5x − 9$
11. $3y$ 13. $8x^4 + 6x^3 + 2x^2 − 3x + 4$ 15. −$2xy^2 − 4xy + 9z$
17. $9h^3 − 9hk + 4k^2$ 19. $2x^3 − 3x^2 + 1$ 21. −$11x^2 + 7x − 3y + 1$

EXERCISE SET 2.5

1. $5x + 4y + z$ 3. −$3x − 2y + 7z$ 5. −$2w − 2x − 2z$ 7. $7x − 3y + 2$
9. −$2x − 3y$ 11. −$2z^2 − y + 13$ 13. $n + 1$ 15. $2x + (7y + 16)$
17. $2y + (−5x − 10)$ 19. $3k + (−6 + 7m)$ 21. −$7x − (2y + 4)$
23. $8 − (2a − 3b)$ 25. $x^2 − (−2xy − y^2)$ 27. $4 − (−6b + 7c)$
29. −$a − (2b + 3c)$ 31. $a + b − [c − (d − e)]$

EXERCISE SET 2.6

1. −$84x^7y^5$ 3. −$21x^5y^2z^2$ 5. −$140x^5y^2z^3$ 7. −$8x + 28y − 12z$
9. −$18z^3 + 24wz^2 + 6z$ 11. $6a^2 + 11a + 4$ 13. $x^3 − x^2 − 8x + 12$
15. $2x^4 − 10x^3y + 6x^2y^2 + x^2y − 5xy^2 + 3y^3$ 17. $x^2 + 4y^2 + 9z^2 − 4xy + 6xz − 12yz$
19. $x^4 − 6x^3 − 20x^2 + 22x + 3$ 21. $4x^2 − 4xy + y^2$
23. $x^2 + y^2 + z^2 + 2xy + 2xz + 2yz$ 25. $x^2 + 4y^2 + 9z^2 − 4xy − 6xz + 12yz$
27. $2x^2$ 29. −$3y^3$ 31. $6x^3 − 4x$ 33. −$4x − 5y + 2z$ 35. −$(y/x) − 4x$
37. $4x$ 39. $[65(a + 2b − 3c)]/9$ 41. $x^2 − x + 5 + [−4/(x + 2)]$
43. $x^2 + 4x + 13 + 56/(x − 4)$ 45. $x + 2y$ 47. −$x + 2 + x/(x^2 + 2x + 4)$
49. $x^2 − 2y^2$ 51. $4y^2 − 16xy + 64x^2 + (−448x^3)/(2y + 8x)$ 53. $x^2 − xy + y^2$
55. $3x − 4$ 57. $3x^2 + 7xy − 9y^2$ 59. $a^5 − a^4b + a^3b^2 − a^2b^3 + ab^4 − b^5$

CHAPTER 3

EXERCISE SET 3.1

1. −2 3. 4 5. −1 7. $\frac{361}{17}$ 9. $m = Fd^2/kM$ 11. $(\tan \theta) = 3$
13. $(\sin \alpha) = \frac{1}{2}$ 15. $(\ln x) = 3$ 17. $(\cos x) = \frac{1}{2}$ 19. $(\log x) = -8$

EXERCISE SET 3.2

1. (a) Yes; (c) no; (e) yes   **2.** (a) $\{-3,-2,-1,0,1\}$; (c) $\{-3,-2,0,4,5\}$ (e) $\{p,r,v,w\}$
3. (a) $\{2,3,4,5,6\}$; (c) $\{-3,-2,-1,1,3\}$; (e) $\{-2,2,3,4\}$ **5.** $\{-2,0,3\}$ **7.** $\{x \mid x \neq 1\}$
9. $\{x \mid x \leq 2\} \cup \{x \mid x \geq 3\}$ **11.** $\{2,4,8\}$ **13.** (a) 0; (c) 4; (e) $a^4 - 3a^2$
14. (a) 13; (c) 1   **15.** (a) 4^0; (c) 4^{-1}; (e) $4^{1/2}$ **16.** (a) 9^1; (c) 9^0
17. $y = x^2$, $x \in \{1,2,3,4\}$ **19.** $y = 1/3^x$

EXERCISE SET 3.3

1. $f^{-1} = \{(-3,0),(3,1),(2,2),(5,3),(2,4)\}$; no   **3.** Yes   **5.** $y = (1 - 3x)/2x$
7. (a) $\{(-2,2),(-1,1),(0,0),(1,1),(2,2)\}$ **8.** (a) $\{(5,5),(5.1,5),(5.4,5),(5.8,5),(5.99,5)\}$ **9.** 0; 0.6

EXERCISE SET 3.4

1. (a) 5; (c) 13   **2.** (a) 13; (c) 10
5. (a) $(x - 4)^2 + (y - 5)^2 = 9$; (b) $(x + 1)^2 + (y + 3)^2 = 4$

EXERCISE SET 3.5

1. $-\sqrt{3}$ **3.** (a) 3; (c) 5   **5.** -1 **7.** $x + y - 1 = 0$ **9.** (a) 2,3; (c) $-\frac{3}{5},\frac{9}{5}$
10. (a) and (c) are parallel   **11.** $3x - 5y + 9 = 0$ **13.** $3x + 2y = 0$

EXERCISE SET 3.6

1. 8,000 cu ft   **3.** 3 in.   **5.** 154 sq ft   **7.** 80 lb/sec   **9.** 36,000 cu ft   **11.** 160 lb

EXERCISE SET 3.7

1. $g = 2s/t^2$ **3.** $C = 5(F - 32)/9$ **5.** $l = (2s - an)/n$ **7.** $n = Ir/(E - IR)$
9. $y = 3x - 4$ **11.** $g = \sqrt{ab}$

CHAPTER 4

EXERCISE SET 4.1

1. x^{7m} **3.** $4a^x$ **5.** $(x + y)^5$ **7.** (30^4) **9.** $1/x^6$ **11.** x^4y^2 **13.** 6
15. x/y^5 **17.** $(x + y)/xy$ **19.** $1/x^{n+2}y$ **21.** $1/(2x + 3y)^2$ **23.** $b^2/(b^2 - a^2)$
25. $(b^2 - a^2)/a^2b^2$ **27.** $(b^{2x} + b^{2y})/b^xb^y$ **29.** $(x^3 - 1)/(x^3 + 1)$ **31.** $3x^2$

EXERCISE SET 4.2

1. 5; 4; 5; 64   **3.** 9; 32; 2; 128; $\frac{1}{7}$; $\frac{7}{9}$ **5.** x^3 **7.** $2x\sqrt{6x}$
9. $7\sqrt{3}$ **11.** 15   **13.** $2\sqrt[3]{7}$ **15.** 3   **17.** 2   **19.** 3   **21.** $\sqrt{3}/3$
23. $\sqrt[3]{50}/5$ **25.** $\sqrt[3]{12x^2y}/2y$ **27.** $3y/x$ **29.** $(16x - 1)/x$ **31.** $x - y$
33. $2(2 + \sqrt{3})$ **35.** $3(\sqrt{5} + 1)/4$ **37.** $a(a + \sqrt{a^2 - 16})/16$
39. $-(\sqrt{a} - \sqrt{a + 1})^2$ **41.** $1 - x$ **43.** $a\sqrt{a - x^2}/(a - x^2)^2$
45. $(2a^2 + b^2)\sqrt{a^2 + b^2}$ **47.** $[(x^4 - 51x^2 + 625)\sqrt{x^2 - 25}]/(x^2 - 25)^2$

EXERCISE SET 4.3

1. 5   **3.** No solution   **5.** $-2\frac{1}{16}$ **7.** $2\frac{1}{16}$ **9.** 6   **11.** 0   **13.** $2\frac{3}{18}$
15. 5   **17.** No solution   **19.** 20   **21.** 225   **23.** 140   **25.** 8   **27.** $-8,1$

EXERCISE SET 4.4

1. (a) 1.7; (c) 1.4; (e) 3; (g) 4.7; (i) 0.2   **2.** (a) 0.6; (c) 1.54; (e) -1.2 **4.** (a) 9
7. $50(\frac{1}{2})^t$ **9.** $0.5\sqrt{2.7}/2.7 = 0.3$ amp (approx.)   **13.** 1   **15.** -4 **17.** -3
19. 3   **21.** -2 **23.** 4   **25.** $\frac{9}{2}$ **27.** -1

CHAPTER 5

EXERCISE SET 5.1

1. (a) $(1,0)$; (c) $(1,0)$; (e) $(0,1)$; (g) $(0,-1)$ **2.** (a) $(\sqrt{2}/2, -\sqrt{2}/2)$; (c) $(-\sqrt{2}/2, -\sqrt{2}/2)$;
(e) $(-\sqrt{2}/2, \sqrt{2}/2)$ **3.** (a) $(\frac{1}{2}, -\sqrt{3}/2)$; (c) $(-\frac{1}{2}, -\sqrt{3}/2)$; (e) $(\frac{1}{2}, -\sqrt{3}/2)$
4. (a) $(-\sqrt{3}/2, \frac{1}{2})$; (c) $(-\sqrt{3}/2, -\frac{1}{2})$; (e) $(\sqrt{3}/2, -\frac{1}{2})$ **5.** $\pi/6$ **6.** (a) $7\pi/4$

EXERCISE SET 5.2

1. (a) $\sin(\pi/2) = 1$, $\cos(\pi/2) = 0$, $\tan(\pi/2)$ is undefined, $\cot(\pi/2) = 0$, $\sec(\pi/2)$ is undefined,
$\csc(\pi/2) = 1$; (c) $\sin(\pi/6) = \frac{1}{2}$, $\cos(\pi/6) = \sqrt{3}/2$, $\tan(\pi/6) = \sqrt{3}/3$, $\cot(\pi/6) = \sqrt{3}$,
$\sec(\pi/6) = 2\sqrt{3}/3$, $\csc(\pi/6) = 2$; (e) $\sin(5\pi/6) = \frac{1}{2}$, $\cos(5\pi/6) = -\sqrt{3}/2$,
$\tan(5\pi/6) = -(\sqrt{3}/3)$, $\cot(5\pi/6) = -\sqrt{3}$, $\sec(5\pi/6) = -(2\sqrt{3}/3)$, $\csc(5\pi/6) = 2$
2. (a) $\sqrt{2}/2$; (c) $\sqrt{2}$; (e) -2; (g) -1; (i) $-\sqrt{3}/2$; (k) 1   **3.** (a) $-\sqrt{3}/2$; (c) $-(\sqrt{3}/3)$
4. (a) $-\frac{1}{2}$; (c) -2 **5.** (a) -1; (c) $\sqrt{2}/2$ **6.** (a) $-\frac{1}{2}$; (c) $2\sqrt{3}/3$
8. (a) III, IV; (c) II, III; (e) I, IV   **9.** $\cos t = \frac{5}{13}$, $\tan t = \frac{12}{5}$, $\cot t = \frac{5}{12}$, $\sec t = \frac{13}{5}$, $\csc t = \frac{13}{12}$
10. (a) $\sin t = \frac{4}{5}$, $\cos t = \frac{3}{5}$, $\tan t = \frac{4}{3}$, $\cot t = \frac{3}{4}$, $\sec t = \frac{5}{3}$, $\csc t = \frac{5}{4}$;
(c) $\sin t = \frac{7}{25}$, $\cos t = \frac{24}{25}$, $\tan t = \frac{7}{24}$, etc.; (e) $\sin t = -\frac{15}{17}$, $\cos t = \frac{8}{17}$, $\tan t = -\frac{15}{8}$, etc.;
(g) $\sin t = -(8\sqrt{164}/164)$, $\cos t = -(10\sqrt{164}/164)$, $\tan t = \frac{8}{10}$, etc.;
(i) $\sin t = -\frac{8}{17}$, $\cos t = \frac{15}{17}$, $\tan t = -\frac{8}{15}$, etc.
11. (a) $\cos t = \frac{5}{13}$, $\tan t = \frac{12}{5}$; (c) $\sin t = -\sqrt{3}/2$, $\cos t = -\frac{1}{2}$; (e) $\cos t = -\sqrt{527}/24$,
$\tan t = -(7\sqrt{527}/527)$; (g) $\cos t = -\sqrt{11}/6$, $\tan t = -(5\sqrt{11}/11)$.
15. (a) True; (c) true; (e) true   **17.** II, IV

EXERCISE SET 5.3

1. (a) 0.8415; (c) -0.3203; (e) 0.8521; (g) 0.6967   **2.** (a) 0.7956; (c) 0; (e) 1; (g) -0.5062;
(i) -0.9001; (k) 0.4954; (m) 5.883; (o) -1.288 **3.** $5\pi/6$ **5.** 5.74   **7.** 0.80

EXERCISE SET 5.4

1. (a) $1 - \sin^2 t$; (c) $\sin t/\pm\sqrt{1 - \sin^2 t}$; (e) $1/\pm\sqrt{1 - \sin^2 t}$; (g) $1/\sin t$; (i) $\sin^2 t/(1 - \sin^2 t)$
2. (a) $\pm\sqrt{1 - \cos^2 \phi}$; (c) $\cos \phi/\pm\sqrt{1 - \cos^2 \phi}$; (e) $1/\pm\sqrt{1 - \cos^2 \phi}$
3. (a) $1/\tan \beta$; (c) $\tan \beta/\pm\sqrt{1 + \tan^2 \beta}$; (e) $1/\pm\sqrt{1 + \tan^2 \beta}$

EXERCISE SET 5.5

1. $45°$ **3.** $30°$ **5.** $72°$ **7.** $-210°$ **9.** $-12°$ **11.** $5°$ **13.** $210°$
15. $-130°$ **17.** $80°$ **19.** $1080°$ **21.** $6480°$ **23.** $-1203.195°$ **25.** $\pi/6$ rad
27. $2\pi/3$ rad   **29.** $-3\pi/2$ rad   **31.** $-11\pi/6$ rad   **33.** $2\pi/5$ rad   **35.** $3\pi/5$ rad
37. $-\pi/18$ rad   **39.** $5\pi/3$ rad   **41.** 3.839 rad   **43.** 6.8055 rad
45. $48°24' = 48.4° = 0.84458$ rad   **47.** $27(0.01745) + 20(0.00029) + 40(0.00000485)$

EXERCISE SET 5.6

1. (a) $25\pi/2$ in.; (c) 5π in.; (e) 100 in.; (g) 2,500 in.   **2.** (a) 1; (c) ½; (e) 2; (g) $\pi/2$
3. (a) 4; (c) 60; (e) $\frac{4}{7}$; (g) $12/7\pi$ **4.** (a) 100; (c) 250; (e) $25\pi/2$; (g) 25π **5.** (a) 18; (c) ½
6. (a) 72; (c) 32   **7.** 3 rad   **9.** 28.8 rad   **11.** $5,000\pi/3$ mi   **13.** $1,550,000\pi/3$ mi

EXERCISE SET 5.7

1. (a) Reasonable; (c) not reasonable; (e) not reasonable; (g) reasonable; (i) not reasonable
3. (a) 0.8; (c) 1.2; (e) -0.9; (g) -0.8; (i) 1.2; (k) 0.9
5. $\sin\theta = \frac{12}{15}$, $\cos\theta = \frac{9}{15}$, $\tan\theta = \frac{12}{9}$, etc.   **7.** $\sin\theta = 8/\sqrt{128}$, $\cos\theta = 8/\sqrt{128}$,
$\tan\theta = 1$, etc.   **9.** $\sin\theta = 4/\sqrt{52}$, $\cos\theta = 6/\sqrt{52}$, $\tan\theta = \frac{4}{6}$, etc.
11. $\sin\theta = \frac{5}{7}$, $\cos\theta = 2\sqrt{6}/7$, $\tan\theta = 5/2\sqrt{6}$ **13.** Quad. I   **15.** Quad. I
17. Quad. I   **19.** $y = 4\sqrt{11}/5$, $r = \frac{24}{5}$

EXERCISE SET 5.8

1. 0.4695   **3.** 0.3839   **5.** 1.046   **7.** 0.9387   **9.** 0.8899   **11.** 0.3314
13. 0.9833   **15.** 0.4074   **17.** 0.4695   **19.** 1.010   **21.** 0.0175   **23.** 0.6041
25. 0.9325   **27.** 0.0145   **29.** 0.0465   **31.** 0.3185   **33.** 2.323   **35.** $37°40'$
37. $36°20'$ **39.** $33°30'$ **41.** $29°30'$ **43.** $4°30'$ **45.** $26°50'$ **47.** $44°40'$
49. $48°20'$ **51.** $55°20'$ **53.** $8°40'$ **55.** $46°30'$ **57.** $37°30'$ **59.** $63°20'$

EXERCISE SET 5.9

1. 0.8599   **3.** 0.6340   **5.** 0.0068   **7.** 1.731   **9.** 1.801   **11.** 0.1583   **13.** 0.8983
15. 30.868   **17.** 0.9008   **19.** 1.074   **21.** 0.9840   **23.** 0.9450   **25.** 1.084
27. 0.2971   **29.** 10.20   **31.** 2.205   **33.** 1.055   **35.** 1.261   **37.** 0.6209
39. 0.3415   **41.** 0.217   **43.** 1.409   **45.** 1.328   **47.** 1.307   **49.** 1.055
51. 0.254   **53.** $17°27'$ **55.** $43°31'$ **57.** $71°24'$ **59.** $49°54'$ **61.** $72°49'$
63. $70°04'$

EXERCISE SET 5.10

1. 0.5000   **3.** -0.1405 **5.** 1.195   **7.** -0.4643 **9.** -0.1132 **11.** 0.5095
13. -0.8158 **15.** 4.915   **17.** -1.039 **19.** -0.9962 **21.** -0.7002
23. -0.0291 **25.** -0.9063 **27.** 0.8391   **29.** $+0.6494$ **31.** -0.4600
33. -1.114 **35.** 0.7513   **37.** $22°18'$, $202°18'$ **39.** $32°21'$, $147°39'$
41. $233°01'$, $306°59'$ **43.** $127°$, $307°$ **45.** $242°36'$, $297°24'$ **47.** $262°20'$, $277°40'$
49. $27°45'$, $332°15'$ **51.** $154°24'$, $334°24'$ **53.** $131°25'$, $228°35'$

EXERCISE SET 5.12

17. (a) 0.9; (c) -1; (e) -0.7 **18.** (a) False; (c) true

CHAPTER 6

EXERCISE SET 6.1

1. $5x(x-2)$ **3.** $9xy(6xy-1)$ **5.** $(2x-y)(3a+4)$ **7.** $(2a-b)(2x+3y)$
9. $(x+y)^3(a^2-3b^2)$ **11.** $(x^2+y^2)(xy+yz+zw)$ **13.** $(x+2y-1)(m^2+2m+3)$
15. $\sin\beta(2\sin\beta-1)(\sin\beta-1)$ **17.** $e^x(4x+1)$ **19.** $\sqrt{3}(2xy+2y+1)$

21. $(2x - \sqrt{7})(7x - 3y)$ **23.** $(x + y)(x + p)$ **25.** $(2x + y)(4x - 3)$

27. $(3 - 2x)(3 + 4x^2)$ **29.** $(2x + 3a)(4y + 5b)$ **31.** $(m + 6)(m^3 - 7)$

33. $(x - y + z)(a - b)$ **35.** $(x + y - z)(a - b + c)$

EXERCISE SET 6.2

1. $(9x + 8)^2$ **3.** $[3x - (y + z)]^2$ **5.** $[5(a - b) + 4c]^2$ **7.** $(7a - 6b)(7a + 6b)$

9. $(5 - x^3)(5 + x^3)$ **11.** $(x/10 - \frac{1}{7})(x/10 + \frac{1}{7})$ **13.** $(a + b - 4)(a + b + 4)$

15. $(x - 2y + 1)(x + 2y - 1)$ **17.** $(3x - 2y - 2)(3x - 2y + 2)$

19. $(x - y - 4)(x - y + 4)$ **21.** $(x + y + z)(x - y - z)$

23. $(4a - b - c - 5d)(4a - b + c + 5d)$ **25.** $(x - 15)(x + 5)$ **27.** $(y^2 - 13)(y^2 - 8)$

29. $(7 - x)(11 + x)$ **31.** $(a - 13b)(a + 7b)$ **33.** $(1 - 2ab)(1 + 7ab)$

35. $(x - 3n)(x - 2m)$ **37.** $(2x + 5)(2x + 9)$ **39.** $(5x - 3a)(5x - 2a)$

41. $(6x - 5)(6x + 7)$ **43.** $[(x - y) - 4][7(x - y) - 2]$

45. $(x^2 - 3y^3z)(x^4 + 3x^2y^3z + 9y^6z^2)$ **47.** $(2a - b^3)(4a^2 + 2ab^3 + b^6)$

49. $(xy + 5z)(x^2y^2 - 5xyz + 25z^2)$ **51.** $[x - (x + y)][x^2 + x(x + y) + (x + y)^2]$

53. $(x - y)(7x^2 + 13xy + 7y^2)$ **55.** $(a^2 - a + 3)(a^2 + a + 3)$

57. $(2x^2 - 5x - 2)(2x^2 + 5x - 2)$ **59.** $(4x^2 - 7x - 4)(4x^2 + 7x - 4)$

61. $(4 - 3x - x^2)(4 + 3x - x^2)$ **63.** $(4a^2 - 8ab - 5b^2)(4a^2 + 8ab - 5b^2)$

65. $(\sin\theta - 3)(2\sin\theta + 1)$ **67.** $(3\cos\alpha - 1)(5\cos\alpha + 2)$

EXERCISE SET 6.3

1. 0,5   **3.** $-5,2,5$ **5.** $-17,-6$ **7.** $-5,22$ **9.** 0,0,2,16   **11.** $-\frac{4}{5},\frac{3}{2}$

13. $-\frac{1}{4},\frac{2}{5}$ **15.** $\frac{1}{2},7$ **17.** -2 is the only real root.   **19.** $\frac{2}{3}$ is the only real root.

21. 23,24   **23.** 13,15   **25.** 25 ft   **27.** $-4,4$ **29.** $0,\pi,3\pi/2$ **31.** $\pi/4, 3\pi/4, 5\pi/4, 7\pi/4$

33. $\pi/6$ **35.** 1.37,π,4.91

EXERCISE SET 6.4

1. $(x - 1)/(x + 1)$ **3.** $-(x^2 + 1)/(x + 3)$ **5.** $(4y - 1)/(2y - 1)$ **7.** $3x + 2y + 1$

9. $4a + 8b + c$ **11.** $(4x^2 + 10x + 25)/[x(x + 3)]$ **13.** $2/(x + 2)$ **15.** 1

17. $(x + 4)/(x^2 + 4)$ **19.** $x^2 + 2xy + 3y^2$

EXERCISE SET 6.5

1. $\dfrac{3x}{(x - 3y)(x + 3y)}$ **3.** $\dfrac{3}{4x - 6}$ **5.** $\dfrac{x - 1}{1 - 2x}$ **7.** $\dfrac{1}{x^2 + y^2}$

9. $\dfrac{34 - 33x}{(2 - 3x)(1 + 2x)(1 - x)}$ **11.** $\dfrac{b^2 + 6b + 7}{(2 + b)(3 + b)(4 + b)}$ **13.** $\dfrac{5(2x^2 + 3xy + 3y^2)}{4(2x - 3y)(2x + 3y)}$

15. $\dfrac{x - xy + y^2}{x^3 - y^3}$ **17.** $\dfrac{x^2 - 8x + 26}{4 - x^2}$ **19.** $\dfrac{6x^2 - 6x - 19}{2x - 3}$

21. $\dfrac{6 - 10n^2 + 40n^4 - 384n^6}{(4n^2 - 1)(64n^6 - 1)}$ **23.** $\dfrac{3 - x - 2x^2}{x + 3}$ **25.** $\dfrac{7y^2 - 5y - 5}{y - 1}$

27. $\dfrac{2(\sec\theta + 3)}{2\sec^2\theta + \sec\theta - 3}$

EXERCISE SET 6.6

1. $(x - 8)/(x + 2)$ **3.** $(x - 3)/(x + 7)$ **5.** $\frac{3}{2}$ **7.** 1   **9.** $a/(a - 1)$

11. $(x + y + 1)/(y - z)$ **13.** $3x^2/4y$ **15.** 1   **17.** $x^2/3(x^2 + y^2)$ **19.** x

EXERCISE SET   6.7

1. 1   **3.** 5   **5.** 1⅗   **7.** No solution   **9.** −3   **11.** $\cos t = 3$ is impossible.
13. 0,⅗   **15.** 4⅘ hr   **17.** 1²⁄₁₉   **19.** $88.89

EXERCISE SET   6.8

1. 4   **3.** No solution   **5.** No solution   **7.** No solution   **9.** No solution
11. $x = (y + 12)/10$ **13.** $y = (x + 10)/6$ **15.** $\pi/6, 5\pi/6$

EXERCISE SET   6.9

1. $x − 1$ **3.** $(b + a)/b$ **5.** $(5y − x)/2$ **7.** $(a^2 + ab + b)/[(a + b)^2 − a]$
9. $a/(a − b)$ **11.** $−3/2y$ **13.** ¾

EXERCISE SET   6.10

1. 64,65   **3.** ⁵⁄₇   **5.** ⁷⁄₁₈   **7.** 5 mi/hr   **9.** 84 min   **11.** 43¹⁷⁄₃₁ min
13. 27 by 9 ft   **15.** 687

CHAPTER 7

EXERCISE SET   7.1

1. 2,2   **3.** 2,2   **5.** 2,4   **7.** ±2,0   **9.** $3 + 2i$ **11.** $3 − 8i$ **13.** $10 + 17i$
15. $18 + 5i$ **17.** $−6 + i$ **19.** $\tfrac{3}{2} + i(\sqrt{3}/2)$ **21.** i **23.** $4 − i(3 + \sqrt{3})$
25. $11 + 16i$ **27.** $−27 − 8i$ **29.** $13 + 11i$ **31.** $31 + 29i$
33. $\sqrt{15} + \sqrt{10} + i(\sqrt{6} − 5)$ **35.** $2i$ **37.** $−4$ **39.** $−3 − 2i$ **41.** $−i$
43. $\tfrac{1}{2} + \tfrac{1}{2} i$ **45.** $−\tfrac{3}{2} − \tfrac{3}{2} i$ **47.** $\tfrac{3}{2} i$ **49.** $^{27}\!/_{41} − \tfrac{3}{41} i$

EXERCISE SET   7.2

1. $\sqrt{5}(\cos 26°30' + i \sin 26°30')$ **3.** $2\sqrt{5}(\cos 206°30' + i \sin 206°30')$
5. $3(\cos 180° + i \sin 180°)$ **7.** $\sqrt{10}(\cos 108°30' + i \sin 108°30')$
9. $\cos 300° + i \sin 300°$ **11.** $\sqrt{3}/2(\cos 54°40' + i \sin 54°40')$
13. $13(\cos 67°20' + i \sin 67°20')$ **15.** $\sqrt{65}/5(\cos 300° + i \sin 300°)$
17. $\sqrt{13}/13(\cos 56°20' + i \sin 56°20')$ **19.** $\sqrt{170}/5(\cos 355°40' + i \sin 355°40')$
21. $3(\cos 90° + i \sin 90°)$ **23.** $−3$ **25.** $−\sqrt{3} + i$ **27.** $−3 − 3i\sqrt{3}$
29. $(5\sqrt{2}/2) − (5\sqrt{2}/2)i$ **31.** $−3 + 3i\sqrt{3}$

EXERCISE SET   7.3

1. $15i$ **3.** $5.909 − 1.042i$ **5.** $27i$ **7.** $2\sqrt{2} + 2i\sqrt{2}$ **9.** $5i$ **11.** $−\tfrac{1}{12} + (\sqrt{3}/12) i$
13. $−2i$ **15.** $−i$ **17.** $−0.1830 + 0.6829i$ **19.** $−\tfrac{1}{2} + \tfrac{1}{2} i$

EXERCISE SET   7.4

1. $(\sqrt{2}/2) + (\sqrt{2}/2)i$ **3.** $(\sqrt{2}/2) − (\sqrt{2}/2)i$ **5.** $−4\sqrt{2} − 4i\sqrt{2}$ **7.** $\tfrac{1}{2} + (\sqrt{3}/2)i$
9. $−4$ **11.** $\sqrt{3}/2 − \tfrac{1}{2} i$ **13.** $−8$ **15.** $−1024i$
21. $2.187 + 0.4649i, −0.4649 + 2.187i, −2.187 − 0.4649i, 0.4649 − 2.187i$
23. $2[\cos(36° + k \cdot 72°) + i \sin(36° + k \cdot 72°)]$, where $k = 0, 1, 2, 3, 4$

25. $-2\sqrt{2} + 2i\sqrt{2}, 2\sqrt{2} - 2i\sqrt{2}$

27. $\cos(6° + k \cdot 120°) + i\sin(6° + k \cdot 120°)$, where $k = 0, 1, 2$

29. $\cos(75° + k \cdot 90°) + i\sin(75° + k \cdot 90°)$, $k = 0, 1, 2, 3$

31. $1.292 + 0.2012i, -0.8203 + 1.018i, 0.4718 - 1.220i$ **33.** $-3, i$

CHAPTER 8

EXERCISE SET   8.1

1. 0,3   **3.** $-9,3$ **5.** $\pi/4 + 2k\pi, 5\pi/4 + 2k\pi, k \in I$

7. $\pi/6 + 2k\pi, 5\pi/6 + 2k\pi, k \in I$ **9.** $-\frac{5}{3},2$ **11.** $k\pi, 153°30' + k\pi, k \in I$

13. 0.0316,10,000   **15.** $-2,10$ **17.** $-\frac{3}{2},3$ **19.** $\tan t = 1 \pm \sqrt{2}$ **21.** $\frac{1}{2},3$

23. $1 \pm \sqrt{3}/3$ **25.** $2i,3i$ **27.** $1 - i, -2 + i$ **29.** -2 **31.** 5   **33.** 0,5

35. 24,25   **37.** 13   **39.** 39 mi/hr   **41.** $-\frac{5}{2},\frac{2}{5}$ **43.** 128 ft

EXERCISE SET   8.2

1. Min. at $x = 2$ **3.** Max. at $x = -1$ **5.** Min. -13 at $x = 4$ **7.** Max. 5 at $x = 2$

9. 20,20   **11.** 40 by 80 rd   **13.** 3 sec; 144 ft

EXERCISE SET   8.3

1. (a) $-1,-6$; (c) $\frac{35}{3},\frac{22}{3}$; (e) $2i,-1-2i$ **2.** (a) $x^2 - 2x - 3 = 0$; (c) $x^2 + 12x + 35 = 0$;

(e) $x^2 - 4x + 13 = 0$; (g) $x^2 + x - 1 = 0$; (i) $x^2 - 2ax + a^2 - b^2 = 0$

4. (a) $x^2 - 8x + 15 = 0$; (c) $x^2 - 2x - 2 = 0$ **5.** $x^2 - 6x + 11 = 0$ **7.** $-4,2$

9. $5,-3$ **11.** -18 **13.** 6   **15.** $4x^2 + 8x - 5 = 0$ **17.** $|k| > \frac{3}{2}$

19. $(kx - 1)(x + k - 1)$

EXERCISE SET   8.4

1. $\pm 2, \pm 3$ **3.** $\pm \frac{1}{3}, \pm \frac{1}{2}$ **5.** 25,16   **7.** $\frac{1}{3}(1 \pm 2i)$ **9.** $\frac{1}{2},-2$ **11.** $\pm 3, \pm \sqrt{14}$

13. $0,4,-6$ **15.** No solution   **17.** 150°   **19.** 135°   **21.** 360°   **23.** $35°16'$

EXERCISE SET   8.5

1. $x > -6$ **3.** $x < 5$ **5.** $x < -2$ **7.** $-\frac{1}{2} < x < \frac{1}{2}$ **9.** $x \le -\frac{1}{4}$

11. $x < -4$, or $x > 4$ **13.** $m - n \le x \le m + n$

15. $x < \dfrac{-3 - \sqrt{41}}{4}$, or $x > \dfrac{-3 + \sqrt{41}}{4}$

17. No real solution   **19.** $x < 2$, or $x > 5$ **21.** $10 \le x \le 60$

23. $x < -3$, or $-2 < x < -1$ **25.** $x < -1$, or $0 < x < 2$

27. $0 < t < \pi/2$, or $3\pi/2 < t < 2\pi$

29. $0 \le t < \pi/3$, or $\pi/2 < t < 3\pi/2$, or $5\pi/3 < t < 2\pi$

31. $0 \le t < \pi/3$, or $5\pi/3 < t < 2\pi$

EXERCISE SET   8.8

1. $x - 1 + [-2/(x - 4)]$ **3.** $x^2 + 3x + 13 + 32/(x - 3)$ **5.** $x^2 - 2 + 1/(x + 3)$

7. $x^2 + 7x + 10 + 25/(x - 5)$ **9.** $2x^2 - 2x + 1/(2x - 1)$

11. $2x^3 + (6 + 2\sqrt{2})x^2 + (10 + 6\sqrt{2})x + 10\sqrt{2}$

EXERCISE SET   8.9

1. -2 **3.** 0 **5.** 64 **7.** $(x-3)(x^2+2x-5)$ **9.** $x-3$ is not a factor.
11. $x+3$ is not a factor.   **13.** $x+1$ is not a factor.   **15.** $(x-5)(x+2)(x+5)$
17. $x+1$ is not a factor.   **19.** $(x+\frac{3}{2})(x^3+x^2+4)(2)$ **29.** $k=1$

EXERCISE SET   8.10

5. $-1<x<9$ **7.** $-3<x<3$ **9.** $f(x)=x^3-10x^2+32x-32$
11. $f(x)=x^5-7x^4+19x^3-25x^2+16x-4$ **13.** $x^3-6x^2+11x-6=0$
15. $x^2+2=0$ **17.** $x^4+2x^2-8=0$ **19.** $-3,-3,-3,1,1$ **23.** $-3,-2,2$

EXERCISE SET   8.11

1. $3,-3\pm 2i$ **3.** $-\frac{1}{4},\frac{3}{2}\pm\frac{1}{2}\sqrt{5}$ **7.** $-2,4,2\pm\sqrt{5}$ **9.** $\frac{1}{3},\frac{3}{2},\frac{5}{2}$
11. $-\frac{3}{4},\frac{1}{5},1,\pm\sqrt{3}$ **13.** $0,\pi$ **15.** $\pi/6,5\pi/6$

EXERCISE SET   8.12

1. $-2<x<-1,-1<x<0,2<x<3$ **3.** $5<x<6$ **5.** 1.15 **7.** -1.23
9. -3.42 **11.** $-1.93,-0.73$ **13.** 1.82 **15.** 1.19

EXERCISE SET   8.13

1. $-2i$ **3.** $1-3i$ **5.** Insufficient information   **7.** $x^3-4x^2+6x-4=0$
9. $x^4-5x^3+10x^2-10x+4=0$ **11.** $\pm i\sqrt{2}$ **13.** 4 **15.** $(1\pm\sqrt{5})/2$
17. $-\frac{3}{2},-i,i$ **21.** $-\frac{1}{2}\sqrt{13}$ **23.** $(2-\sqrt{3})/2$ **25.** $-3-5\sqrt{2}$
27. $f(x)=2x^3-5x^2+10x-4$ **29.** $f(x)=x^4-2x^3-10x^2-2x-11$
31. $\pi/3,\pi,5\pi/3$ **33.** $0,1.231,5.052$

CHAPTER 9

EXERCISE SET   9.1

1. (a) $\log_{10}100=2$; (c) $\log_7(\frac{1}{7})=-1$; (e) $\log_8(\frac{1}{4})=-\frac{2}{3}$; (g) $\log_4 x=y$; (i) $\log_{10}x=\log_{10}y$
2. (a) $5^3=125$; (c) $8^{2/3}=4$; (e) $4^{3/2}=8$; (g) $10^{-3}=0.001$ **3.** (a) 3; (c) 1; (e) 4
4. (a) 4; (c) 4; (e) $\frac{1}{32}$; (g) no solution   **5.** (a) 0.9542; (c) 1.2552; (e) 0.1761; (g) 0.0791;
(i) 2; (k) -0.0485; (m) 0.8495; (o) 3.2814; (q) 0.2845 **7.** $\log_b(4\pi r^3/3)$ **9.** $\log_b(3,200)^{1/3}$
11. $\log_b[9^{1/3}(7^{1/2})]/16$

EXERCISE SET   9.2

1. (a) 1; (c) 2; (e) 4; (g) 4; (i) 9 -10 (k) 9 -10; (m) 9 -10; (o) 0; (q) 1; (s) 0; (u) 0
2. (a) 1.7604; (c) $9.7604-10$; (e) 6.7604 **3.** (a) 42.65; (c) 0.04265; (e) 0.4265;
(g) 0.00004265; (i) 0.000004265 **4.** (a) 2.4362; (c) $9.9782-10$; (e) 1.9420; (g) $7.7316-10$;
(i) 2.0043; (k) 0.0000; (m) 4.0112; (o) 5.0163; (q) 0.4816 **5.** (a) 4; (c) 527; (e) 0.358;
(g) 0.0772; (i) 0.627; (k) 0.0776; (m) $1,040$; (o) 1.05 **6.** (a) 0.6990; (c) 0.2014; (e) 2.51;
(g) 540; (i) 25.1; (k) 0.9031; (m) 0.0862; (o) -14.6181

EXERCISE SET 9.3

1. (a) 3.0990; (c) 1.7606; (e) 8.6836 − 10   2. (a) 3.146; (c) 0.2475; (e) 0.9474
3. (a) 1.403(10¹³); (c) 8.424(10⁸); (e) 2.79   4. (a) 26.83; (c) 0.3077; (e) 2.654; (g) 1.478;
(i) 0.5224; (k) 2.658; (m) 81,360   5. (a) 166.8; (c) 0.5028; (e) 0.6702; (g) 19.56
7. 71,900 cu ft   9. 1,241 sq ft   11. $3,065

EXERCISE SET 9.4

1. 3.114   3. 3.635   5. 0.870   7. 0.903   9. −2.405   11. 0   13. 0.8330
15. −2.578   17. 0.799   19. 1.404   21. 1.431   23. 6.69   25. 3.02

EXERCISE SET 9.5

1. 6   3. 0   5. −0.2386   7. 2.978 9. 2.874   11. 7 or −13 13. 8⁷⁄₂
15. ²⁰⁴⁄₁₉₇   17. 96   19. 3   21. No solution   23. $x < -0.585$ 25. 10, 0.1
27. $x = 10y^2$ 29. $x = y^3 e^{2y}$ 31. 0   33. 5   35. 3, 2.676

CHAPTER 10

EXERCISE SET 10.1

1. $\beta = 75°$, $a = 7.4$, $c = 8.9$ 3. $\gamma = 42°$, $b = 32$, $c = 70$
5. $\beta = 36°30'$, $a = 216$, $b = 130$ 7. $\gamma = 20°30'$, $a = 0.988$, $b = 1.60$
9. $\beta = 67°$, $\gamma = 53°$, $c = 69$ 11. No solution   13. 73 ft (approx.)
15. 460 ft (approx.)

EXERCISE SET 10.2

1. $b = 33$, $c = 41$ 3. $a = 5.72$, $b = 7.41$ 5. $\alpha = 27°$, $c = 70$
7. $\alpha = 42°50'$, $a = 329$ 9. $\alpha = 42°40'$, $b = 46.8$ 11. $b = 1.79$, $c = 4.67$
13. $a = 6.449$, $c = 9.707$ 15. 67°   17. 38°10′   19. 9.28 in.   21. 4,230 ft
23. 3,147 mi   25. N40°W   27. 76.2 ft

EXERCISE SET 10.3

1. $v_x = 31$ lb, $v_y = 40$ lb   3. $v_x = 80$ lb   5. Slightly greater than 34.2 lb
7. 38° south of east   9. 520 mi/hr; S18°E   11. N2°W   13. 6.8(10⁷)
14. (a) [3,10], $\sqrt{109}$; (c) [1,0], 1; (e) [8,−2], $\sqrt{68}$ 17. (a) 5[⅘,⅗]; (c) 5[−⅗,⅘]
18. (a) ($\sqrt{10}/10$, $3\sqrt{10}/10$); (c) ($-3\sqrt{13}/13$, $2\sqrt{13}/13$)

EXERCISE SET 10.4

1. $c = 143$, $\alpha = 30°50'$ 3. $a = 112$, $\gamma = 35°10'$ 5. $b = 363$, $\alpha = 68°20'$
7. No solution   9. $\alpha = 51°$, $\beta = 42°$ 11. $\alpha = 36°40'$, $\beta = 88°10'$ 13. 328 ft
15. 340 lb; the resultant makes an angle of 46°10′ with the 370-lb force.   17. 10.9 in.

EXERCISE SET 10.5

1. 25 sq units (approx.)   3. 1,720 sq units (approx.)   5. 18,800 units (approx.)
7. 27.7 sq units (approx.)   9. 570 sq units (approx.)   17. 37 mi/hr, N35°E

EXERCISE SET   10.6

1. 31   **3.** 17   **5.** $13/\sqrt{338}$ **9.** -14 **11.** $30°$

CHAPTER 11

EXERCISE SET   11.1

1. (a) $(5\sqrt{91} + 3\sqrt{75})/100$; (c) $(15 + 5\sqrt{273})/100$; (e) $\frac{3}{10}$; (g) $\sqrt{91}/10$ **3.** $(\sqrt{6} + \sqrt{2})/4$
5. (a) $\cos 0$; (c) $-\cot \pi/12$; (e) $\cot 0.65$ **6.** (a) $\sin 15°$; (c) $\cot 12°$; (e) $-\tan 38°$
7. (a) 0.9801; (c) 0.9553 **8.** (a) $(\sqrt{6} + \sqrt{2})/4$; (c) $-(\sqrt{6} + \sqrt{2})/4$; (e) $(\sqrt{6} - \sqrt{2})/4$
9. (a) 0; (c) 1 **11.** $\frac{84}{85}$ **13.** 1 **15.** $\frac{56}{65}$ **17.** $\cos \pi/6$ **19.** $\sin \frac{3}{2}$
21. $\cos 5\pi/12$ **23.** $\cos u$ **25.** $\pi/30, \pi/6, 13\pi/30$

EXERCISE SET   11.2

1. $\frac{1}{2}(\sqrt{2 + \sqrt{3}})$ **3.** $\frac{1}{2}(\sqrt{2 - \sqrt{2}})$ **5.** $\frac{1}{2}(\sqrt{2 + \sqrt{3}})$ **7.** $\frac{1}{2}(\sqrt{2 - \sqrt{3}})$
9. $\sqrt{2} - 1$ **11.** $-\sqrt{3}/2$ **13.** $2/\sqrt{2 - \sqrt{2}}$ **15.** $\frac{1}{2}(\sqrt{2 - \sqrt{2 + \sqrt{3}}})$
16. (a) $-\frac{120}{169}$; (c) $\frac{120}{119}$; (e) $2\sqrt{13}/13$; (g) $-\frac{169}{119}$; (i) $\frac{119}{120}$ **17.** (a) $720/1,681$;
(c) $1,519/1,681$; (e) $720/1,519$ **18.** (a) $\frac{3}{4}$; (c) $-\frac{3}{5}$; (e) $-\frac{4}{5}$ **19.** (a) $2,035/2,197$;
(c) $828/2,197$; (e) $2,035/828$ **33.** $0, \pi/3$ **35.** $\pi/6, \pi/2$ **37.** $0, \pi$ **39.** π

EXERCISE SET   11.3

1. $\frac{1}{2}(\sin 8t - \sin 2t)$ **3.** $\sin 6t + \sin 2t$ **5.** $\sin 9 - \sin 3$ **7.** $\frac{1}{2}(\cos 4 + \cos 2)$
9. $(\sqrt{3} - 1)/4$ **11.** $\sqrt{2}/4$ **13.** $\sqrt{6}/2$ **15.** $\sqrt{2}/2$ **23.** $0, \pi/4$ **25.** $0, \pi/2$
27. $0, \pi/3, \pi/2, \pi, 3\pi/2, 5\pi/3$

CHAPTER 12

EXERCISE SET   12.1

1. $f^{-1}(x) = (3x + 1)/5$, domain $-2 \leq x \leq 8$; range $-1 \leq y \leq 5$
3. $f^{-1}(x) = -\sqrt{25 - x}$, domain $0 \leq x \leq 25$; range $-5 \leq y \leq 0$
5. $f^{-1}(x) = (x + \sqrt{x^2 - 4})/2$, domain $x \geq 2$; range $y \geq 1$
7. For $x > 1, f^{-1}(x) = \sqrt{(x + 1)/(x - 1)}$, domain $1 < x < \infty$; range $1 < y < \infty$.
For $0 \leq x < 1, f^{-1}(x) = \sqrt{(x + 1)/(x - 1)}$, domain $-\infty < x \leq -1$; range $0 \leq y < 1$.
For $-1 < x < 0, f^{-1}(x) = -\sqrt{(x + 1)/(x - 1)}$, domain $-\infty < x < -1$; range $-1 < y < 0$.
For $x < -1, f^{-1}(x) = \sqrt{(x + 1)/(x - 1)}$, domain $1 < x < \infty$; range $-\infty < y < 1$.

EXERCISE SET   12.2

1. $\text{Cot } x = \cot x, 0 < x < \pi$ **3.** (a) $\pi/2$; (c) $-\pi/4$; (e) $\pi/4$; (g) $-\pi/3$; (i) π; (k) $-\pi/4$;
(m) $\pi/4$; (o) $3\pi/4$ **4.** (a) 0.70; (c) 0.52; (e) 1.01; (g) 0.61; (i) -0.90; (k) 1.76; (m) -1.40
5. (a) x; (c) x, if $-\pi/2 \leq x \leq \pi/2$; $\pi - x$, if $\pi/2 < x < 3\pi/2$; $x - 2\pi$, if $3\pi/2 \leq x \leq 5\pi/2$;
$3\pi - x$, if $5\pi/2 < x < 7\pi/2$. In general,
$\text{Sin}^{-1}(\sin x) = x - 2k\pi$, if $(4k - 1)\pi/2 \leq x \leq (4k + 1)\pi/2$
$\qquad\qquad = (2k + 1)\pi - x$, if $(4k + 1)\pi/2 < x < (4k + 3)\pi/2$
k an integer; (e) t; (g) x, if $-\pi/2 < x < \pi/2$; $x - \pi$, if $\pi/2 < x < 3\pi/2$;
$x - 2\pi$, if $3\pi/2 < x < 5\pi/2$, etc.   **6.** (a) $1/\sqrt{1 + x^2}$; (c) $t/\pm\sqrt{1 + t^2}$; (e) t

7. (a) $\frac{4}{5}$; (c) 0.6; (e) $\frac{3}{5}$; (g) $\frac{3}{5}$ **8.** (a) $\sqrt{3}/2$; (c) $\pi/2$
9. (a) $2t/(1-t^2)$; (c) $(t^2-1)/(t^2+1)$ **21.** $\frac{16}{65}$ **23.** 0   **31.** $\sqrt{2}/2$ **33.** $\frac{1}{2}$
35. $0, \frac{4}{5}$

CHAPTER 13

EXERCISE SET   13.1

1. (0.8,2.8)   **3.** (1,2)   **5.** $(-3,-10)$ **7.** $(\frac{7}{2},1)$ **9.** $(-2,11)$ **11.** $(-7,-3)$
13. $5x - 20 = 0$ **15.** $27x - 29 = 0$ **17.** $8y + 8 = 0$
      $2x + y - 8 = 0$ $3x + 3y - 7 = 0$ $2x - 5y - 16 = 0$
19. $11y + 41 = 0$
      $2x + 3y - 9 = 0$
21. $(3,-1)$ **23.** $(1,-2)$ **25.** $(-\frac{3}{11},\frac{9}{11})$ **27.** $(-\frac{2}{3},\frac{3}{5})$ **29.** $(-4,2)$ **31.** $(0,-i/2)$
33. $(\frac{1}{2},1)$ **35.** (3,1)   **37.** $(\frac{5}{2},-\frac{2}{3})$ **39.** (3,4)   **41.** (10,100)   **43.** (7,11)
45. 20 men, 32 women   **47.** 5 mi/hr; $\frac{1}{2}$ mi/hr   **49.** 24   **51.** 32 by 24 ft

EXERCISE SET   13.2

1. Inconsistent   **3.** Inconsistent   **5.** Inconsistent   **7.** Consistent and dependent
9. Consistent and dependent   **11.** Any $k \neq -1$ **13.** Any $k \neq -\frac{5}{3}$ **15.** Any $k \neq -3$
17. Any $k \neq -2$ **19.** $k = 32$ **21.** None

EXERCISE SET   13.3

1. (3,1,2)   **3.** $(-1,-3,2)$ **5.** $(5,-4,6)$ **7.** (3,2,1)   **9.** (9,12,8)   **11.** $(\frac{1}{2},1,-\frac{1}{3})$
13. $(2,1,-2,3)$ **15.** \$5,000   **17.** (25,28,31)   **19.** 734

EXERCISE SET   13.4

1. $(-1,2),(-3,-2)$ **3.** $(-\frac{5}{2},-\frac{3}{5}),(-9,2)$ **5.** $(2,1),(-\frac{5}{2},-2)$
7. $(1,2),(1,-2),(-1,2),(-1,-2)$
9. $(\sqrt{5}/2,\sqrt{3}/3),(\sqrt{5}/2,-\sqrt{3}/3),(-\sqrt{5}/2,\sqrt{3}/3),(-\sqrt{5}/2,-\sqrt{3}/3)$
11. $(\sqrt{329/34},\pm i\sqrt{12/17})$ **13.** $(2,1),(-2,-1)$ **15.** $(2,3),(-2,-3),(2,-1),(-2,1)$
17. $(3,6),(-3,-6),(4\sqrt{3},-5\sqrt{3}),(-4\sqrt{3},5\sqrt{3})$ **19.** 4,9   **21.** 24 by 32 in.   **23.** 3,4
25. $-8,-4$ **27.** 20 by 15 ft   **29.** 27   **31.** 30 lb   **33.** 32 by 24 yd; 48 by 16 yd

EXERCISE SET   13.5

13. $(-4,-7), (2.5,-3.8)$ **15.** (2.75,1.25), (3.5,2.5)

EXERCISE SET   13.6

1. $(-2\frac{1}{5},\frac{7}{5},-2\frac{2}{5})$ **3.** $(\frac{16}{11},-2\frac{1}{11},-\frac{9}{2})$ **5.** $(\frac{9}{5},-\frac{1}{5},1)$ **7.** (1,0,8)   **9.** $(2,0,-1,3)$
11. 369

EXERCISE SET   13.7

1. $x = 3, y = 5$ **3.** $x = y = 1$ **5.** $x = 3, y = 1, z = -2$

7. $\begin{pmatrix} 5 & 3 & 3 \\ 7 & 7 & 8 \end{pmatrix}$ **9.** $\begin{pmatrix} 3x - 2y & -x & 14 \\ 2y & 2x + 15y & -3y \end{pmatrix}$ **11.** $\begin{pmatrix} 3 & -5 & 4 \\ -1 & -5 & 9 \\ 4 & -3 & 3 \end{pmatrix}$

13. $\begin{pmatrix} +13 & +4 & -9 \\ -14 & +18 & -15 \\ -2 & -19 & -4 \end{pmatrix}$ **15.** -5 **17.** $\begin{pmatrix} 16 & 4 & 8 \\ 14 & 3 & 9 \\ 7 & 6 & 7 \end{pmatrix}$ **19.** $\begin{pmatrix} 2 & 0 & 3 & 11 \\ 0 & 0 & 3 & -15 \end{pmatrix}$

23. $\begin{pmatrix} 1 & 2 & 5 \\ 3 & 4 & 0 \\ 2 & 5 & 7 \end{pmatrix}$ **27.** $2x + 3y = 6$ **29.** $\begin{pmatrix} \frac{7}{29} & \frac{5}{29} \\ \frac{3}{29} & -\frac{2}{29} \end{pmatrix}$
  $5x - 2y = -4$

EXERCISE SET   13.8

1. $-3,2$ **3.** $-\begin{vmatrix} 3 & -5 \\ 1 & 7 \end{vmatrix}, \begin{vmatrix} 1 & -5 \\ 3 & 7 \end{vmatrix}, -\begin{vmatrix} 1 & 3 \\ 3 & 1 \end{vmatrix}$ **5.** $-4,1$

7. $-\begin{vmatrix} 1 & 2 \\ 2 & 5 \end{vmatrix}, \begin{vmatrix} 0 & 1 \\ 2 & 5 \end{vmatrix}, -\begin{vmatrix} 0 & 1 \\ 1 & 2 \end{vmatrix}$ **9.** 22 **11.** -19 **13.** $2x^2 + x - 12$ **15.** -4

17. -35 **19.** 9 **21.** 183 **23.** 3 **25.** 1 **27.** 1

EXERCISE SET   13.9

1. $(3,-2)$ **3.** $(\frac{3}{4},\frac{1}{2})$ **5.** $(\frac{7}{2},\frac{5}{2})$ **7.** $(3,1,2)$ **9.** $(-1,-3,2)$ **11.** $(\frac{1}{2},1,-\frac{1}{3})$
13. $(-4,3,6,7)$

EXERCISE SET   13.10

1. $10, -6$ **2.** 17 **3.** 480 **4.** 400 **5.** $0, -30$ **6.** $x = 4, y = 2$ **7.** $\$17.45$
8. Two A items and three B items yield the maximum profit of $36.

CHAPTER 14

EXERCISE SET   14.1

1. $3,14,47,146$ **3.** $5,-\frac{1}{5},5,-\frac{1}{5}$ **5.** $2,3,5\frac{5}{3},14\frac{4}{9}$ **7.** $\frac{1}{2},\frac{1}{4},\frac{1}{6},\frac{1}{8}$ **9.** $\frac{1}{2},\frac{1}{6},\frac{1}{12},\frac{1}{20}$
11. $1,4,27,256$ **13.** $1 - \frac{1}{8} + \frac{1}{27} - \frac{1}{64}$ **15.** $1,3,7,15$ **17.** $1,1,1,1$

19. $\displaystyle\sum_{n=1}^{\infty} \frac{1}{2^{n-1}}$ **21.** $\displaystyle\sum_{n=1}^{\infty} \frac{2n-1}{2n}$ **23.** $\displaystyle\sum_{n=1}^{\infty}(-1)^{n-1}2n$ **25.** $\displaystyle\sum_{n=1}^{\infty}\frac{x^n}{(1)(2)(3)\cdots(n)}$
27. $2 + 0 + 2$ **29.** $1 + x + x^2 + x^3 + x^4$ **31.** $\frac{1}{2} + \frac{2}{9} + \frac{3}{64} + \frac{4}{625} + \cdots$

EXERCISE SET   14.3

1. -35 **3.** $-\frac{34}{3}$ **5.** 37th   **7.** $-1,360$ **9.** $1,275$ **11.** $1,044$ **13.** 658
15. $9\frac{1}{2},6,15\frac{1}{12},9,21\frac{1}{2},12,27\frac{1}{2}$ **17.** 10 **19.** $3,14,25$ **21.** 512 **23.** 0.0000003 **25.** 9
27. $\sqrt{2}/3, \sqrt{6}/3, \sqrt{2}, \sqrt{6}, 3\sqrt{2}$ **29.** $1,456$ **31.** $255\frac{5}{8}$ **33.** $511\frac{1}{16}$ **37.** $20,10$
39. ± 12

EXERCISE SET   14.4

1. 24 **3.** 14 **5.** $25\frac{1}{4}$ **7.** $13\frac{9}{99}$ **9.** $\frac{7}{9}$ **11.** $124\frac{4}{999}$ **13.** $392\frac{2}{55}$ **15.** 60 ft
17. 81 sq in.

EXERCISE SET 14.5

1. $a^6 + 6a^5b + 15a^4b^2 + 20a^3b^3 + 15a^2b^4 + 6ab^5 + b^6$
3. $x^6 + 6x^4 + 15x^2 + 20 + 15/x^2 + 6/x^4 + 1/x^6$
5. $y^{8/3} - 4y^2x^{2/3} + 6y^{4/3}x^{4/3} - 4y^{2/3}x^2 + x^{8/3}$
7. $y^3 + 6y^{5/2}x^{1/2} + 15y^2x + 20y^{3/2}x^{3/2} + 15yx^2 + 6y^{1/2}x^{5/2} + x^3$
9. $2a^7 + 42a^5b^2 + 70a^3b^4 + 14ab^6$ 11. $64x^{12} + 96x^{19/2} + 60x^7$ 13. $280a^3b^4$
15. $126x^5y^5$ 17. -252 19. $960x^{12}$ 21. $1{,}030{,}301$

EXERCISE SET 14.6

1. $\frac{1}{2} + m/4 + m^2/8 + m^3/16 + \cdots$ 3. $1/y^2 - 4/y^3 + 12/y^4 - 32/y^5 + \cdots$
5. $4 - x/3 - x^2/144 - x^3/2{,}592 - \cdots$ 7. $1 + x/2 + 3x^2/8 + 5x^3/16 + \cdots$
9. $2 + y/4 - y^2/64 + y^3/512 - \cdots$ 11. 1.051 13. 1.059 15. 1.020 17. 1.987
19. 1.01 21. -3.04

CHAPTER 15

EXERCISE SET 15.1

1. $\frac{1}{110}$ 3. $\frac{1}{1260}$ 5. $k + 1$ 7. $(n + 1)/n$ 9. $k(k - 1)$ 11. $n/(n + 2)$
13. $9{,}000$ 15. 216 17. $60{,}000$

EXERCISE SET 15.2

1. (a) $2{,}520$; (c) $4{,}896$ 5. 4 7. $6{,}840$ 9. $1{,}680$ 11. $5{,}040$ 13. $22{,}050$
15. $1{,}260$ 17. 48 19. $10{,}368$ 21. $3{,}628{,}800$

EXERCISE SET 15.3

4. (a) 10; (c) 120 5. (a) $1{,}140$; (c) $11{,}480$; (e) $7{,}140$ 6. (a) 5 7. 126
9. $C(13,5)C(11,4)$ 11. $3{,}003$ 15. 2^{10}

EXERCISE SET 15.4

7. $4{,}680$ 9. 255

EXERCISE SET 15.5

1. (a) $\frac{1}{7}$; (c) $180/1{,}001$; (e) $72/1{,}001$ 2. (a) $1/2{,}024$ 3. $\frac{5}{18}$ 5. $\frac{1}{8}$ 7. $44/4{,}165$
9. $[C(10,3)C(8,2)]/C(18,5)$ 11. \$5   13. $\frac{1}{36}$

EXERCISE SET 15.6

1. $\frac{7}{15}$ 3. $\frac{1}{3}$ 5. $\frac{1}{10}$ 7. $\frac{3}{4}$ 9. $193\frac{3}{512}$ 11. $37\frac{1}{120}$ 13. $11/5{,}184$
15. $2{,}197/20{,}825$ 17. $52\frac{2}{77}$

INDEX

Page numbers in *italics* indicate marginal notes.